Low-Carbon Cementitious Materials with 100% Solid Wastes

Changwang Yan · Ru Bai · Ju Zhang

Low-Carbon Cementitious Materials with 100% Solid Wastes

Springer

Changwang Yan
Inner Mongolia University of Technology
Hohhot, China

Ju Zhang
Inner Mongolia University of Technology
Hohhot, China

Ru Bai
Inner Mongolia Academy of Science and Technology
Hohhot, China

ISBN 978-981-96-8741-1 ISBN 978-981-96-8742-8 (eBook)
https://doi.org/10.1007/978-981-96-8742-8

This Springer imprint is published by the registered company Springer Nature Singapore Pte Ltd.
The registered company address is: 152 Beach Road, #21-01/04 Gateway East, Singapore 189721, Singapore

Foreword

Under the global strategic backdrop of addressing climate change and achieving sustainable development, the cement industry, as a representative high-energy-consuming and high-carbon-emission basic materials sector, is confronted with an urgent requirement for green and low-carbon transformation. Specifically, how to overcome the excessive dependence of traditional cement production on natural mineral resources while simultaneously enabling large-scale and high-value utilization of industrial solid waste has emerged as a critical scientific challenge and technical bottleneck that urgently demands resolution within the current field of materials science and engineering.

The core innovation of this book resides in the proposition of utilizing "100% industrial solid waste as a substitute for natural raw materials" in the preparation of cementitious materials. By precisely controlling the phase composition of multiple solid wastes, including coal gangue, calcium carbide slag, and desulfurization gypsum, low-carbon cementitious materials dominated by $3CaO{\cdot}3Al_2O_3{\cdot}CaSO_4$ ($C_4A_3\hat{S}$), and $2CaO{\cdot}SiO_2$ (C_2S) as the primary mineral phases were successfully developed. This material exhibits substantial advantages in terms of resource utilization efficiency and carbon emission reduction. The author's team has systematically conducted research on the phase reconstruction mechanism of multi-component solid waste, the hydration reaction characteristics of low-carbon cementitious materials, and the mechanical properties of hardened slurries. In addition, low-carbon cementitious materials have been utilized in related technologies, including seawater mixing, grouting materials, and the impact-resistant concrete. These innovative advancements have offered significant theoretical support and practical technical solutions, contributing to the achievement of the "dual carbon" strategic goals within the building materials industry.

This book encompasses disciplines including civil engineering, materials engineering, chemical engineering, and mechanics, demonstrating significant academic value and substantial engineering application potential. The development of low-carbon cementitious materials not only paves a novel pathway for the resourceful utilization of solid waste but also delivers a practical and viable technical approach for the green transition of the building materials industry. It is anticipated that the release of this book will provide a valuable reference for scientific researchers and engineering professionals in relevant fields, and significantly contribute to advancing the development of a circular economy and the construction of ecological civilization.

Shilang Xu
Academician of Chinese Academy of Sciences
Professor of Zhejiang University
Hangzhou, China

Preface

The cement production process is characterized by high resource consumption and environmental pollution. It extensively utilizes natural resources such as limestone, clay, gypsum, and bauxite, whose extraction generates significant environmental contamination while simultaneously releasing substantial quantities of carbon dioxide. Concurrently, the industry annually produces millions of tons of industrial solid wastes, including coal gangue, fly ash, desulfurization gypsum, steel slag, and carbide slag, which collectively impose severe ecological burdens. The optimal solution lies in substituting natural raw materials with these industrial solid wastes for cement production.

This book demonstrates an innovative approach to synthesizing low-carbon cementitious materials (LCCM) containing $3CaO{\cdot}3Al_2O_3{\cdot}CaSO_4$ ($C_4A_3\hat{S}$), and $2CaO{\cdot}SiO_2$ (C_2S) through 100% utilization of industrial solid wastes, achieving dual environmental benefits by eliminating bulk solid waste accumulation and significantly reducing CO_2 emissions associated with conventional cement production derived from natural resource extraction. The research is organized into ten chapters. Chapter 1 establishes the research background by discussing the current state of investigations into solid waste-based low-carbon cementitious materials and examining their feasibility. Chapter 2 discusses the formation and characteristics of the silica-alumina phase in LCCM, mainly exploring the influence of the calcination system on the formation of the main mineral components in LCCM. Chapter 3 analyzes the hydration characteristics of LCCM, emphasizing hydration degree, the development of hydration products, and the evolution of associated pore structures of LCCM at varying curing ages. Chapter 4 explores the AC impedance spectra during the hydration process of LCCM, mainly clarifying the relationship between the hydration degree and the AC impedance characteristics, as well as the relationship between the pore structure and the impedance behavior. Moreover, the ARIMA model is utilized to predict the impedance parameters. Chapter 5 examines the regulation of the hydration process of LCCM by desulfurization gypsum, detailing its effects on hydration evolution, mechanical strength, and microstructural development. Chapter 6 investigates the mechanical properties of hardened pastes of seawater-mixed LCCM, including the free chloride ions content, compressive strength, and pore structure. Chapter 7 focuses on the properties of grouting materials prepared by LCCM, concentrating on the influence of steel slag and fly ash content on the flowability, expansion ratio, compressive strength, microstructure, and anchoring capability of grouting material. Chapter 8 presents the preparation and performance analysis of the concrete containing LCCM, with particular attention to early-age compressive strength, pore structure, and the influence of hydration products on mechanical performance. Chapter 9 addresses the mechanical behavior of the concrete containing LCCM under impact loading, evaluating impact resistance, the aggregate-matrix transition zone, and impact life predictions. Chapter 10 investigates the environmental impact assessment of LCCM, covering aspects such as resource input, energy input, pollutant emissions, and environmental impact analysis.

This book presents an innovative paradigm for synthesizing LCCM through 100% utilization of industrial solid wastes, incorporating $C_4A_3\check{S}$ and C_2S, which simultaneously addresses the dual challenges of bulk solid waste disposal and CO_2 emission reduction in conventional cement production. The research systematically explores the phase reconstruction mechanisms of multi-component solid wastes under controlled calcination conditions, hydration behavior evolution across curing ages with characterization of reaction products and pore structure development, and AC impedance spectroscopy-based hydration monitoring enhanced by ARIMA modeling. It further investigates the regulatory effects of desulfurization gypsum on hydration kinetics and microstructure, seawater-activated hydration pathways yielding hardened pastes with compressive strength. Develop optimized grouting materials with specific fluidity and interface bonding strength, concrete formulations with early strength development, and impact-resistant composite materials capable of absorbing energy. Environmental verification through life cycle assessment has confirmed the reduction of resource extraction and carbon dioxide equivalent emissions, establishing a comprehensive framework for the development of sustainable cement materials from solid waste valorization to engineering applications.

The key innovation of this book is the 100% utilization of solid waste for the preparation of a cementitious material containing $C_4A_3\check{S}$ and C_2S. Specifically, solid wastes such as coal gangue, fly ash, carbide slag, and desulfurization gypsum are employed. Through high-temperature calcination, the chemical bonds in these solid wastes are broken, and their crystal structures are reorganized. The adoption of LCCM enables effective enhancement of seawater mixing performance, fulfillment of grouting material functional requirements, improvement of concrete's early-age properties, and significant enhancement of concrete impact resistance. The LCCM is capable of fully substituting for Portland cement in functionality.

Hohhot, China

Changwang Yan
Ru Bai
Ju Zhang

Acknowledgments

The author has received following research grants to sponsor the related research programs:

- National Natural Science Foundation of China (NO. 52068059; NO. 52368036)
- Science and Technology Program Project of Inner Mongolia Autonomous Region (NO. 2022YFDZ0023; NO. 2023YFDZ0049; NO. 2024JBGS0017)
- Youth Leading Project of the Inner Mongolia Autonomous Region (NO. QNLJ012008)
- Special Training Program for Outstanding Young Talents in Inner Mongolia Autonomous Region
- Science and Technology Project of Huaneng Group Co., Ltd. of China (NO. HNKJ24-H117).

Graduate students in the author's research group contributed a great deal to the research work and completion of this book, including Mr. Cunyong Sun, Mr. Chang Liu, Mr. Jie Liu, Ms. Rongrong Feng, Ms. Xiaowei Shen, Mr. Fei Cao, Mr. Jiaxin Gao, Mr. Linghao Zhang, Mr. Lizhi Xia.

Special thanks are extended to Mr. Rammohan Krishnamurth, the production editor, for his significant contributions in editing throughout the entire compilation process of this book.

Contents

1 Introduction 1
1.1 Research Background 1
1.2 Research Status 3
1.2.1 Low-Carbon Cementitious Materials Prepared from Single-Source Industrial Solid Waste 3
1.2.2 Low-Carbon Cementitious Materials Prepared from Multi-Source Industrial Solid Wastes 5
1.3 Book's Outline 8
References 9

2 Characterization of Aluminosilicate Minerals in Low-Carbon Cementitious Materials 13
2.1 Introduction 13
2.2 Raw Materials 13
2.3 Materials Proportion 15
2.4 Cement Clinker Preparation 16
2.5 Mineral Phase Analysis 17
2.6 Functional Group Analysis 18
2.7 Chemical Binding Energy Analysis 21
2.8 Chemical Shift Analysis 23
2.9 Micro Morphology Analysis 25
2.10 Conclusions 27
References 29

3 Hydration Characteristics of Low-Carbon Cementitious Material 31
3.1 Introduction 31
3.2 Raw Materials and Test Methods 31
3.2.1 Raw Materials 31
3.2.2 Treatment of Raw Materials 32

3.2.3 Mineral Composition Design and Preparation Process ... 33
3.2.4 Test Method ... 35
3.3 Formation and Characterization of Clinker Minerals ... 39
3.3.1 Appearance Characteristics ... 39
3.3.2 Maturity Analysis ... 40
3.3.3 Mineral Phase Analysis ... 41
3.3.4 Mineral Molecular Structure Analysis ... 44
3.3.5 Thermogravimetric Comprehensive Analysis ... 47
3.3.6 Microstructural Analysis ... 48
3.4 Optimal Calcination System ... 51
3.4.1 Calcination Temperature ... 51
3.4.2 Calcination Time ... 53
3.4.3 Mineral Composition Analysis ... 55
3.4.4 Workability Analysis ... 57
3.5 Hydration Characteristics of Low Carbon Cementitious Materials ... 59
3.5.1 Hydration Heat ... 59
3.5.2 Chemically Bound Water ... 60
3.5.3 Hydration Products ... 61
3.5.4 Micro-morphology of Hydration Products ... 68
3.6 Conclusions ... 69
References ... 70

4 AC Impedance Spectrum of Low-Carbon Cementitious Materials During the Hydration Process ... 73
4.1 Introduction ... 73
4.2 Test Survey ... 74
4.3 Hydration and AC Impedance Characteristics of LCCM ... 75
4.3.1 Hydration Characteristics Analysis ... 75
4.3.2 AC Impedance Spectrum Analysis ... 80
4.3.3 AC Impedance Parameters Analysis ... 84
4.3.4 Relationship Between AC Impedance Characteristics and Hydration Degree ... 90
4.4 Microstructure Evolution and AC Impedance Characteristics of LCCM ... 93
4.4.1 Equivalent Circuit Model Based on Microstructure ... 93
4.4.2 Analysis of AC Impedance Parameters Based on Microstructure ... 97
4.4.3 Pore Structure Analysis ... 102
4.4.4 Relationship Between AC Impedance Parameters and Pore Structure ... 109

4.5 Prediction of AC Impedance Parameters of LCCM Based on ARIMA Model ... 112
4.5.1 Basic Principle of ARIMA Model ... 112
4.5.2 Steps for ARIMA Model ... 113
4.5.3 Prediction of AC Impedance Parameter ... 115
4.6 Conclusions ... 120
References ... 121

5 Desulfurization Gypsum Regulates Hydration Process of Low-Carbon Cementitious Material ... 123
5.1 Introduction ... 123
5.2 Experimental Overview ... 123
5.2.1 Raw Materials ... 123
5.2.2 Preparation Process of LCCM ... 124
5.2.3 Specimen ... 126
5.2.4 Test Methods ... 127
5.3 Desulfurization Gypsum Regulates the Strength Development of LCCM ... 130
5.3.1 Setting Time ... 130
5.3.2 Strength Development ... 130
5.4 Desulphurization Gypsum Regulates the Microstructure of LCCM ... 132
5.4.1 Phase Transition ... 132
5.4.2 Comprehensive Thermal Analysis ... 133
5.4.3 Molecular Characteristics Analysis ... 136
5.4.4 Quantitative Analysis of Phase Change ... 138
5.4.5 Microscopic Morphology Development Process ... 139
5.5 Desulfurization Gypsum Regulates the Pore Structure of LCCM ... 141
5.5.1 Porosity ... 141
5.5.2 Pore Size Distribution ... 143
5.6 AC Impedance Characteristics of LCCM Regulated by Desulfurization Gypsum ... 144
5.6.1 Nyquist Curves ... 144
5.6.2 An Equivalent Circuit Model Based on Microstructure Change ... 146
5.6.3 Relationship Between Impedance Parameters and Compressive Strength ... 150
5.7 Hydration Kinetics of LCCM Regulated by Desulfurization Gypsum ... 154
5.7.1 Hydration Heat ... 154
5.7.2 Hydration Kinetic Model of LCCM ... 156

5.8 Hydration Mechanism of LCCM Regulated by Desulfurization Gypsum 160
5.9 Conclusions 162
References 163

6 Mechanical Properties of Hardened Pastes of Seawater-Mixed Low-Carbon Cementitious Materials 165
6.1 Introduction 165
6.2 Specimen Preparation and Research Methods 165
6.2.1 Raw Materials 165
6.2.2 Preparation Process 167
6.2.3 Test Methods 168
6.3 Free Chloride Ions of Hardened Pastes of Seawater-Mixed LCCM 169
6.4 Hydration Process of Seawater-Mixed LCCM 171
6.4.1 Hydration Products 171
6.4.2 Thermogravimetric Analysis 174
6.4.3 Molecular Structure of Hydration Products 177
6.4.4 Microstructural of Hydration Products 178
6.5 Compressive Strength of LCCM Mixed with Seawater 180
6.5.1 Compressive Strength Under Different Mixing and Curing Conditions 180
6.5.2 Mechanism Analysis Chloride Ion Influence on Compressive Strength in Seawater 182
6.6 Pore Structure of LCCM Mixed with Seawater 185
6.6.1 Porosity 185
6.6.2 T_2 Spectrum 186
6.6.3 Pore Size Distribution 188
6.6.4 Relationship Between Pore Structure and Compressive Strength 190
6.7 Conclusions 193
References 195

7 Mechanical Properties of Grouting Materials Prepared by Low-Carbon Cementitious Materials 197
7.1 Introduction 197
7.2 Specimen Preparation and Test Methods 197
7.2.1 Raw Materials 197
7.2.2 Preparation Process 201
7.2.3 Test Methods 202
7.3 Flowability of Grouting Materials Prepared by LCCM 206
7.3.1 Effect of Steel Slag Content on Flowability 206
7.3.2 Effect of Fly Ash Content on Flowability 207
7.4 Vertical Expansion Rate of Grouting Materials Prepared by LCCM 209

7.4.1 Effect of Steel Slag Content on Vertical Expansion Rate ... 209
7.4.2 Effect of Fly Ash Content on Vertical Expansion Rate ... 210
7.5 Compressive Strength of Grouting Materials Prepared by LCCM ... 212
7.5.1 Effect of Steel Slag Content on Compressive Strength ... 212
7.5.2 Effect of Fly Ash Content on Compressive Strength ... 214
7.6 Content of Ettringite of Grouting Materials Prepared by LCCM ... 215
7.6.1 Effect of Steel Slag Content on Ettringite Content ... 215
7.6.2 Effect of Fly Ash Content on Ettringite Content ... 217
7.7 Uniaxial Tension Test of Reinforcement Sleeve with Grouting Material ... 219
7.7.1 Bond Stress Analysis ... 219
7.7.2 Effect of Anchorage Length on Bond Stress ... 224
7.7.3 Strain Analysis of Grouted Rebar Sleeves ... 228
7.8 Conclusions ... 234

8 Natural Curing Properties of Concrete Containing Low-Carbon Cementitious Materials ... 237
8.1 Introduction ... 237
8.2 Experimental Overview ... 237
8.2.1 Materials ... 237
8.2.2 Preparation of LCCM ... 241
8.2.3 Preparation and Mix Proportion of Concrete Containing LCCM ... 242
8.2.4 Test Methods ... 243
8.3 Early Age Strength Formation Mechanism of Concrete Containing LCCM ... 245
8.3.1 Hydration Products at Early Age ... 245
8.3.2 Thermal Analysis of Early Age Hydration Products ... 248
8.3.3 Molecular Structure Analysis of Early Age Hydration Products ... 254
8.3.4 Micro-morphology of Early-Age Hydration Products ... 257
8.4 Pore Structure of Concrete Containing LCCM ... 264
8.4.1 Evolution Law of Pore Structure ... 264
8.4.2 Distribution of Pore Structure ... 267
8.4.3 Porosity ... 268
8.4.4 Permeability Evolution Law Based on Pore Structure ... 268

8.5 Early Strength Performance of Concrete Containing LCCM 270
8.5.1 Compressive Strength 270
8.5.2 Failure Mode 272
8.5.3 Relationship Between Pore Structure and Compressive Strength 274
8.5.4 Relationship Between Hydration Product and Compressive Strength 278
8.6 Conclusions 280
References 281

9 Impact Properties of Concrete Containing Low-Carbon Cementitious Materials 283
9.1 Introduction 283
9.2 Raw Material and Test Methods 283
9.2.1 Raw Material 283
9.2.2 Preparation of LCCM 284
9.2.3 Mix Proportion Design of Concrete Containing LCCM 286
9.2.4 Sample Preparation of Concrete Containing LCCM 287
9.2.5 Test Methods 288
9.3 Mechanical Properties of Concrete Containing LCCM 289
9.3.1 Compressive Strength 289
9.3.2 Impact Resistance Analysis 293
9.4 Microstructure of Concrete Containing LCCM 302
9.4.1 Microstructure Characteristics of Concrete 302
9.4.2 Interface Transition Zone of Concrete Aggregate 304
9.4.3 Evolution of Micro-pore Structure in Concrete 308
9.5 Prediction of Impact Life for Concrete Containing LCCM 313
9.5.1 Weibull Distribution Model 313
9.5.2 Impact Times Analysis 313
9.5.3 Prediction of Impact Life of Concrete Under Different Failure Rates 315
9.6 Conclusions 320
References 323

10 Environmental Impact Assessment of Low-Carbon Cementitious Materials 325
10.1 Introduction 325
10.2 Research Objectives and System Boundary 326
10.2.1 Research Objectives 326
10.2.2 System Boundary 326
10.3 List Analysis 327
10.3.1 Resource Input List 327
10.3.2 Energy Input List 328
10.3.3 Pollutant Output List 330

10.4 Environmental Impact Analysis 330
10.4.1 Characterization Analysis 330
10.4.2 Standardization Analysis 335
10.5 Comparative Analysis of Products 335
10.6 Conclusions 338

Chapter 1
Introduction

1.1 Research Background

As the world's largest cement producer, China has dominated global cement output since 2009, contributing to over 50% of the total production (Fig. 1.1a). The country's cement production surged to its first historical peak in 2014, reaching approximately 2.49 billion tons. However, between 2014 and 2018, output experienced a noticeable decline before gradually rebounding in subsequent years (Fig. 1.1b). As a cornerstone material for national economic development, cement production trends closely mirror broader economic activity. Given its indispensable role in infrastructure and construction, cement is expected to remain a vital building material for the foreseeable future.

The conventional production of Portland cement relies heavily on non-renewable raw materials. During the calcination process, limestone provides the CaO component, while clay contributes SiO_2, Fe_2O_3, and Al_2O_3. These raw materials undergo high-temperature calcination to form mineral components such as $2CaO{\cdot}SiO_2$ (C_2S), $3CaO{\cdot}SiO_2$ (C_3S), $3CaO{\cdot}Al_2O_3$ (C_3A), and $4CaO{\cdot}Al_2O_3{\cdot}Fe_2O_3$ (C_4AF) (Jiang & Zhang, 1995). Among these phases, C_3S is the predominant mineral phase in Portland cement, with a CaO content as high as 73.7%. Its calcination temperature is approximately 1450 °C, and the production of each ton of C_3S results in the emission of approximately 0.86 tons of CO_2 (Barcelo et al., 2014). Evidently, the high content of C_3S is the root cause of the substantial energy consumption and carbon emissions associated with Portland cement production.

Future perspectives in sustainable cement production point toward the development of innovative low-carbon cementitious materials featuring reduced firing temperatures and lower carbon emissions (Cao et al., 2020; Olsson et al., 2023; Watari et al., 2022). Promising mineral phases such as $4CaO{\cdot}2SiO_2{\cdot}CaSO_4$ ($C_5S_2\hat{S}$), $3CaO{\cdot}3Al_2O_3{\cdot}CaSO_4$ ($C_4A_3\hat{S}$), and C_2S offer distinct advantages including decreased limestone requirements, lower calcination temperatures (1300–1350 °C), and reduced CO_2 emissions (Chi et al., 2023; Cui et al., 2023; Zhang et al., 2023).

C. Yan et al., *Low-Carbon Cementitious Materials with 100% Solid Wastes*,
https://doi.org/10.1007/978-981-96-8742-8_1

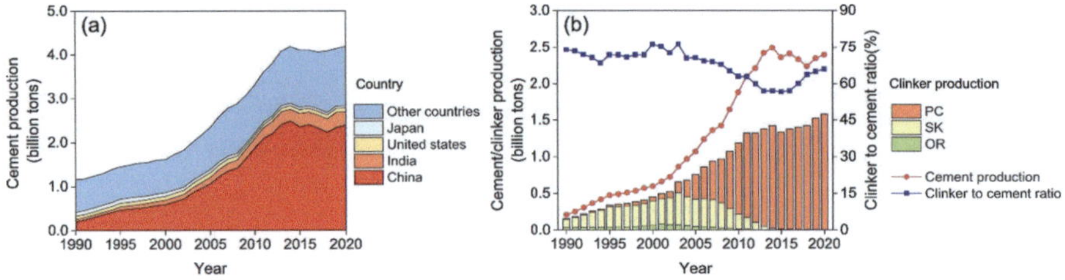

Fig. 1.1 Historical development of cement industry from 1990 to 2020. **a** Historical cement production in the world, **b** historical cement production in China (Wang et al., 2023)

These phases are typically synthesized from conventional raw materials (limestone, bauxite, and gypsum) through optimized low-temperature calcination processes, exhibiting favorable engineering properties such as rapid setting characteristics, enhanced early-age strength development, and stable long-term performance (Huang et al., 2022; Burris & Kurtis, 2018). The combination of environmental benefits and satisfactory performance metrics suggests that cementitious systems based on $C_5S_2\hat{S}$, $C_4A_3\hat{S}$, and C_2S represent viable and sustainable alternatives to conventional Portland cement.

Despite their environmental advantages, current production methods for low-carbon cementitious materials based on $C_5S_2\hat{S}$, $C_4A_3\hat{S}$, and C_2S still predominantly depend on non-renewable resources. A critical limiting factor is the requirement for high-purity bauxite, which not only incurs substantial production costs but also faces stringent quality specifications, thereby severely restricting the industrial-scale manufacturing of these materials. Consequently, current production volumes remain insufficient to satisfy increasing market demands. In response to these challenges—encompassing cost reduction, resource efficiency, and environmental sustainability—researchers have increasingly focused on utilizing industrial solid wastes as alternative raw materials for synthesizing these low-carbon cementitious systems. This approach has emerged as a key strategy to overcome existing limitations while advancing circular economy principles in construction materials development.

This research is situated at the intersection of carbon mitigation in cement production and industrial solid waste valorization. The study investigates the synthesis of low-carbon cementitious materials incorporating $C_5S_2\hat{S}$, $C_4A_3\hat{S}$, and C_2S phases through the strategic substitution of conventional raw materials with industrial solid wastes. This dual-focused approach simultaneously promotes the high-value utilization of industrial byproducts while providing a viable technological pathway for the cement industry to achieve its carbon peak and neutrality objectives.

1.2 Research Status

1.2.1 Low-Carbon Cementitious Materials Prepared from Single-Source Industrial Solid Waste

Calcium sulfoaluminate cement represents a prominent class of low-carbon cementitious materials, with its primary mineral phases comprising $C_5S_2\hat{S}$, $C_4A_3\hat{S}$, and C_2S. Conventionally, calcium sulfoaluminate cement is synthesized by proportioning non-renewable raw materials—limestone, bauxite, and gypsum—followed by high-temperature calcination at 1350 ± 50 °C (Chen et al., 2020; Gao et al., 2020; Shen et al., 2018). However, the reliance on high-cost bauxite has impeded industrial-scale production, thereby restricting the broader commercialization and application of these sustainable cement alternatives. In response, recent research has prioritized the recycling and resource utilization of industrial solid wastes in construction materials. Global studies have investigated the partial or complete substitution of bauxite and limestone with mining or industrial byproducts. Initial methodologies predominantly employed single-source solid wastes as partial raw material replacements, often supplemented with analytical-grade corrective additives such as $CaSO_4$ and Al_2O_3 to achieve desired compositions.

Wu et al. (2011a) investigated the feasibility of synthesizing calcium sulfoaluminate cement from municipal solid waste incineration ash with a chemical composition of CaO-SiO_2-Al_2O_3. The study found that using incineration ash as the primary raw material, supplemented with analytical-grade $CaCO_3$, $CaSO_4$, and Al_2O_3, calcium sulfoaluminate cement dominated by $C_4A_3\hat{S}$ and C_2S could be synthesized at 1250 °C, provided the incineration ash substitution rate did not exceed 30%.

Red mud, a solid waste generated during alumina extraction, is rich in Al, Fe, Ca, and Si. Substituting bauxite with red mud and blending it with limestone and gypsum enables the production of calcium sulfoaluminate cement with high early-age strength and dense hardened paste structures (Jia et al., 2023; Singh et al., 1996; Zhao et al., 2006). Singh et al. (1996) synthesized calcium sulfoaluminate cement clinker using red mud, limestone, bauxite, and gypsum, systematically investigating the effects of red mud substitution ratio, calcination duration, and temperature on cementitious properties. Results demonstrated that the 28-day compressive strength of the cementitious material prepared with red mud replacing bauxite was comparable to that of ordinary Portland cement. Zhao et al. (2006) increased the red mud substitution ratio to 40% and produced calcium sulfoaluminate cement composed of $C_4A_3\hat{S}$, C_2S, and C_4AF under calcination at 1300 °C. Jia et al. (2023) synthesized iron-rich calcium sulfoaluminate cement by calcining red mud, limestone, bauxite, and anhydrite, further analyzing the effects of anhydrite and flue gas desulfurization gypsum on hydration and mechanical properties. The results indicated that anhydrite minimally enhanced early compressive strength, yet the system achieved 94.61 MPa at 3 days. In contrast, desulfurization gypsum significantly improved early compressive strength, elevating the 4-h compressive strength of the hardened paste to 30 MPa.

Phosphogypsum, a solid waste generated during wet-process phosphoric acid production, primarily consists of $CaSO_4 \cdot 2H_2O$. Shen (2015) investigated the calcination process of calcium sulfoaluminate cement using phosphogypsum, analyzing the effects of calcination parameters on mineral phase formation and mechanical properties of hardened paste. Results revealed that the optimal calcination regime for phosphogypsum-based calcium sulfoaluminate cement involves: a calcination temperature of 1270 °C, a holding time of 30 min, and rapid air quenching for cooling. Liu (2017) further studied staged calcination protocols for phosphogypsum-derived calcium sulfoaluminate cement. The findings demonstrated that a multi-stage calcination process (CO_2 atmosphere at 950 °C for 1 h, air atmosphere at 1100 °C for 1 h, and 1200 °C for 30 min) achieved optimal performance, confirming the feasibility of fully replacing limestone with phosphogypsum in staged cement production.

Converter slag, a by-product of steelmaking, primarily contains CaO, SiO_2, and Fe_2O_3. Xue et al. (2016) utilized converter slag as a main raw material supplemented with chemical-grade reagents to synthesize calcium sulfoaluminate cement, investigating the sintering process under varying substitution ratios. Results indicated that incorporating converter slag improved the clinker's melt phase, enhanced sintering efficiency, and yielded hardened pastes with superior early-age performance compared to ordinary Portland cement at a 14% substitution rate.

Lepidolite residue, an industrial waste from lithium extraction via sulfuric acid treatment of lepidolite, is rich in SiO_2 and Al_2O_3. Li (2012) produced calcium sulfoaluminate cement clinker using lepidolite residue, bauxite, limestone, and gypsum, optimizing the calcination parameters. The study identified an optimal substitution ratio of 16% for lepidolite residue, a calcination temperature range of 1270–1320 °C, and achieved mechanical strengths of 42.4 MPa (compressive strength at 28 d) and 6.8 MPa (flexural strength at 28 d).

Sulfur-fixation slag, a coal combustion by-product from fluidized bed boilers with sulfur-fixing agents, mainly comprises CaO, SiO_2, and SO_3. Liu et al. (2008) synthesized calcium sulfoaluminate cement clinker by substituting 30% of conventional calcium/silica sources with sulfur-fixation slag, combined with limestone, bauxite, and analytical-grade gypsum. Calcination at 1250 °C for 30 min successfully yielded cement clinker containing $C_4A_3\hat{S}$, C_2S, and C_4AF phases.

Marble sludge, dominated by CaO, was evaluated by El-Alfi and Gado (2016) for producing early-strength cementitious materials under varying calcination temperatures. A raw mix comprising 25% kaolin, 20% gypsum, and 55% marble sludge calcined at 1250 °C significantly increased $C_4A_3\hat{S}$ and C_2S phases in the clinker, substantially enhancing the mechanical properties of the hydrated cementitious material.

Matos et al. (2022) evaluated the feasibility of producing calcium sulfoaluminate cement using waste fluid catalytic cracking catalysts. Results demonstrated that waste fluid catalytic cracking catalysts could replace 10.24–20.81% of conventional raw materials as silica-alumina sources. Furthermore, the Fe_2O_3 present in the waste fluid catalytic cracking catalysts not only facilitated the formation of C_4AF but also stabilized the crystalline structure of $C_4A_3\hat{S}$ phases.

Wang et al. (2023) synthesized calcium sulfoaluminate cement using electrolytic manganese residue as a silica and iron source. Analysis of hydration products revealed that short rod-like ettringite formed within 3 days, providing rigid skeletal support, while gel phases filled interstitial pores, contributing flexible reinforcement. At 28 days of hydration, a dense microstructure emerged from the synergistic interaction of rigid crystalline frameworks (AFm, AFt, and strätlingite) and flexible matrices (iron-alumina gels and C-S-H gels), achieving a compressive strength of up to 91 MPa.

These studies collectively confirm the feasibility of utilizing single-source industrial solid wastes, combined with natural mineral resources and chemical additives, to produce low-carbon cementitious materials dominated by $C_4A_3\hat{S}$, calcium $C_5S_2\hat{S}$, and C_2S.

1.2.2 Low-Carbon Cementitious Materials Prepared from Multi-Source Industrial Solid Wastes

With the deepening of research, scholars have conducted extensive studies on synthesizing calcium sulfoaluminate cement using multi-source industrial solid wastes to improve waste substitution rates.

Wu et al. (2020) successfully prepared calcium sulfoaluminate cement using phosphogypsum as the sole CaO source, supplemented with minor aluminum ash and fly ash. In follow-up work, Wu et al. (2019) reported that calcium sulfoaluminate cement hardened paste achieved compressive strengths of 63.2, 88.5, and 100.2 MPa at 1 d, 3 d, and 28 d, respectively, when the Al_2O_3/SiO_2 ratio was controlled at 2.0 or 2.5. Beyond phosphogypsum, Wu et al. (2019) explored replacing natural gypsum and limestone with desulfurization gypsum. Calcium sulfoaluminate cement synthesized from desulfurization gypsum, aluminum slag, red mud, and fly ash exhibited optimal performance at a calcination temperature of 1270–1310 °C and holding time of 60–100 min, yielding compressive strengths of 63.3, 85.3, and 102.8 MPa at 1 d, 3 d, and 28 d.

Tao et al. (2023) investigated the effects of calcination parameters on calcium sulfoaluminate cement phase composition using limestone, bauxite, fly ash, and phosphogypsum. Results indicated that calcination temperature significantly influenced phase formation, with the optimal conditions being 1200 °C for 60–120 min. Hydration products included AFt, AFm, C-S-H, C_4AH_X, and AH_3.

He et al. (2022a) developed belite-sulfoaluminate cement via co-processing of electrolytic manganese residue, barium slag, limestone, and bauxite. The optimal calcination conditions (1300 °C, 30 min) produced compressive strengths of 37.9, 47.3, and 60 MPa at 3 d, 7 d, and 28 d. Microstructural analysis revealed that low-melting-point compounds from electrolytic manganese residue reduced the eutectic temperature of target minerals and enhanced liquid-phase sintering. Barium slag further promoted the crystal transition of calcium sulfoaluminate from orthorhombic to cubic systems. In subsequent research, He et al. (2022b) demonstrated that early hydration generated metastable ettringite, which gradually transformed into AFm, while C_2S hydration remained suppressed initially.

Canbek et al. (2020) synthesized calcium sulfoaluminate cement using red mud, fly ash, and desulfurization gypsum, achieving a 200 °C reduction in calcination temperature compared to Portland cement. By reducing bauxite content from 40 to 20% without compromising 28-day strength, they validated industrial feasibility. At a 38% solid waste substitution rate, mortar specimens attained compressive strengths >15 MPa (1d) and >40 MPa (28d). However, increasing substitution to 55% degraded mechanical performance.

Isteri et al. (2022) produced iron-rich sulfoaluminate cement using ladle furnace slag, iron slag, and phosphogypsum, successfully scaling up the process from laboratory trials to pilot-scale production in a 7-m cement kiln.

Wang et al. (2022) synthesized high-belite sulfoaluminate cement at reduced temperatures using petroleum coke desulfurization residue, fly ash, calcium carbide slag, and bauxite. Under calcination at 1225–1300 °C, cement clinker containing $C_4A_3\hat{S}$, C_2S, and C_4AF was produced, achieving an 80% industrial solid waste utilization rate. Optimal calcination conditions (10 °C/min heating rate to 1300 °C with a 45-min hold) yielded hardened paste compressive strengths of 27.7 MPa (1 d) and 51.4 MPa (28 d). Early hydration products included ettringite, aluminum gel, unhydrated $C_4A_3\hat{S}$, C_2S, gypsum, and $Ca(OH)_2$, where the aluminum gel-filled ettringite network significantly enhanced mechanical strength.

Ren et al. (2017a) prepared calcium sulfoaluminate cement using 8–15% coal gangue, 25–35% desulfurization gypsum, 30–35% aluminum slag, and 30–35% calcium carbide slag, calcined at 1150–1300 °C. The clinker primarily consisted of $C_4A_3\hat{S}$, C_2S, and C_4AF, achieving a 28-day compressive strength of 75 MPa. In subsequent studies, Ren et al. (2017b) conducted a life cycle assessment comparing calcium sulfoaluminate cement production from industrial solid wastes (e.g., red mud, desulfurization gypsum) versus conventional raw materials, demonstrating a 38.62% reduction in environmental burden. Ren et al. (2020) further developed calcium sulfoaluminate cement 3D-printable materials using tailings, coal gangue, aluminum slag, and desulfurization gypsum. With optimized additives, the materials exhibited adjustable setting times (15–80 min) and 2-h compressive strengths of 15–20 MPa.

Wu et al. (2011b) synthesized calcium sulfoaluminate cement using municipal solid waste incineration ash and evaluated its environmental safety via three leaching test protocols. Results confirmed that 30% incineration ash substitution was feasible, with clinker quality optimized by controlling the alkalinity modulus (1.05), Al/S ratio (2.5), and Al/Si ratio (2.0–3.0). Hazardous elements were effectively immobilized within the clinker mineral phases and hydration products, mitigating environmental risks.

Hou et al. (2012) synthesized calcium sulfoaluminate cement by calcining electrolytic manganese residue, limestone, and kaolin at 1200 °C. The cement system achieved final setting within 30 min and exhibited 28-day compressive strengths of

35–65 MPa. Adding gypsum enhanced early and long-term strengths by 50 and 30%, respectively.

Yao et al. (2020) developed high-performance calcium sulfoaluminate cement using desulfurization gypsum, coal gangue, limestone tailings, and aluminum ash. The clinker primarily comprised $C_4A_3\hat{S}$ and C_2S, with minor gabbro phases. Mortar specimens demonstrated compressive strengths of 45.7 MPa (3 d) and 67.8 MPa (28 d).

Rungchet et al. (2016) produced calcium sulfoaluminate cement via hydrothermal calcination of fly ash, aluminum-rich sludge, and desulfurization gypsum. Calcination at 1050 °C maximized $C_4A_3\hat{S}$ and C_2S formation, yielding a 28-day compressive strength of 30 MPa.

Dong et al. (2020) explored the feasibility of using pyrite-rich cyanidation tailings as Fe_2O_3 and SO_3 sources for calcium sulfoaluminate cement clinker. Clinker calcined at 1300 °C for 40 min contained $C_4A_3\hat{S}$, C_4AF, and $C_{12}A_7$ as primary phases.

Mao et al. (2020) prepared calcium sulfoaluminate cement clinker using 35% washed municipal solid waste incineration fly ash, desulfurization gypsum, and aluminum ash, effectively immobilizing eight heavy metals (Cu, Zn, Cd, Pb, Cr, Ni, Ba, As) within the hydration products.

Shen et al. (2014) investigated the feasibility of utilizing phosphogypsum for calcium sulfoaluminate cement production and analyzed its decomposition behavior during processing. Results demonstrated that phosphogypsum can serve as a raw material for calcium sulfoaluminate cement synthesis, with an optimal calcination temperature range of 1250–1300 °C. However, the decomposition of phosphogypsum during calcination was found to reduce the compressive strength of hardened paste. In subsequent research, Shen et al. (2015) developed a $C_5S_2\hat{S}$-$C_4A_3\hat{S}$-C_2S system via secondary thermal treatment of phosphogypsum. While the presence of $C_5S_2\hat{S}$ significantly suppressed early-age compressive strength, hydration products derived from $C_5S_2\hat{S}$ after 56 days of curing enhanced the long-term strength, demonstrating a trade-off between early performance and durability.

Li et al. (2021) synthesized calcium sulfoaluminate cement from aluminum-rich sludge and phosphogypsum, evaluating its hydration behavior through setting time and hydration product analysis. The optimal calcination conditions (1150 °C for 1 h) yielded calcium sulfoaluminate cement with a 28-day compressive strength of 39.4 MPa.

In summary, the partial replacement of non-renewable resources with multi-source industrial solid wastes to produce low-carbon cementitious materials dominated by $C_4A_3\hat{S}$, $C_5S_2\hat{S}$, and C_2S is technically feasible. However, current studies generally exhibit low substitution rates for non-renewable resources and still require supplementation with natural minerals (e.g., limestone, bauxite, natural gypsum) and analytical-grade reagents (e.g., $CaCO_3$, $CaSO_4$, Al_2O_3) as corrective additives.

1.3 Book's Outline

This book demonstrates an innovative approach to synthesizing low-carbon cementitious materials (LCCM) containing calcium sulphoaluminate ($C_4A_3\check{S}$) and dicalcium silicate (C_2S) through 100% utilization of industrial solid wastes, achieving dual environmental benefits by eliminating bulk solid waste accumulation and significantly reducing CO_2 emissions associated with conventional cement production derived from natural resource extraction. The research is organized into ten chapters.

This Chapter establishes the research background by discussing the current state of investigations into solid waste-based low-carbon cementitious materials and examining their feasibility.

Chapter 2 discusses the formation and characteristics of the silica-alumina phase in LCCM, mainly exploring the influence of the calcination system on the formation of the main mineral components in LCCM.

Chapter 3 analyzes the hydration characteristics of LCCM, emphasizing hydration degree, the development of hydration products, and the evolution of associated pore structures of LCCM at varying curing ages.

Chapter 4 explores the AC impedance spectra during the hydration process of LCCM, mainly clarifying the relationship between the hydration degree and the AC impedance characteristics, as well as the relationship between the pore structure and the impedance behavior. Moreover, the ARIMA model is utilized to predict the impedance parameters.

Chapter 5 examines the regulation of the hydration process of LCCM by desulfurization gypsum, detailing its effects on hydration evolution, mechanical strength, and microstructural development.

Chapter 6 investigates the mechanical properties of hardened pastes of seawater-mixed LCCM, including the free chloride ions content, compressive strength, and pore structure.

Chapter 7 focuses on the properties of grouting materials prepared by LCCM, concentrating on the influence of steel slag and fly ash content on the flowability, expansion ratio, compressive strength, microstructure, and anchoring capability of grouting material.

Chapter 8 presents the preparation and performance analysis of the concrete containing LCCM, with particular attention to early-age compressive strength, pore structure, and the influence of hydration products on mechanical performance.

Chapter 9 addresses the mechanical behavior of the concrete containing LCCM under impact loading, evaluating impact resistance, the aggregate-matrix transition zone, and impact life predictions.

Chapter 10 investigates the environmental impact assessment of LCCM, covering aspects such as resource input, energy input, pollutant emissions, and environmental impact analysis.

References

Barcelo, L., Kline, J., Walenta, G., & Gartner, E. (2014). Cement and carbon emissions. *Materials and Structures, 47*(6), 1055–1065.

Burris, L. E., & Kurtis, K. E. (2018). Influence of set retarding admixtures on calcium sulfoaluminate cement hydration and property development. *Cement and Concrete Research, 104*, 105–113.

Canbek, O., Shakouri, S., & Erdoğan, S. T. (2020). Laboratory production of calcium sulfoaluminate cements with high industrial waste content. *Cement and Concrete Composites, 106*, 103475.

Cao, Z., Myers, R. J., Lupton, R. C., Duan, H., Sacchi, R., Zhou, N., Miler, T., Cullen, J., Ge, Q., & Liu, G. (2020). The sponge effect and carbon emission mitigation potentials of the global cement cycle. *Nature Communications, 11*(1), 3777.

Chi, L., Li, M., Zhang, Q., Liang, X., Huang, C., Peng, B., & Sun, H. (2023). Cobalt immobilization performance and mechanism analysis of low carbon belite calcium sulfoaluminate cement. *Construction and Building Materials, 386*, 131545.

Chen, X., Wang, Y., Chong, H., & Huang, J. (2020). Use of sulphoaluminate cement in grouted macadam as sustainable pavement material. *Journal of Transportation Engineering, Part b: Pavements, 146*(2), 04020018.

Cui, K., Lu, D., Jiang, T., Zhang, J., Jiang, Z., Zhang, G., Chang, J., & Lau, D. (2023). Understanding the role of carbon nanotubes in low carbon sulfoaluminate cement-based composite. *Journal of Cleaner Production, 416*, 137843.

Dong, K., Xie, F., Wang, W., Chang, Y., Chen, C., & Gu, X. (2020). Calcination of calcium sulphoaluminate cement using pyrite-rich cyanide tailings. *Crystals, 10*(11), 971.

El-Alfi, E. A., & Gado, R. A. (2016). Preparation of calcium sulfoaluminate-belite cement from marble sludge waste. *Construction and Building Materials, 113*, 764–772.

Gao, Y., Li, Z., Zhang, J., Zhang, Q., & Wang, Y. (2020). Synergistic use of industrial solid wastes to prepare belite-rich sulphoaluminate cement and its feasibility use in repairing materials. *Construction and Building Materials, 264*, 120201.

He, W., Li, R., Zhang, Y., & Nie, D. (2022a). Synergistic use of electrolytic manganese residue and barium slag to prepare belite-sulphoaluminate cement study. *Construction and Building Materials, 326*, 126672.

He, W., Li, R., Nie, D., Zhang, J., Wang, Y., Zhang, Y., & Chen, Q. (2022b). Belite-calcium sulphoaluminate cement prepared by EMR and BS: Hydration characteristics and microstructure evolution behavior. *Construction and Building Materials, 333*, 127415.

Hou, P. K., Qian, J. S., Wang, Z., & Deng, C. (2012). Production of quasi-sulfoaluminate cementitious materials with electrolytic manganese residue. *Cement and Concrete Composites, 34*(2), 248–254.

Huang, G., Zhao, J., Gupta, R., & Liu, W. V. (2022). Influence of tartaric acid dosage on the early-age and long-term properties of calcium sulfoaluminate belite cement composites. *Construction and Building Materials, 356*, 129257.

Isteri, V., Ohenoja, K., Hanein, T., Kinoshita, H., Kletti, H., Rößler, C., Tanskanen, P., Illikainen, M., & Fabritius, T. (2022). Ferritic calcium sulfoaluminate belite cement from metallurgical industry residues and phosphogypsum: Clinker production, scale-up, and microstructural characterisation. *Cement and Concrete Research, 154*, 106715.

Jia, F., Sun, H., Yu, L., Lu, X., Zhou, Y., Elchalakani, M., & Du, P. (2023). Effects of different gypsum varieties on the hydration and mechanical properties of red mud-based ferrite-rich sulfoaluminate cement. *Journal of Materials Research and Technology, 24*, 3086–3097.

Jiang, Y., & Zhang, L. (1995). (1995) Combustion of high quality Portland cement with low grade limestone clay. *Cement, 03*, 11–13.

Li, F. (2012). Study on preparation of sulfoaluminate cement from leucite slag and low-grade bauxite. Nanchang University.

Li, Z., Huang, L., Wang, S., Yang, Z., & Yu, L. (2021). Using alumina-rich sludge and phosphogypsum manufactures low-CO2 cement. *Construction and Building Materials, 288*, 123016.

Liu, N. (2017). *Study on preparation of Belite-sulfoaluminate cement by stage decomposition of phosphogypsum*. Chongqing University.

Liu, R., Hua, W., & Li, F. (2008). Research on firing Belite sulfoaluminate cement with sulfur fixing slag. *Comprehensive Utilization of Fly Ash, 05*, 19–20.

Matos, S. R., de Matos, P. R., Neto, J. S. A., Campos, C. E., Kirchheim, A. P., & Gonçalves, J. P. (2022). Synthesis and hydration of ye'elimite-containing cement (YCC) produced with spent fluid catalytic cracking catalyst. *Construction and Building Materials, 359*, 129364.

Mao, Y., Wu, H., Wang, W., Jia, M., & Che, X. (2020). Pretreatment of municipal solid waste incineration fly ash and preparation of solid waste source sulphoaluminate cementitious material. *Journal of Hazardous Materials, 385*, 121580.

Olsson, J. A., Miller, S. A., & Alexander, M. G. (2023). Near-term pathways for decarbonizing global concrete production. *Nature Communications, 14*(1), 4574.

Ren, C., Wang, W., & Li, G. (2017a). Preparation of high-performance cementitious materials from industrial solid waste. *Construction and Building Materials, 152*, 39–47.

Ren, C., Wang, W., Mao, Y., Yuan, X., Song, Z., Sun, J., & Zhao, X. (2017b). Comparative life cycle assessment of sulfoaluminate clinker production derived from industrial solid wastes and conventional raw materials. *Journal of Cleaner Production, 167*, 1314–1324.

Ren, C., Wang, W., Yao, Y., Wu, S., & Yao, X. (2020). Complementary use of industrial solid wastes to produce green materials and their role in CO_2 reduction. *Journal of Cleaner Production, 252*, 119840.

Rungchet, A., Chindaprasirt, P., Wansom, S., & Pimraksa, K. (2016). Hydrothermal synthesis of calcium sulfoaluminate–belite cement from industrial waste materials. *Journal of Cleaner Production, 115*, 273–283.

Shen Y. (2015). *Preparation of high sulfur Belite sulfoaluminate cement by phosphogypsum*. Chongqing University.

Shen, Y., Li, X., Chen, X., Zhang, W., & Yang, D. (2018). Synthesis and calorimetric study of hydration behavior of sulfate-rich belite sulfoaluminate cements with different phase compositions. *Journal of Thermal Analysis and Calorimetry, 133*, 1281–1289.

Shen, Y., Qian, J., Chai, J., & Fan, Y. (2014). Calcium sulphoaluminate cements made with phosphogypsum: Production issues and material properties. *Cement and Concrete Composites, 48*, 67–74.

Shen, Y., Qian, J., Huang, Y., & Yang, D. (2015). Synthesis of belite sulfoaluminate-ternesite cements with phosphogypsum. *Cement and Concrete Composites, 63*, 67–75.

Singh, M., Upadhayay, S. N., & Prasad, P. M. (1996). Preparation of special cements from red mud. *Waste Management, 16*(8), 665–670.

Tao, X., Li, P., Wang, G., Li, W., & Ma, S. (2023). Preparation and hydration of Belite-Ye'elimite-Ternesite clinker based on industrial solid waste. *Case Studies in Construction Materials, 18*, e01922.

Wang, F., Long, G., Bai, M., Wang, J., Shi, Y., Zhou, X., & Zhou, J. L. (2023). A new perspective on Belite-ye'elimite-ferrite cement manufactured from electrolytic manganese residue: Production, properties, and environmental analysis. *Cement and Concrete Research, 163*, 107019.

Wang, X., Guo, M. Z., Yue, G., Li, Q., & Ling, T. C. (2022). Synthesis of high belite sulfoaluminate cement with high volume of mixed solid wastes. *Cement and Concrete Research, 158*, 106845.

Watari, T., Cao, Z., Hata, S., & Nansai, K. (2022). Efficient use of cement and concrete to reduce reliance on supply-side technologies for net-zero emissions. *Nature Communications, 13*(1), 4158.

Wu, K., Shi, H., & Guo, X. (2011a). Preparation of sulphoaluminate cement from waste inclination fly ash and its hydration heat release characteristics. *Journal of the Chinese Ceramics, 39*(04), 670–675.

Wu, K., Shi, H., & Guo, X. (2011b). Utilization of municipal solid waste incineration fly ash for sulfoaluminate cement clinker production. *Waste Management, 31*(9–10), 2001–2008.

Wu, S., Wang, W., Ren, C., Yao, X., Yao, Y., Zhang, Q., & Li, Z. (2019). Calcination of calcium sulphoaluminate cement using flue gas desulfurization gypsum as whole calcium oxide source. *Construction and Building Materials, 228*, 116676.

Wu, S., Yao, X., Ren, C., Yao, Y., & Wang, W. (2020). Recycling phosphogypsum as a sole calcium oxide source in calcium sulfoaluminate cement and its environmental effects. *Journal of Environmental Management, 271*, 110986.

Xue, P., Xu, A., He, D., Yang, Q., Liu, G., Engström, F., & Björkman, B. (2016). Research on the sintering process and characteristics of belite sulphoaluminate cement produced by BOF slag. *Construction and Building Materials, 122*, 567–576.

Yao, Y., Wang, W., Ge, Z., Ren, C., Yao, X., & Wu, S. (2020). Hydration study and characteristic analysis of a sulfoaluminate high-performance cementitious material made with industrial solid wastes. *Cement and Concrete Composites, 112*, 103687.

Zhang, W., Cao, L., Zhi, X., Liu, L., Ren, X., & Ye, J. (2023). Development of mineral hydration activity, properties and clinker preparation of low calcium cement clinker. *Journal of the Chinese Academy of Ceramics, 51*(09), 2465–2477.

Zhao, H., Li, J., & Liu, H. (2006). Phase composition and hydration properties of sulphoaluminate cement clinker prepared from red mud. *Nonferrous Metals, 04*, 119–123.

Chapter 2
Characterization of Aluminosilicate Minerals in Low-Carbon Cementitious Materials

2.1 Introduction

In Inner Mongolia, a significant amount of coal gangue, desulfurization gypsum, and carbide slag are discharged during the operation of coal power and calcium carbide industries. Coal gangue, rich in SiO_2 (40–60%) and Al_2O_3 (15–30%), exhibits chemical composition comparable to bauxite (Zhang et al., 2023; Zhou et al., 2023). Carbide slag primarily consists of $Ca(OH)_2$ (54–72%), demonstrating composition similarity to natural limestone (Ren et al., 2022; Yuan et al., 2017). Desulfurization gypsum, composed mainly of $CaSO_4 \cdot 2H_2O$ (92–95%), shows superior quality to natural gypsum in terms of both chemical composition and physical properties (Caillahua & Moura, 2018; Liu et al., 2021). This study presents a comprehensive methodology for synthesizing low-carbon cementitious materials containing $C_5S_2\hat{S}$, $C_4A_3\hat{S}$, and C_2S using these industrial byproducts as raw materials. The mineral phase evolution was systematically investigated through multi-technique characterization at varying calcination temperatures.

2.2 Raw Materials

The raw materials utilized for low-carbon cementitious materials synthesis comprised coal gangue, carbide slag, and desulfurization gypsum. Coal gangue was sourced from the Inner Mongolia Yitai Coal Co., Ltd. Carbide slag was obtained from a chemical plant in Erdos, Inner Mongolia, China. Desulfurization gypsum originated from a power plant in Hohhot, Inner Mongolia, China. The main chemical compositions of raw materials analyzed by XRF are given in Table 2.1.

It is evident that coal gangue primarily consists of SiO_2 and Al_2O_3. Carbide slag exhibits a high content of CaO while desulfurization gypsum mainly contains CaO

C. Yan et al., *Low-Carbon Cementitious Materials with 100% Solid Wastes*,
https://doi.org/10.1007/978-981-96-8742-8_2

Table 2.1 Chemical compositions of raw materials and the raw mixes employed for the synthesis of low-carbon cementitious materials (wt.%)

Material	CaO	SiO_2	Al_2O_3	SO_3	Fe_2O_3	MgO	Na_2O	K_2O	ZrO_2	TiO_2	Others	LOI	AR
CG	2.32	47.8	44.90	1.31	1.46	0.11	0.05	0.49	0.07	1.36	0.13	31.94	42.57
CS	99.20	0.41	0.14	0.06	0.04	0.02	0.08	0.00	0.00	0.00	0.05	24.10	37.82
DG	41.30	1.35	0.56	54.50	0.31	0.46	0.05	0.08	0.00	0.00	1.36	5.65	19.61

Note CG means coal gangue; CS represents carbide slag; DG stands for desulfurization gypsum; LOI denotes loss on ignition calcined at 1050 °C for 2 h; AR means addition ratio (wt.%)

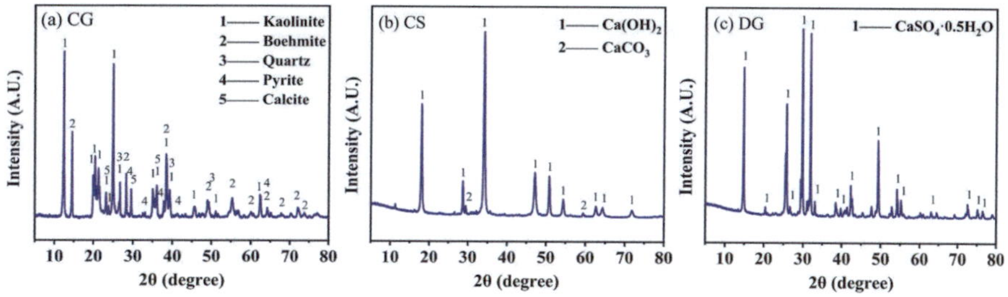

Fig. 2.1 XRD patterns of raw materials. **a** Coal gangue, **b** carbide slag, and **c** desulfurization gypsum

and SO_3. Furthermore, trace components such as MgO, Na_2O, K_2O, and Ti_2O are also present in coal gangue, carbide slag, and desulfurization gypsum.

The mineral compositions of raw materials determined by XRD are shown in Fig. 2.1. The main crystalline phases of coal gangue are kaolinite, boehmite, quartz, pyrite, and calcite. Carbide slag predominantly contains $Ca(OH)_2$ with minor amounts of $CaCO_3$. Desulfurization gypsum is mainly composed of $CaSO_4{\cdot}0.5H_2O$.

2.3 Materials Proportion

In order to ensure the adequate formation of alumino-silicate minerals, the ingredient calculation was based on two fundamental principles. The CaO content was designed to compensate for any insufficiency. The SO_3 content was designed to account for any excess (Majling et al., 1993). The three-modulus method of alkalinity (C_m), aluminum/sulfur ratio (P), and aluminum/silicon ratio (N) were employed to calculate the materials proportion, as shown in Eqs. 2.1 and 2.3 (Wang et al., 1999). Based on previous experiments and relevant literature (Huang et al., 2022; Shen et al., 2015; Wu et al., 2011), it has been determined that the appropriate materials proportions are listed in Table 2.1.

$$Cm = \frac{\omega(\mathrm{CaO}) - 0.70\omega(\mathrm{TiO_2})}{0.73[\omega(\mathrm{Al_2O_3}) - 0.64\omega(\mathrm{Fe_2O_3})] + 1.40\omega(\mathrm{Fe_2O_3}) + 1.87\omega(\mathrm{SiO_2})} \tag{2.1}$$

$$P = \frac{\omega(\mathrm{Al_2O_3}) - 0.64\omega(\mathrm{Fe_2O_3})}{\omega(\mathrm{SO_3})} \tag{2.2}$$

$$N = \frac{\omega(\mathrm{Al_2O_3}) - 0.64\omega(\mathrm{Fe_2O_3})}{\omega(\mathrm{SiO_2})} \tag{2.3}$$

2.4 Cement Clinker Preparation

The preparation of cement clinker using coal gangue, carbide slag, and desulfurization gypsum was conducted in three stages: raw material treatment, mixture preparation, and cement clinker calcination, as elaborated in Fig. 2.2.

(1) The stage of raw materials preparation. The treatment of coal gangue involved the processes of crushing, grinding, and screening to achieve a particle size smaller than 0.074 mm. The carbide slag was dried in an oven at 100 °C. Subsequently, the dried carbide slag underwent grinding and sieving to obtain particles with a size smaller than 0.074 mm. Desulfurization gypsum with an initial particle size below 90 μm can be directly utilized after screening. Physical diagrams of coal gangue, carbide slag, and desulfurization gypsum are presented in Fig. 2.3a–e.

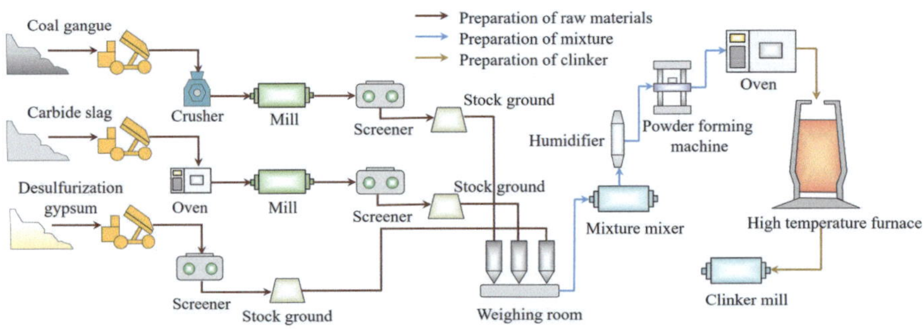

Fig. 2.2 Schematic diagram of preparation for cement clinker

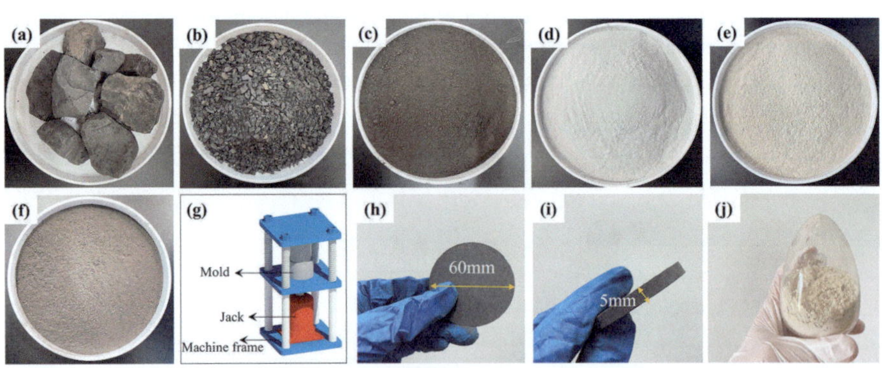

Fig. 2.3 Physical diagram of raw materials, mixture, and cement clinker. **a–c** The coal gangue is available in bulk, granular, and powdered forms, **d** carbide slag, **e** desulfurization gypsum, **f** mixture, **g** powder forming machine, **h–i** cylindrical sample of mixture, and **j** powdered form of cement clinker

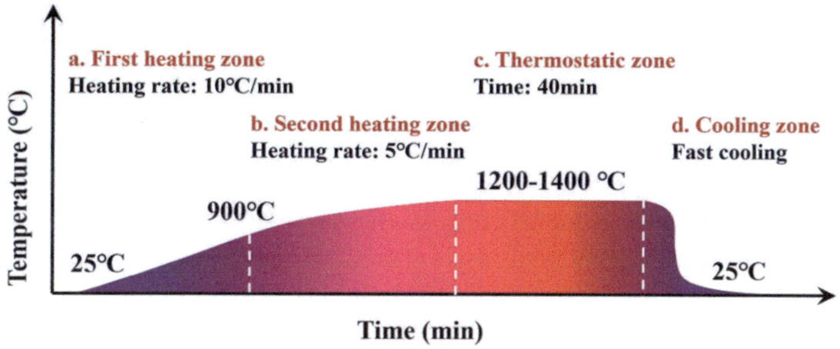

Fig. 2.4 Temperature rise curve for the synthesis of cement clinker

(2) The stage of mixture preparation. The raw materials in predetermined proportion in Table 2.1 were taken and mixed thoroughly. Following complete homogenization, the mixtures containing 10% water were pressed and molded into cylindrical samples with a diameter of 60 mm and a height of 5 mm using a powder forming machine, as depicted in Fig. 2.3h–i. Subsequently, these cylindrical samples were subjected to drying at 105 °C for a duration of 24 h.

(3) The stage of cement clinker calcination. The cylindrical samples of mixture were subjected to firing in a high-temperature furnace. The temperature profile of the calcination process is depicted in Fig. 2.4. It is manifest that the clinker's calcination can be divided into four distinct stages, namely, the initial heating zone, secondary heating zone, thermostatic zone, and the cooling zone. The heating rate of the first and second heating zone were set as 10 and 5 °C/min respectively. In the thermostatic zone, temperatures were maintained at predetermined level (1200, 1250, 1300, 1350, and 1400 °C) for a duration of 40 min. Subsequently, rapid cooling of the clinkers was achieved by exposing them to ambient temperature conditions. Finally, the clinkers underwent crushing, grinding, and sieving processes to obtain particles with size smaller than 0.074 mm, as shown in Fig. 2.3j.

2.5 Mineral Phase Analysis

In the calcination temperature range of 1200–1250 °C (Fig. 2.5a and b), $C_5S_2\hat{S}$, $C_4A_3\hat{S}$, and C_2S coexist within the system. However, at higher temperatures of 1300–1400 °C (Fig. 2.5c–e), $C_4A_3\hat{S}$ and C_2S become predominant species. Quantitative analysis results demonstrate that at temperatures of 1200 and 1250 °C, the content of $C_5S_2\hat{S}$ is determined to be 49.03% and 15.89%, respectively (Figs. 2.5 and 2.6). As the calcination temperature is raised to 1300–1400 °C, the diffraction peak corresponding to $C_5S_2\hat{S}$ becomes undetectable. It is possible that within the temperature range of 1300–1400 °C, $C_5S_2\hat{S}$ undergoes decomposition reaction according to Eq. 2.4, leading to its transformation into C_2S and $CaSO_4$. The cement clinker primarily consists of orthogonal $C_4A_3\hat{S}$ with a minor presence of cubic $C_4A_3\hat{S}$. At a temperature of 1200 °C in the $C_5S_2\hat{S}$-$C_4A_3\hat{S}$-C_2S system, $C_4A_3\hat{S}$ accounts for approximately 20.64%. As the calcination temperature increases, there is a marginal decline observed in the $C_4A_3\hat{S}$ content, stabilizing at around 14.5% within the range of 1300–1400 °C. The content of C_2S changed significantly with the increase of calcination temperature, and the mineral component is only 23.21% at 1200 °C. However, when the calcination temperature rises to 1400 °C, the content of C_2S increases significantly to 85.33%. The diffraction peak of the trace transitional mineral component ($Ca_2Al_2SiO_7$), calcined at 1200 °C, is exclusively detected with a content of only 3.44%. With an increase in the calcination temperature, $Ca_2Al_2SiO_7$ may undergo reactions with $CaSO_4$ and CaO to transform into $C_4A_3\hat{S}$ and C_2S, as illustrated in Eq. 2.5. The $CaSO_4$ content exhibits an initial increase followed by a decrease within the temperature range of 1200−1400 °C. At 1200 °C, the $CaSO_4$ content is measured at 1.89%, which may be attributed to excessive amount of $CaSO_4$ in the mix ratio. As the calcination temperature rises, the $CaSO_4$ content reaches 13.13% at 1250 °C and further increases to 15.68% at 1300 °C. This analysis suggests that the significant presence of $CaSO_4$ in cement clinker could be due to $C_5S_2\hat{S}$ decomposition into C_2S and $CaSO_4$ at this temperature range. However, when the calcination temperature reaches 1350 °C, there is a decrease in $CaSO_4$ content, and at 1400 °C, it becomes difficult to detect any diffraction peak associated with $CaSO_4$ due to its decomposition reaction.

$$4CaO \cdot 2SiO_2 \cdot CaSO_4 \rightarrow 2(2CaO \cdot SiO_2) + CaSO_4 \tag{2.4}$$

$$\begin{aligned} 3(2CaO \cdot Al_2O_3 \cdot SiO_2) + 3CaO + CaSO_4 \rightarrow 3(2CaO \cdot SiO_2) + \\ 3CaO \cdot 3Al_2O_3 \cdot CaSO_4 \end{aligned} \tag{2.5}$$

where: $4CaO{\cdot}2SiO_2{\cdot}CaSO_4$ is $C_5S_2\hat{S}$; $2CaO{\cdot}SiO_2$ is C_2S, $3CaO{\cdot}3Al_2O_3{\cdot}CaSO_4$ is $C_4A_3\hat{S}$.

2.6 Functional Group Analysis

The XRD pattern of cement clinker reveals that the predominant mineral groups at different calcination temperatures can be categorized as $C_5S_2\hat{S}$, $C_4A_3\hat{S}$, and C_2S. The $C_4A_3\hat{S}$ compound contains functional groups of [SO_4] and [AlO_4], while the C_2S compound consists of functional groups of [SiO_4]. The $C_5S_2\hat{S}$ compound exhibits functional groups of both [SO_4] and [SiO_4]. The results from FTIR spectroscopy of the cement clinker at different calcination temperatures are present in Fig. 2.7.

Based on the findings depicted in Fig. 2.7, it can be inferred that the bands in close proximity to 411, 616, 640, 690, 801, 822, and 882 cm^{-1} correspond to the [AlO_4] in $C_4A_3\hat{S}$. Among these absorption bands, the bands at 411 cm^{-1} are attributed to the bending vibrations of the [AlO_4] (Black et al., 2003; Wang & Yang, 1990). The bands

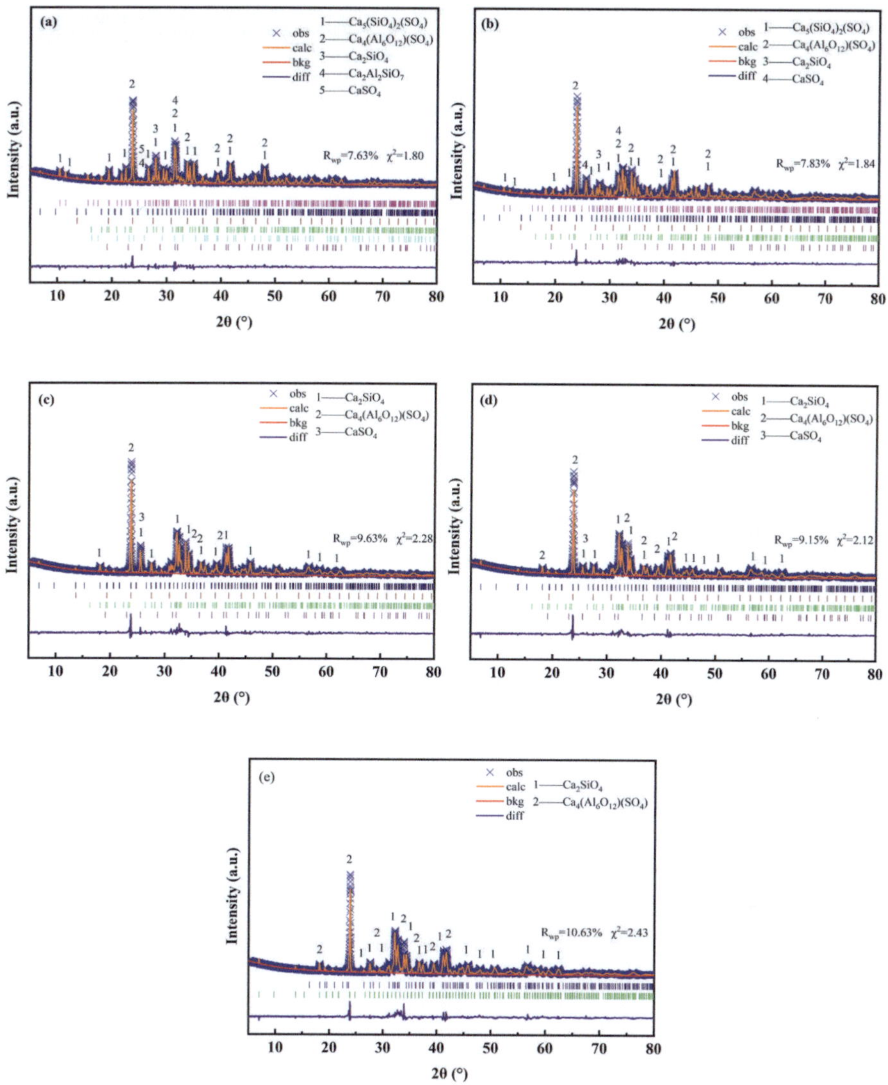

Fig. 2.5 Rietveld refinement of XRD pattern for the cement clinker at different calcination temperatures. **a** 1200 °C, **b** 1250 °C, **c** 1300 °C, **d** 1350 °C, and **e** 1400 °C The blue symbol indicates the observation scan, the origin line denotes the calculated profile, the red line means the background, and the purple line at the bottom shows the difference plot

within the range of 600–700 cm^{-1} represent the vibrational coupling of the [AlO_4] and [SO_4] (Wang & Yang, 1990). The symmetric stretching vibration of [AlO_4] is observed at wavenumbers of approximately 801, 822, and 882 cm^{-1} (HuaJian et al., 2008). The FTIR absorption bands of the [SiO_4] may be assigned near 520, 545, 845, 900, and 994 cm^{-1}. Among them, the bands at 994, 900, and 845 cm^{-1} are attributed

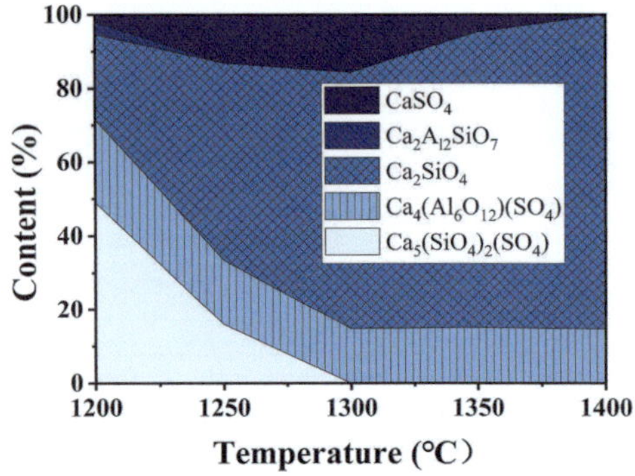

Fig. 2.6 Quantitative results of Rietveld analysis of cement clinker at different calcination temperatures (wt.%)

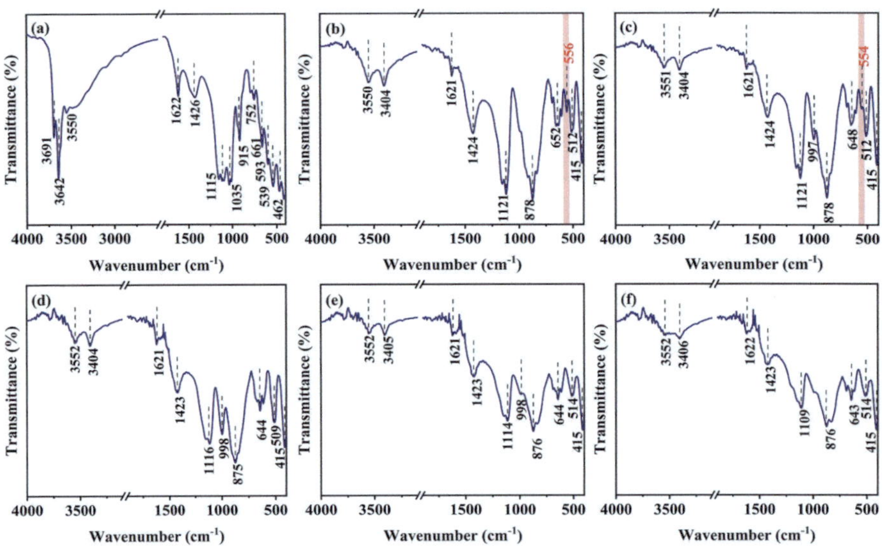

Fig. 2.7 FTIR patterns of cement clinker at different calcination temperatures. **a** Mixture, **b** 1200 °C, **c** 1250 °C, **d** 1300 °C, **e** 1350 °C, **f** 1400 °C

to the asymmetric stretching vibration of the [SiO_4] (Chen et al., 2015). The bands at 545 and 520 cm^{-1} are caused by the bending vibration of the [SiO_4] (Chen et al., 2015). Considering the presence of [SiO_4] in both $C_5S_2\hat{S}$ and C_2S, it is plausible that the aforementioned infrared characteristic wave represents an overlapping group derived from these two compounds. However, meticulous examination reveals a distinct peak at 556 cm^{-1} as highlighted in Fig. 2.7b and c. This particular band diminishes in the FTIR spectrum of the clinker above 1300 °C. Hence, it can be inferred that the sharp peak observed in Fig. 2.7b and c potentially corresponds to the existence of $C_5S_2\hat{S}$. The asymmetric stretching vibration of [SO_4] give rise to peaks at wave numbers around 1098 and 1190 cm^{-1} (Isobe et al., 2003; Okada et al., 1998), potentially corresponding to overlapping peaks attributed to $C_5S_2\hat{S}$ and $C_4A_3\hat{S}$.

2.7 Chemical Binding Energy Analysis

The XPS survey scan spectra of the cement clinker at different calcination temperatures are shown in Fig. 2.8a. The survey scan spectra reveal prominent binding energy peaks corresponding to O 1s, Ca 2s, Ca 2p, C 1s, S 2p, Si 2s, Al 2s, Si 2p, and Al 2p. Additionally, distinct spectral peaks of Ca LMM and O KLL can also be observed.

To further investigate the precise locations of chemical binding energies for Ca, Al, Si, and S elements at various calcination temperatures, high-resolution scan spectra were acquired as illustrated in Fig. 2.8b–e. The photoelectron spectrum of Ca 2p exhibits a characteristic double-peaked structure attributed to the splitting of energy levels caused by spin–orbit coupling. The peak binding energies for Ca $2p_{1/2}$ and Ca $2p_{3/2}$ are approximately 350.85 and 347.29 eV, respectively, as depicted in Fig. 2.8b. Notably, the spectral peaks of Ca 2p remain unchanged following calcination, indicating that the chemical bonding state of Ca atoms in both the mixture and the cement clinker is unaffected by variations in calcination temperature.

The high-resolution scan spectra of Al 2p are depicted in Fig. 2.8c. The photoelectron spectrum of Al 2p exhibits a single-peak structure with a binding energy envelope ranging from 76.27 to 71.44 eV. It is evident from the figure that the peak

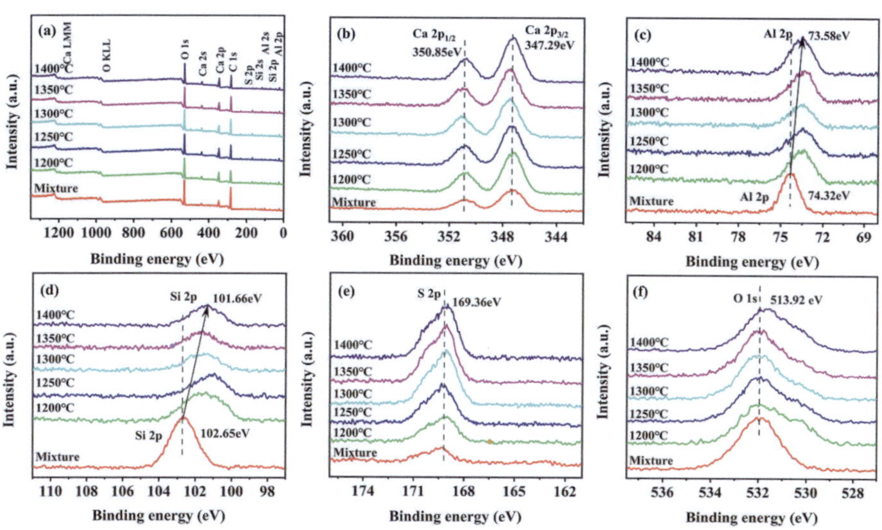

Fig. 2.8 XPS spectrum of the cement clinker at different calcination temperatures **a** survey scan spectra, **b** Ca 2p, **c** Al 2p, **d** Si 2p, **e** S 2p, and **f** O 1s

binding energy of Al 2p undergoes a shift following calcination, transitioning from 74.32 eV in the mixture to 73.58 eV in the cement clinker. The analysis reveals that the presence of Al element in the mixture originates from kaolinite present of the coal gangue. Within the aluminum oxide octahedral structure of kaolinite, each Al atom is coordinated with six O atoms. Previous research has demonstrated that the chemical binding energy of six-coordinated Al ranges between 74.1 and 75.0 eV (Barr et al., 1997; Black et al., 2003; Li et al., 2010; Wagner & Joshi, 1988). After calcination, the mixture undergoes continuous heat absorption and conversion into internal energy, leading to enhanced molecular thermal motion. Alongside the evaporation of adsorbed water due to thermal motion, crystal water and bound water are also released from the mixture, impeding the polymerization of long chains composed of –Si-O- and -Al-O- octahedrons in kaolinite. With further increase in calcination temperature, the structure of kaolinite within the mixture is deconstructed while Al begins to recombine with other atoms to form $C_4A_3\hat{S}$. The crystal structure of $C_4A_3\hat{S}$ reveals that each Al atom coordinates with four O atoms within a single cell. Previous studies have demonstrated that tetragonal-coordinated Al exhibits binding energies ranging between 73.4 and 74.55 eV (Barr et al., 1997; Black et al., 2003; Li et al., 2010; Wagner & Joshi, 1988). Consequently, after calcination, there is a decrease in the chemical binding energy of Al in cement clinker.

Similarly, the photoelectron spectrum of Si 2p exhibits a singular peak structure with the envelope of the peaks ranging from 104.87 to 99.20 eV, as illustrated in Fig. 2.8d. The peak binding energy of Si 2p is reduced from 102.65 eV in the mixture to 101.57 eV in the cement clinker. The analysis reveals that the Si-containing minerals in the mixture originate from kaolinite of coal gangue, which possesses a 1:1 type layered silicate structure. After calcination at temperatures ranging from 1200 to 1250 °C, kaolinite undergoes thermal decomposition and transforms into isolated $C_5S_2\hat{S}$ and C_2S through solid-phase reactions with other elements at elevated temperatures. With the increase of calcination temperature, when it exceeds 1300 °C, the island $C_5S_2\hat{S}$ are decomposed into belite, so the Si element only exists in the belite belonging to the island structure. Some studies have shown that the Si 2p chemical binding energy in island, ring, chain, layered and shelf silicate minerals increases from 101.3 to 103.4 eV (Okada et al., 1998; Wu et al., 2018). Therefore, when the Si-containing elements in the mixture are transformed from kaolinite to $C_5S_2\hat{S}$ and C_2S, the chemical binding energy is reduced.

The high-resolution scanning spectra of S 2p and O 1 s are depicted in Fig. 2.8e–f. It is evident that the photoelectron spectra of S 2p and O 1s exhibit a unimodal structure, characterized by peak binding energies of 169.36 and 513.92 eV, respectively.

2.8 Chemical Shift Analysis

The NMR spectra of cement clinkers at various calcination temperatures are presented in Fig. 2.9. As depicted in Fig. 2.9a prior to calcination, the ^{27}Al NMR spectra exhibit a prominent broad signal peak around the chemical shift of 2.45 ppm. Through calcination at temperatures ranging from 1200 °C to 1400 °C, the NMR spectra of cement clinker exhibited a noticeable shift, with its chemical shift moving towards approximately 68.59 ppm. The chemical shift has been demonstrated in related studies to be associated with the coordination environment of Al in silicoaluminate minerals. Generally, Al exists in three coordination states: $[AlO_4]$, $[AlO_5]$, and $[AlO_6]$. The chemical shifts of $[AlO_4]$, $[AlO_5]$, and $[AlO_6]$ correspond to signal positions at 50–100, 20–30, and 0–20 ppm respectively (Isobe et al., 2003). In this study, kaolinite is the main Al-containing phase in mixture, and its aluminum oxide polyhedron mainly exists in the form of $[AlO_6]$. After calcination at temperatures ranging from 1200 °C to 1400 °C, the predominant mineral component in cement clinker containing Al is $C_4A_3\hat{S}$. Additionally, the aluminum oxide polyhedron predominantly exists in the form of $[AlO_4]$. Therefore, the Al element changes from the $[AlO_6]$ in the mixture to the $[AlO_4]$ in the cement clinker, which is reflected in the increase of chemical shift in the ^{27}Al NMR spectrogram.

The ^{29}Si NMR spectrum in Fig. 2.9b exhibits a distinct peak positioned within the chemical shift range of −87.46 to −94.15 ppm. Upon calcination at temperatures ranging from 1200 to 1400 °C, the chemical shift of the cement clinker undergoes alteration, resulting in a peak shift towards −72.91 ppm. Previous studies have demonstrated that the polymerization degree of silica tetrahedron can be reflected by chemical shift. Generally, there exist five polymerization degrees of silica tetrahedron in silicate mineral components, namely Q_0, Q_1, Q_2, Q_3 and Q_4. Q_0 represents the isolated SiO^{4-} in monosilicate with a corresponding chemical shift ranging from −68 to −76 ppm. Low polymerization degrees of silica tetrahedron are denoted as Q_1 and Q_2 with a corresponding chemical shift ranging from −76 to −82 ppm. Q_1 encompasses the siloxane group with a terminal moiety, while Q_2 corresponds to

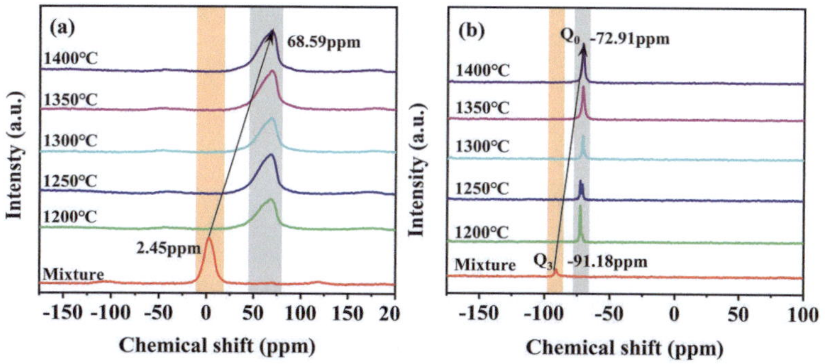

Fig. 2.9 NMR spectrums of the cement clinker at different calcination temperatures **a** ^{27}Al and **b** ^{29}Si

an intermediate group of the chain structure. High polymerization degrees of silica tetrahedron are denoted as Q_3 and Q_4 with a corresponding chemical shift ranging from −88 to −120 ppm. Q_3 denotes groups located at branch chains or layer positions, whereas Q_4 refers to groups possessing a three-dimensional skeleton structure. The polymerization degree of Si in the layered structure of kaolinite in the mixture is predominantly Q_3 or Q_4, as revealed by this study. Upon calcination at temperatures ranging from 1200 to 1400 °C, the polymerization degree of Si in C_2S and $C_4A_3\hat{S}$ within the cement clinker transforms into a Q_0 structure. Consequently, there is a discernible increase in chemical shift observed in the ^{29}Si NMR spectrogram, indicating a transition from Q_3 to Q_0 structure during the formation of the cement clinker.

The above analysis reveals that both C_2S and $C_5S_2\hat{S}$ are Si-containing mineral components in cement clinker when calcined at 1200 and 1250 °C. Consequently, the absorption bands of ^{29}Si NMR spectra under the calcination conditions of 1200 and 1250 °C arise from the overlapping resonance absorption signals originating from C_2S and $C_5S_2\hat{S}$. Relevant literature suggests that C_2S and $C_5S_2\hat{S}$ exhibit peak chemical shifts at approximately −71.4 and −72.6 ppm, respectively (Grimmer et al., 1985; Huang, 2018). Therefore, the evolution process of mineral components of C_2S and $C_5S_2\hat{S}$ can be further assessed by performing peak fitting analysis on the ^{29}Si NMR spectrogram using PeakFit software, as depicted in Fig. 2.10.

The results demonstrate that the ^{29}Si NMR spectrum of the mixture exhibits a single peak structure, with the chemical shift of the peak centered around −91.18 ppm. Upon calcination at 1200 °C, the ^{29}Si NMR spectrum of the cement

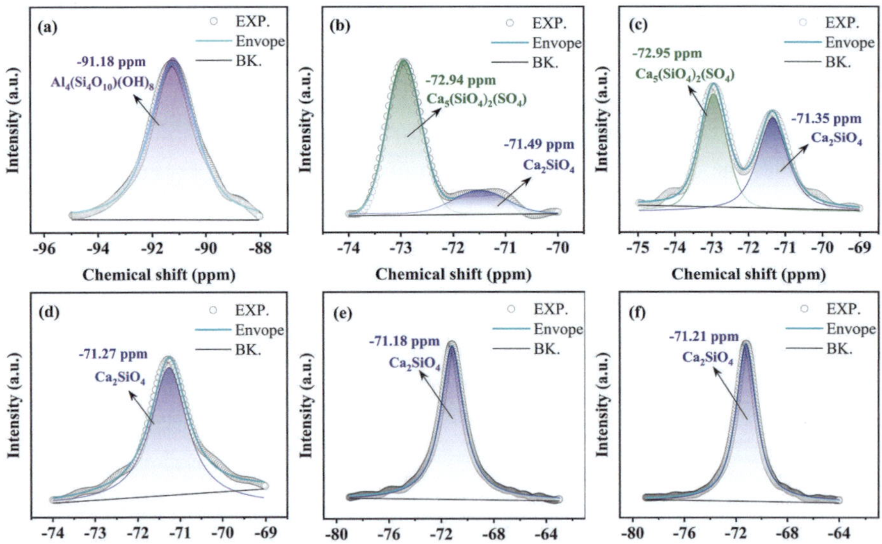

Fig. 2.10 The fitting result of ^{29}Si NMR spectra at different calcination temperatures **a** mixture, **b** 1200 °C, **c** 1250 °C, **d** 1300 °C, **e** 1350 °C, **f** 1400 °C

clinker displays a double peak structure, with chemical shifts assigned to peak 1 and peak 2 at approximately −72.94 and −71.49 ppm, respectively. Notably, the area encompassed by peak 1 is significantly larger than that of peak 2. Similarly, when subjected to a calcination temperature of 1250 °C, the cement clinker's ^{29}Si NMR curve also manifests as a double peak structure, with chemical shifts corresponding to peak 1 and peak 2 observed at approximately −72.95 and −71.35 ppm, respectively. However, it is worth mentioning that at this calcination temperature, the area surrounded by peak 2 is notably larger compared to that observed at 1200 °C. Furthermore, when the calcination temperature is within the range of 1300–1400 °C, the ^{29}Si NMR spectrum of cement clinker exhibits a singular peak structure, with chemical shifts observed at approximately −71.27, −71.18, and −71.21 ppm respectively. It can be seen that the peak fitting results of the ^{29}Si NMR spectrogram demonstrate a more comprehensive depiction of the evolution process of mineral components in C_2S and $C_5S_2\hat{S}$ at various calcination temperatures. Notably, the obtained peak fitting results from Peakfit software align consistently with the mineral component derived through Rietveld refinement method.

2.9 Micro Morphology Analysis

The surface morphology, grain size, and grain stacking of cement clinker were observed under different calcination temperatures, as depicted in Fig. 2.11. Consistent with previous investigations (Londono-Zuluaga et al., 2017; Shen et ål., 2021), $C_5S_2\hat{S}$ predominantly exhibited angular granular or plate shapes; $C_4A_3\hat{S}$ appeared as hexagonal plates with a grain size ranging between 1 and 2 μm; C_2S manifested as large ellipsoids or pebbles with a grain size below 10 μm. Based on the aforementioned research findings, an analysis was conducted to elucidate the evolutionary patterns of mineral component micro-morphology in cement clinker concerning calcination temperature.

The microstructure of cement clinker at different magnification ratios when the calcination temperature is 1200 °C is illustrated in Fig. 2.11a–c. As depicted in Fig. 2.11a, grains with diverse morphologies and sizes aggregate into a cluster-like structure, characterized by noticeable gaps between the clusters. Consequently, the cement clinker prepared under this specific calcination temperature exhibits favorable grindability. Upon further magnification, as illustrated in Fig. 2.11b–c, three distinct types of grains are observed within the cement clinker: ovoid grains measuring less than 0.5 μm, hexagonal lamellar grains ranging from 0.5 to 1 μm, and ellipsoidal grains measuring between 1 and 2 μm.

The microstructure of cement clinker at 1250 °C is depicted in Fig. 2.11d–f at various magnification ratios. The grains in the cement clinker are observed to be distributed in clusters with significant interstitial spaces, indicating favorable grindability. Comparing Fig. 2.11a–c with the cement clinker calcined at 1200 °C, it can be noted that the grain boundary becomes more distinct and crystal development is enhanced at 1250 °C. Furthermore, there is a notable reduction in ovoid grains

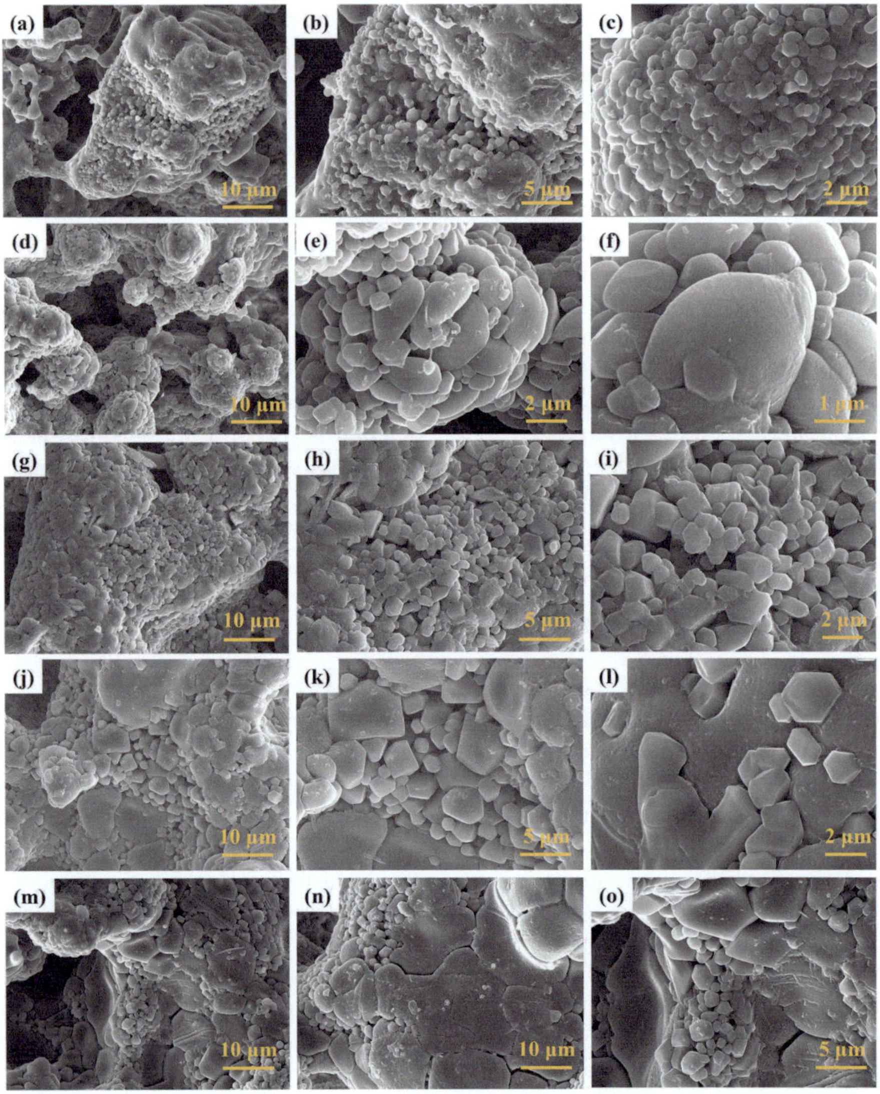

Fig. 2.11 SEM micrographs of the cement clinker at different calcination temperatures **a–c** 1200 °C, **d–f** 1250 °C, **g–i** 1300 °C, **j–l** 1350 °C, **m–o** 1400 °C

while ellipsoidal grains exhibit a substantial increase both in terms of quantity and size (3–4 μm).

The microstructure of cement clinker at 1300 °C is depicted in Fig. 2.11g–i. At this calcination temperature, various types of grains are interconnected to form blocks, with noticeably reduced interstitial gaps. Consequently, compared to the cement clinker at calcination temperatures of 1200 and 1250 °C, the grindability of the cement clinker produced under these conditions is diminished. Furthermore, only ellipsoidal and hexagon plate-like grains are observed at 1300 °C.

The microstructure of cement clinker at a calcination temperature of 1350 °C is presented in Fig. 2.11j–l. At this temperature, various types of grains exhibit close interconnections with well-defined grain boundaries and distinct edges and corners. In comparison to the cement clinker at 1300 °C, the ellipsoidal grains increase in size from 3 to 4 μm to over 10 μm, while the hexagonal flake grains enlarge from 1 to 2 μm to more than 4 μm.

The microstructure of cement clinker at 1400 °C is depicted in Fig. 2.11m–o. At this temperature, the grain boundary between particles appears blurred due to partial melting. Furthermore, the average grain size among ellipsoidal grains increases to a range of 15–20 μm.

The EDS element mapping of the cement clinker at different calcination temperatures is illustrated in Fig. 2.12. The scan area for EDS mapping is depicted in Fig. 2.12a, c, e, and g. The distribution of Ca, Si, Al, and S elements acquired in the EDS element mapping scan area is presented in Fig. 2.12b, d, f, h, and i. From Fig. 2.12b, it can be observed that Ca, Si, Al, and S elements are uniformly distributed within the EDS mapping scan area at 1200 °C. Combining these findings with XRD results reveals that $C_4A_3\hat{S}$, $C_5S_2\hat{S}$, and C_2S are the main minerals present in the cement clinker at 1200 °C. It can be inferred that three distinct grain sizes shown in Fig. 2.12a correspond to stacked layers of $C_4A_3\hat{S}$, $C_5S_2\hat{S}$, and C_2S. From Fig. 2.12d, f, h, to i, the distribution results of Ca, Si, Al, and S elements further support the conclusion that hexagonal plate grains correspond to $C_4A_3\hat{S}$ while ellipsoidal grains represent C_2S.

2.10 Conclusions

Coal gangue, carbide slag, and desulfurization gypsum were utilized as raw materials to completely substitute non-renewable resources for the preparation of low-carbon cementitious materials containing $C_4A_3\hat{S}$, $C_5S_2\hat{S}$, and C_2S. The evolution of phase types, functional groups, chemical binding energy, chemical shift, grain morphology, and grain size of the cement clinker at different calcination temperatures was characterized using XRD, FTIR, XPS, SS-NMR and SEM-EDS techniques. Based on the experimental results, the following major conclusions can be drawn:

(1) When calcined at 1200–1250 °C, the cement clinker undergoes a coexistence phenomenon involving three mineral components: $C_4A_3\hat{S}$, $C_5S_2\hat{S}$, and C_2S. The proportion of $C_5S_2\hat{S}$ decreases from 49.03% at 1200 °C to 15.89% at 1250 °C, while the content of $C_4A_3\hat{S}$ varies between 16.33% and 20.64%. Simultaneously, there is an increase in the C_2S content from 23.21% at 1200 °C to reach a value of 53.48% at the higher temperature of 1250 °C. At calcination temperatures ranging from 1300 to 1400 °C, the predominant mineral components of cement

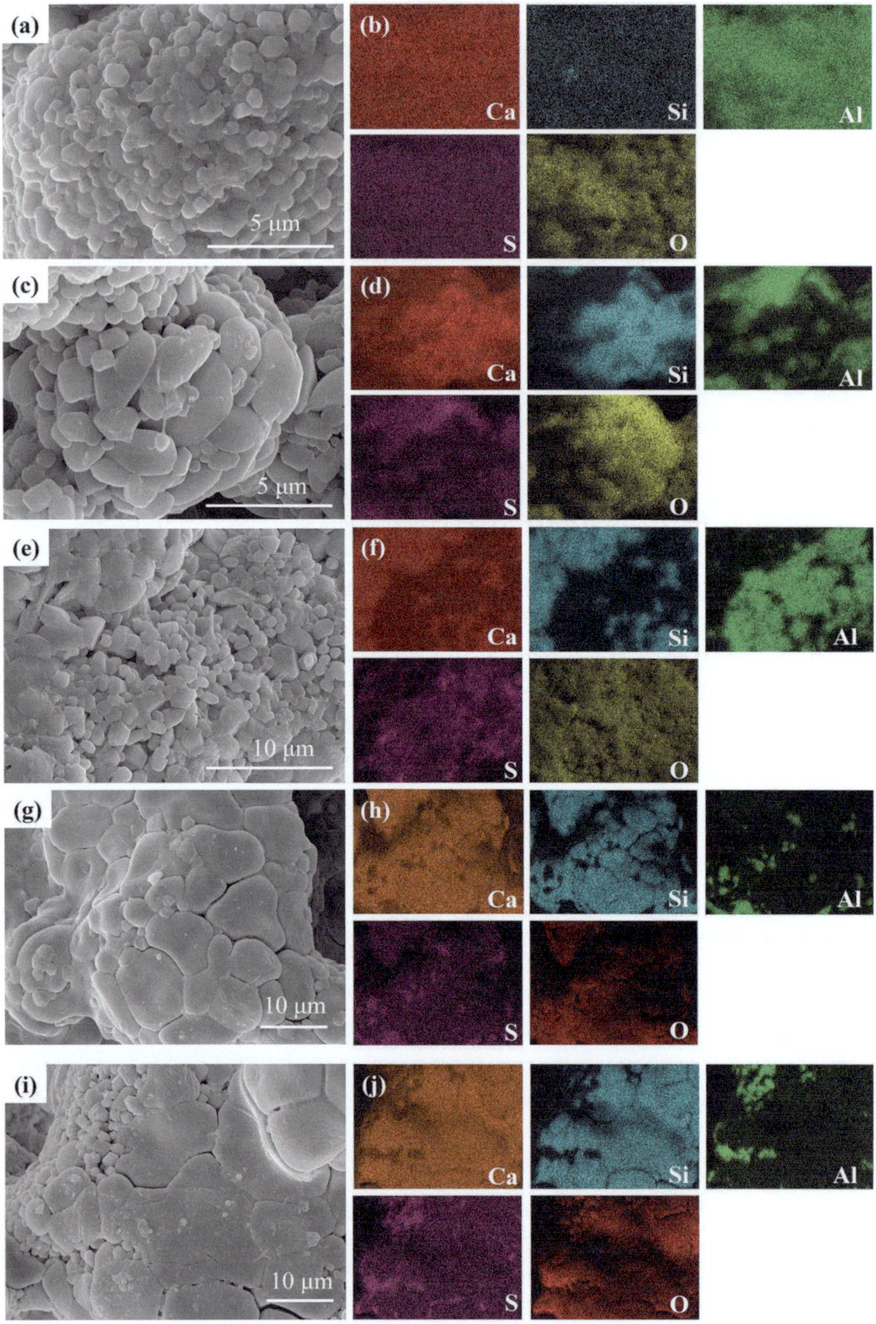

Fig. 2.12 EDS element mapping of cement clinker at different calcination temperatures **a–c** 1200 °C, **d–f** 1250 °C; **g–i** 1300 °C, **j–l** 1350 °C, **m–o** 1400 °C

clinker are C_2S and $C_4A_3\hat{S}$, with belite constituting as much as 85.33% at 1400 °C.

(2) After calcination, the chemical binding energy of the Al 2p and the chemical shift of ^{27}Al both exhibit a significant decrease, indicating that kaolinite undergoes a transformation from a high binding energy state with $[AlO_6]$ structure to $C_4A_3\hat{S}$ with $[AlO_4]$ structure. Similarly, the chemical binding energy of Si 2p and the chemical shift of ^{29}Si also decreased after calcination, suggesting a conversion from kaolinite with Q_3 polymerization degree to $C_5S_2\hat{S}$ and C_2S with Q_0 polymerization degree. Furthermore, the fitting analysis of ^{29}Si revealed valuable insights into the evolution pattern of $C_5S_2\hat{S}$ and C_2S content in cement clinker at different calcination temperatures.

(3) At calcination temperatures of 1200 and 1250 °C, the cement clinker predominantly comprises spherical $C_4A_3\hat{S}$, hexagonal $C_5S_2\hat{S}$, and ellipsoidal C_2S with a fine grain size. However, when the calcination temperature is increased to 1300–1400 °C, the grain size of both $C_4A_3\hat{S}$ and C_2S in the cement clinker increases proportionally.

References

Barr, T. L., Seal, S., Wozniak, K., & Klinowski, J. (1997). ESCA studies of the coordination state of aluminium in oxide environments. *Journal of the Chemical Society, Faraday Transactions, 93*(1), 181–186.

Black, L., Garbev, K., Stemmermann, P., Hallam, K. R., & Allen, G. C. (2003). Characterisation of crystalline C-S-H phases by X-ray photoelectron spectroscopy. *Cement and Concrete Research, 33*(6), 899–911.

Caillahua, M. C., & Moura, F. J. (2018). Technical feasibility for use of FGD gypsum as an additive setting time retarder for Portland cement. *Journal of Materials Research and Technology, 7*(2), 190–197.

Chen, L., Wang, X., Jia, Z., Luo, Y., & Jia, D. (2015). Use of precipitated silica with silanol groups as an inorganic chain extender in polyurethane. *Materials & Design, 87*, 324–330.

Grimmer, A. R., von Lampe, F., Mägi, M., & Lippmaa, E. (1985). High-resolution solid-state ^{29}Si NMR of polymorphs of Ca_2SiO_4. *Cement and Concrete Research, 15*(3), 467–473.

HuaJian, L., HengHu, S., Xuchu, T., & XueJun, X. (2008). A new method to evaluate the hydraulic activity of Al-Si materials. *Science in China Series E Technological Sciences, 51*(2), 113–120.

Huang, Y. (2018) Utilization of phosphogypsum for the preparation of ternesite-calcium sulfoaluminate cement. Chongqing University.

Huang, Y., Dong, D., Wang, X., Zhang, Z., Zhao, P., Cui, N., & Lu, L. (2022). Ternesite-calcium sulfoaluminate cement: Preparation and hydration. *Construction and Building Materials, 344*, 128187.

Isobe, T., Watanabe, T., de la Caillerie, J. B. D., Legrand, A. P., & Massiot, D. (2003). Solid-state ^{1}H and ^{27}Al NMR studies of amorphous aluminum hydroxides. *Journal of Colloid and Interface Science, 261*(2), 320–324.

Li, C., Wan, J., Sun, H., & Li, L. (2010). Investigation on the activation of coal gangue by a new compound method. *Journal of Hazardous Materials, 179*(1–3), 515–520.

Liu, S., Liu, W., Jiao, F., Qin, W., & Yang, C. (2021). Production and resource utilization of flue gas desulfurized gypsum in China—A review. *Environmental Pollution, 288*, 117799.

Londono-Zuluaga, D., Tobón, J. I., Aranda, M. A. G., Santacruz, I., & De la Torre, A. G. (2017). Clinkering and hydration of belite-alite-yeĪelimite cement. *Cement and Concrete Composites, 80*, 333–341.

Majling, J., Sahu, S., Vlna, M., & Roy, D. M. (1993). Relationship between raw mixture and mineralogical composition of sulphoaluminate belite clinkers in the system $CaO-SiO_2-Al_2O_3-Fe_2O_3-SO_3$. *Cement and Concrete Research, 23*(6), 1351–1356.

Okada, K., Kameshima, Y., & Yasumori, A. (1998). Chemical shifts of silicon X-ray photoelectron spectra by polymerization structures of silicates. *Journal of the American Ceramic Society, 81*(7), 1970–1972.

Ren, Y., Wang, Z., Qu, G., Kuang, L., Ning, P., Ren, N., Zhang, C., Chen, X., Wu, F., Hu, Y., Wang, J., Lu, P., & Tong, J. (2022). Comprehensive performance study of aluminum ash and calcium carbide slag for brick making under ultra-high pressure. *Construction and Building Materials, 359*, 129526.

Shen, Y., Qian, J., Huang, Y., & Yang, D. (2015). Synthesis of belite sulfoaluminate-ternesite cements with phosphogypsum. *Cement and Concrete Composites, 63*, 67–75.

Shen, Y., Wang, P., Chen, X., Zhang, W., & Qian, J. (2021). Synthesis, characterisation and hydration of ternesite. *Construction and Building Materials, 270*, 121392.

Wagner, C. D., & Joshi, A. (1988). The auger parameter, its utility and advantages: A review. *Journal of Electron Spectroscopy and Related Phenomena, 47*, 283–313.

Wang, D. F., & Yang, Z. X. (1990). Auger parameters and configurations of silicon in silicate minerals. *Journal of the Chinese Ceramic Society, 18*, 282–287.

Wang, Y. M., Su, M. Z., & Zhang, L. (1999). *Sulphoaluminate cement*. Beijing Industry University Press.

Wu, C.-L., Wang, B.-B., Tao, R., Fang, L.-W., & Li, H.-X. (2018). Study of mineral structure transformation of coal ash with high ash melting temperature by XPS. *Spectroscopy and Spectral Analysis., 38*(7), 2296–2301.

Wu, K., Shi, H., & Guo, X. (2011). Utilization of municipal solid waste incineration fly ash for sulfoaluminate cement clinker production. *Waste Management, 31*(9), 2001–2008.

Yuan, B., Straub, C., Segers, S., Yu, Q. L., & Brouwers, H. J. H. (2017). Sodium carbonate activated slag as cement replacement in autoclaved aerated concrete. *Ceramics International, 43*(8), 6039–6047.

Zhang, K., Liu, L., Liu, L., Yang, Y., & Liu, Q. (2023). Iron removal from kaolinitic coal gangue via magnetic separation after oxidizing calcination with the crystal structure of kaolinite protected. *Materials Today Communications, 37*, 107175.

Zhou, A., Wang, X., Yu, S., Deng, S., Tan, H., & Mikulčić, H. (2023). Process design and optimization on self-sustaining pyrolysis and carbonization of municipal sewage sludge. *Waste Management, 159*, 125–133.

Chapter 3
Hydration Characteristics of Low-Carbon Cementitious Material

3.1 Introduction

A low-carbon cementitious material (LCCM) containing $C_4A_3\check{S}$, C_4AF, and C_2S was designed and synthesized using coal gangue, carbide slag, steel slag as raw materials. The mineral composition of the clinker was optimized, and the calcination system (calcination temperature and holding time) was systematically investigated to determine the optimal parameters. The phase evolution, f-CaO content, and microstructural characteristics of the clinker under different calcination conditions were analyzed. Furthermore, the hydration characteristics of the LCCM was comprehensively characterized through hydration heat analysis, chemically bound water measurement, X-ray diffraction (XRD), Fourier transform infrared spectroscopy (FTIR), and scanning electron microscopy (SEM). The results revealed the hydration product formation, molecular structure evolution, and microstructural development during the hydration process.

3.2 Raw Materials and Test Methods

3.2.1 Raw Materials

The raw materials used for preparing low-carbon cementitious material (LCCM) include coal gangue, carbide slag, steel slag, and desulfurization gypsum. The coal gangue was sourced from the Ordos mining area in Inner Mongolia. The carbide slag was obtained from industrial park of Ulanqab City, Inner Mongolia. The steel slag was provided by Baotou Iron and Steel Group. The desulfurization gypsum was collected from a thermal power plant in the Jungar region. The admixtures used were $CaSO_4 \cdot 2H_2O$ and $CaCO_3$, both supplied by Shandong Yousuo Chemical Technology Co., Ltd. For the control experiment, the cement used was P.O. 42.5R cement.

C. Yan et al., *Low-Carbon Cementitious Materials with 100% Solid Wastes*,
https://doi.org/10.1007/978-981-96-8742-8_3

Table 3.1 Chemical composition of raw materials (wt. %)

Materials	Al_2O_3	SiO_2	CaO	Fe_2O_3	SO_3	CO_2	TiO_2	MgO	Cr_2O_3	MnO	Loss
CG	38.84	42.75	0.09	0.11	0.27	/	0.70	0.03	/	/	17.01
CS	0.22	0.63	74.96	0.05	/	/	/	/	/	/	23.97
SS	7.11	16.93	33.76	26.30	0.83	/	0.55	3.74	2.18	2.55	4.37
DG	0.31	7.62	25.62		47.62	/	/	/	/	/	14.62
$CaSO_4 \cdot 2H_2O$			32.49	/	46.42	/	/	/	/	/	20.89
$CaCO_3$	/	/	54.98	/	/	43.17	/	/	/	/	/

Note CG: coal gangue, CS: carbide slag, SS: steel slag, DG: desulfurization gypsum, Loss: ignition loss

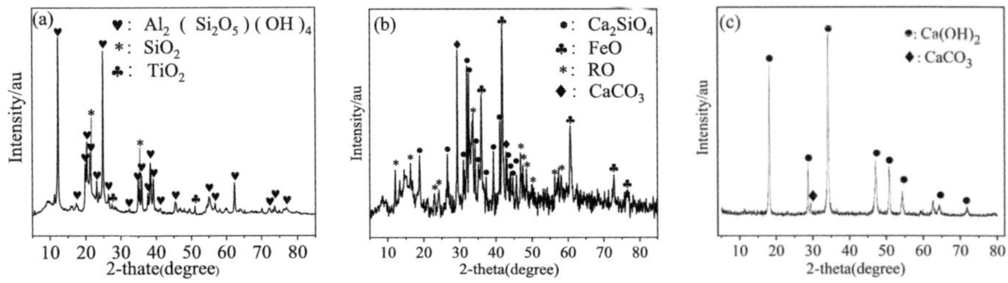

Fig. 3.1 XRD spectra of raw materials. **a** Coal gangue, **b** steel slag, **c** carbide slag

The chemical and mineral composition of the raw materials was presented in Table 3.1 and Fig. 3.1. Coal gangue is primarily composed of kaolinite, with its main chemical constituents being SiO_2 and Al_2O_3, accounting for approximately 80% of the total composition. Carbide slag consists mainly of $Ca(OH)_2$, with CaO content reaching up to 75%. Desulfurization gypsum is dominated by SO_3, which constitutes 47.3% of its composition. Steel slag contains dicalcium silicate and iron phases as its principal mineral components, with Fe_2O_3 content measured at 26.3%. Additionally, $CaCO_3$ was incorporated to regulate the hydration reaction of the low-carbon cementitious material.

3.2.2 Treatment of Raw Materials

Coal gangue

The coal gangue used in this study was obtained from the Jungar mining area. The pretreatment process consisted of the following steps: First, the raw coal gangue was crushed into 5–10 mm particles using a jaw crusher. Subsequently, the crushed material was finely ground in a cement mill and sieved through a 200-mesh vibrating screen to ensure a sieve residue of less than 3%. The powdered coal gangue was then dried at 105 °C for 6 h in a blast drying oven. Finally, the dried product was sealed and stored for further experimentation.

Carbide slag

The carbide slag, obtained from the Ulanqab region, was pretreated through the following procedure: First, the raw material was dried at 105 °C for 6 h in a blast drying oven to eliminate moisture. The dried carbide slag was then passed through a 200-mesh sieve, maintaining a sieve residue below 3%. Finally, the processed material was sealed and stored for subsequent use.

Steel slag

The steel slag, obtained from the Baotou region, was pretreated through a systematic process. Initially, the raw material was crushed to 5–10 mm particles using a jaw crusher. The crushed slag was then subjected to fine grinding in a cement mill. Following grinding, the material was passed through a 200-mesh vibrating screen to achieve a sieve residue below 3%. Subsequently, the processed slag was dried at 105 °C for 6 h in a blast drying oven. The final product was sealed in airtight containers for storage and further experimentation.

Desulfurization gypsum

The desulfurization gypsum used in this study was obtained from the Jungar region. The pretreatment process consisted of three key steps: (1) moisture removal through drying at 45 °C for 6 h in a blast drying oven to eliminate free water content; (2) particle size control by sieving through a 200-mesh screen with less than 3% residue; and (3) proper storage of the processed material in sealed containers to maintain sample integrity for subsequent experimental procedures.

3.2.3 Mineral Composition Design and Preparation Process

The low-carbon cementitious material (LCCM) was designed based on the following mineralogical principles:

Phase transformation: Silica from coal gangue was designed to form C_2S, while alumina was converted into $C_4A_3\check{S}$.

Burnability enhancement: Steel slag addition (2–5 wt.%) was incorporated to improve raw meal reactivity, where the iron oxide component reacted with Al_2O_3 and CaO to form C_4AF.

Mineral system: The final clinker composition was designed to primarily consist of $C_4A_3\check{S}$ (35–45%), C_4AF (15–25%), and C_2S (30–40%), forming a ternary mineral system.

The raw material proportions were calculated using the modified Bogue method (Zhang & Guolong, 2011), with detailed calculation procedures provided in Eqs. 3.1–3.5.

To ensure that the Fe_2O_3, Al_2O_3, and CaO in the raw materials fully react and form the target mineral phases, the basicity coefficient (C_m) was set to 1.01. Additionally, to ensure that the Al_2O_3 in the raw materials fully reacts with $CaSO_4 \cdot 2H_2O$ to form $C_4A_3\check{S}$, an extra 5% of desulfurization gypsum was added. The appropriate addition of Fe_2O_3 not only effectively improves the burnability of the LCCM but also enhances the activity of $C_4A_3\check{S}$. However, excessive addition of Fe_2O_3 increases the content of less reactive C_4AF while reducing the content of more reactive $C_4A_3\check{S}$, thereby weakening the overall activity of the cement. Therefore, the dosage of steel slag must be strictly controlled.

The clinker's Al_2O_3 and Fe_2O_3 components were primarily derived from coal gangue and steel slag, respectively. This material sourcing strategy enabled precise control over the $C_4A_3\check{S}/C_4AF$ ratio through systematic adjustment of the coal gangue-to-steel slag proportion. Four distinct mix ratios were investigated in this study: 1:0.14, 1:0.20, 1:0.26, and 1:0.32 (coal gangue: steel slag). The CaO was sourced from carbide slag, while the $CaSO_4 \cdot 2H_2O$ was derived from desulfurization gypsum. The specific raw material ratios and oxide contents are shown in Table 3.2.

$$W(CaO) = 0.6512W(C_2S) + 0.3762W\left(C_4A_3\check{S}\right) + 0.4616W(C_4AF) \tag{3.1}$$

$$W(SiO_2) = 0.3488W(C_2S) \tag{3.2}$$

$$W(Al_2O_3) = 0.5106W(C_4A_3\check{S}) + 0.2098W(C_4AF) \tag{3.3}$$

$$W(Fe_2O_3) = 0.33W(C_4AF) \tag{3.4}$$

$$W(SO_3) = 0.1311W(C_4A_3\check{S}) \tag{3.5}$$

The LCCM was prepared through a three-stage process:

1. **Sample Preparation**

   - Pre-treated raw materials (coal gangue, steel slag, carbide slag, and desulfurization gypsum) were homogenized according to the predetermined ratio.

Table 3.2 Ratio of raw materials and chemical components (wt.%)

NO	Chemical components					Raw material			
	Al_2O_3	SiO_2	CaO	Fe_2O_3	SO_3	CG	SS	CS	DG
LCCM-A	17.39	19.7	53.89	1.66	7.06	34.26	4.8	49.03	11.91
LCCM-B	17.06	19.55	53.94	2.28	6.87	33.29	6.66	48.45	11.6
LCCM-C	16.75	19.41	53.99	2.86	6.69	32.38	8.42	47.9	11.31
LCCM-D	16.46	19.28	54.04	3.41	6.52	31.51	10.08	47.37	11.04

- 10% water was added to improve workability and facilitate molding.
- The moist mixture was compacted using a tablet press to form cylindrical specimens (Ø60 mm × 5 mm).
- Specimens were oven-dried at 105 °C for 6 h to remove moisture.

2. **Controlled Calcination**
   - Dried specimens were loaded into a high-temperature electric furnace.
   - Temperature was ramped at 5 °C/min to the target calcination temperature.
   - Isothermal holding was maintained for the specified duration.
   - Calcined samples were immediately air-quenched to room temperature.

3. **Final Product Processing**
   - The clinker was blended with $CaSO_4 \cdot 2H_2O$ and $CaCO_3$ (87:8:5 mass ratio).
   - The mixture was ground and passed through a 200-mesh sieve.
   - Final product specifications: <3% sieve residue (square-hole sieve).

3.2.4 Test Method

f-CaO content

The content of f-CaO in clinker was quantitatively analyzed using the standardized glycol extraction method, with the calculation procedure detailed in Eq. 3.6. The experimental protocol consisted of the following steps:

1. **Sample Preparation**
   - A representative clinker sample was finely ground and homogenized.
   - Precisely weighed the clinker were combined with ethylene glycol extraction solution.

2. **Thermal Extraction**
   - The mixture was immediately transferred to a Ca-5 Free Calcium Oxide Analyzer.
   - The extraction process was carried out at 90 ± 2 °C for 5.0 ± 0.1 min with constant magnetic stirring.

3. **Titration Procedure**
   - The hot extract was titrated against benzoic acid–ethanol standard solution.
   - The titration endpoint was determined by complete decolorization of the phenolphthalein indicator.
   - The extraction-titration cycle was repeated until reproducible endpoints were obtained (typically 2–3 repetitions)

$$\omega_{\text{fCaO}} = \frac{T''_{\text{CaO}} \times V_{33}}{m_{49} \times 1000} \times 100 \tag{3.6}$$

where: ω_{fCaO} is mass fraction of f-CaO, %; T''_{CaO} is the titration value of CaO by benzoic acid solution, mg/ml; V_{33} is volume of benzoic acid solution consumed during the titration of sample, ml; m_{49} is the mass of the sample, mg.

Water requirement of normal consistency and setting time

The water requirement of normal consistency and setting time of LCCM were tested in accordance with GB/T 1346-2011.

Compressive Strength

1. **Mixing**

   - The LCCM paste was prepared at a fixed water-to-binder ratio of 0.3 (by mass).
   - Homogenization was achieved using an NJ-160A standardized paste mixer operated at low speed (140 ± 5 rpm) for 90 s and high speed (285 ± 5 rpm) for an additional 90 s.
   - Mixing was paused at 90 s for 15 s of manual scraping to ensure uniform consistency.

2. **Molding and Curing**

   - The homogenized paste was cast into 20 mm^3 steel molds in two equal layers.
   - Each layer was compacted using a vibrating table (50 Hz frequency) for 30 ± 5 s to eliminate entrapped air.
   - Specimens were covered with plastic sheets and maintained at 23 ± 2 °C for 6 h prior to demolding.
   - Curing regime: Temperature: 20 ± 1 °C (controlled water bath). Relative humidity: ≥99% (saturated limewater environment).

3. **Mechanical Test**

   - Compressive strength testing was performed using an Instron 5985 universal testing machine.
   - Loading parameters: Loading rate: 0.8 MPa/s. Load cell capacity: 300 kN (0.5% accuracy). Six replicate specimens were tested at each curing interval to ensure statistical significance.

Hydration Heat

The hydration heat of LCCM was measured according to the dissolution method in GB/T 12959-2008. The calculation formula of the hydration heat of LCCM is shown in Eqs. 3.7–3.13.

$$\mathrm{C} = \frac{G0[1072.0 + 0.4(30 - ta) + 0.5(T - t_a)]}{R0} \quad (3.7)$$

where: C is heat capacity of the calorimeter, j/°C. G_0 is the weight of ZnO, g. T is the temperature at which ZnO is added to the calorimeter. t_a is the first measured value during the dissolution phase, °C. R_0 is the upward appreciation of temperature after corrected. Coefficient 1072.0 represents the dissolution heat value of zinc oxide at 30 °C, J/g. Coefficient 0.4 means the specific heat capacity of negative temperature of the dissolution heat, J/(g·°C). Coefficient 0.5 denotes the specific heat capacity of zinc oxide, J/(g·°C).

$$R_0 = (\theta_a - \theta_0) - \frac{a}{b-a}(\theta_b - \theta_0) \tag{3.8}$$

where: θ_0 is the value of the Baylor or calorimetric thermometer at the end of the initial testing period (i.e., when zinc oxide is added), °C. θ_a is the value of the Baylor or calorimetric thermometer at the first measured during the dissolution period, °C. θ_b is the value of the Baylor or calorimetric thermometer at the end of the dissolution period, °C. a represents the intervals between the test times of θ_a and θ_0, min. b represents the intervals between the test times of θ_b and θ_0, min.

$$q_1 = \frac{R_1 C}{G_1} - 0.8(T' - \mathrm{t}_{a'}) \tag{3.9}$$

where: q_1 is hydration heat of unhydrated sample, J/g. C is the calorimeter heat capacity. G_1 is mass of unhydrated sample after calcination, g. T' is the room temperature when the unhydrated sample is added to the thermos, °C. t'_a is the temperature in Celsius that is the sum of the first reading of the unhydrated cement sample during the dissolution period and the Baylor thermometer at 0 °C, °C. R_1 is the upward appreciation of the corrected temperature, °C. Coefficient 0.8 is the specific heat capacity of unhydrated cement sample, J/(g·°C).

R_1 is calculated according to Eq. 3.10, and the calculation result is retained to 0.001 °C.

$$R_1 = (\theta'_a - \theta'_0) - \frac{a'}{b' - a'}(\theta'_b - \theta'_a) \tag{3.10}$$

where: θ_0', θ_a', and θ_b' are the Baylor thermometer readings at the end of the initial test period, the first and second test readings at the dissolution period of the unhydrated cement sample, °C. a' and b' are the elapsed times from the initial reading θ_0 to the first measurement θ_1 and second measurement θ_2, min.

$$q_2 = \frac{R_2 C}{G_2} - 1.7(T'' - T_a) + 1.3 + (\mathrm{t}''_a - \mathrm{t}'_a) \tag{3.11}$$

where: q_2 is the heat of solution of hydrated cement sample after hydration for a certain age, J/g. C is the calorimeter heat capacity corresponding to the reading time, J/°C. G_2 is the mass of hydrated cement sample after burning at a certain age, g. T'' is the room temperature at which the hydrated cement sample is loaded into the calorimeter, °C. t_a'' is the temperature in Celsius that is the sum of the first reading of the hydrated cement sample during the dissolution period and the Baylor thermometer at 0 °C, °C. t_a' is the temperature in Celsius that is the sum of the first reading of the unhydrated cement sample during the dissolution period and the Baylor thermometer at 0 °C, °C. R_2 is the upward appreciation of the corrected temperature, °C. Coefficient 1.7 is the specific heat capacity of hydrated cement sample, J/(g·°C). Coefficient 1.3 is the temperature corrected specific heat capacity, J/(g·°C).

The value of R_2 is calculated according to Eq. 3.12, and the calculation result is retained to 0.001 °C.

$$R_2 = (\theta_a'' - \theta_0'') - \frac{a''}{b'' - a''}(\theta_b'' - \theta_a'') \tag{3.12}$$

where: the definition of θ_0'', θ_a'', θ_b'' a'', and b'' is the same as above, but here it represents the hydrated cement sample.

$$q = q_1 - q_2 + 0.4(20 - t_a') \tag{3.13}$$

where: q is the hydration heat released by the cement sample at a certain hydration age, J/g. q_1 is the solution heat of the unhydrated cement sample, J/g. q_2 is the solution heat of hydrated cement sample at a certain hydration age, J/g. t_a' is the temperature in Celsius that is the sum of the first reading of the unhydrated cement sample during the dissolution period and the Baylor thermometer at 0 °C, °C. Coefficient 0.4 is the negative temperature specific heat capacity of the heat of solution, J/(g·°C).

Chemical bound water

(1) Sample Preparation: Grind the hydration-terminated samples to achieve a particle size smaller than 200 mesh.
(2) Drying Treatment: Place the powdered samples in a forced-air drying oven and dry at 45 °C.
(3) Calcination and Weighing: First, precisely weigh 1–2 g of sample (accuracy: ±0.0001 g) into a crucible. Then, calcinate the sample in a muffle furnace at 950 °C for 60 min. Next, remove the sample and cool in a desiccator to room temperature. Final, record the mass of the sample. Repeat the calcination-cooling-weighing cycle until constant mass is achieved.

$$W = \frac{m_1 - m_2}{m_2} - \frac{R}{1 - R} \tag{3.14}$$

where: W is the content of the chemical bound water, %. m_1 is the mass of the sample before calcination, g. m_2 is the mass of the sample after calcination, g. R is the loss on ignition of unhydrated cement %.

X-ray diffractometer

The phase composition of both clinker and hydrated samples was analyzed using an X-ray diffractometer (XRD). The particle size of the test sample should be controlled under 200 mesh. The XRD measurements were performed with a scanning rate of 0.02°/s over a 2θ range of 5–90°.

Fourier-transform infrared

The functional group composition of both clinker and hydration products was characterized by Fourier-transform infrared (FTIR) spectroscopy. The particle size of the test sample should be controlled under 200 mesh and homogenously mixed with KBr at a mass ratio of 100:1. FTIR spectra were acquired in the range of 340–4700 cm^{-1} with a resolution of 2 cm^{-1}, accumulating 40 scans per measurement to ensure signal-to-noise ratio.

Thermogravimetric and differential scanning calorimetry

The endothermic and exthermic behavior of samples in different temperature ranges was measured by a comprehensive thermal analyzer. The particle size of the tested sample is controlled under 200 mesh. The heating rate was set at 10 °C/min, the nitrogen flow rate was set at 50 mL/min, and the data collection frequency was set at 1 °C/time.

Scanning electron microscope

The microstructural morphology of both clinker and hydration products was examined using scanning electron microscopy (SEM). The samples were trimmed to dimensions below 1 cm and sputter-coated with a thin gold layer to enhance surface conductivity and improve image resolution.

3.3 Formation and Characterization of Clinker Minerals

3.3.1 Appearance Characteristics

The appearance characteristics of clinker at different calcination temperatures are shown in Fig. 3.2. It can be observed that when the calcination temperature ranges from 700 to 1100 °C, the clinker appears light gray in color with no significant volume change. Within this temperature range, the clinker has a soft texture and is easily crushed. This is mainly due to the low calcination temperature, where no liquid phase forms in the clinker, and minerals are primarily generated through solid-state reactions. At 1100 °C, the clinker exhibits slight pulverization. This may be attributed to the formation of a small amount of C_2S at 1100 °C. During cooling, C_2S undergoes a polymorphic transformation accompanied by crystal volume expansion (Koumpouri & Angelopoulos, 2016), leading to slight pulverization of the clinker.

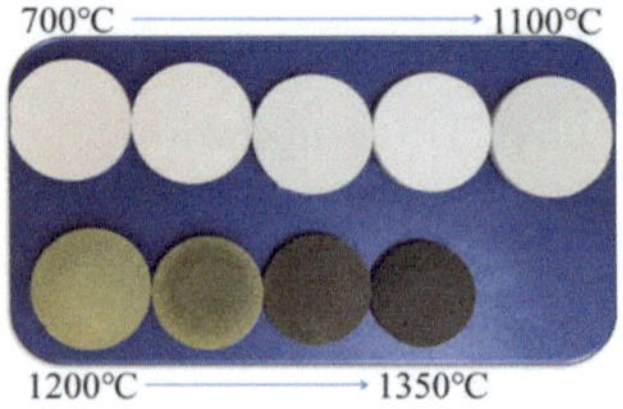

Fig. 3.2 Appearance characteristics of clinker at different calcination temperatures

When the calcination temperature ranges from 1200 to 1320 °C, significant changes occur in the color, volume, and texture of the clinker. At 1200 °C, the clinker appears light green, likely due to the presence of the transition mineral calcium sulphate aluminate. Additionally, the clinker volume slightly contracts, and its texture becomes noticeably harder, indicating the formation of a small amount of liquid phase at this temperature. The presence of the liquid phase promotes ion diffusion, facilitating mineral nucleation and crystallization. As the calcination temperature increases, the clinker color gradually changes from light green to dark brown, accompanied by further volume contraction. This is because the amount of liquid phase increases with temperature, promoting the formation of C_4AF, C_2S, and $C_4A_3\check{S}$, while the transition minerals gradually decompose and disappear.

When the calcination temperature reaches 1350 °C, the clinker exhibits severe shrinkage, a metallic luster on the surface, and a hard texture. This is due to the excessive liquid phase generated in the clinker at such high temperatures. Excessive liquid phase increases the viscosity and surface tension of the liquid, thereby reducing ion diffusion rates. This is unfavorable for the formation and growth of mineral crystals (Li, 2013). Additionally, excessively high temperatures may lead to the decomposition of active minerals in the clinker, affecting its performance.

3.3.2 *Maturity Analysis*

The content of f-CaO in clinker directly reflects the mineral formation rate and clinker maturity at different calcination temperature ranges. The variation trend of the content of f-CaO in clinker with calcination temperature is shown in Fig. 3.3. It can be observed that the content of f-CaO in clinker generally decreases with increasing calcination temperature, indicating that the amount of mineral formation increases with temperature. Based on the mineral formation rate, the calcination temperature range can be divided into three stages: 700–1000, 1000–1200, and 1200–1350 °C.

In the temperature range of 700–1000 °C, the content of f-CaO changes slightly, decreasing by only 8.8%. This indicates that the mineral formation rate is slow in this range, with the generated minerals mainly being transition minerals. When the calcination temperature rises to the range of 1000–1200 °C, the content of f-CaO decreases significantly, dropping from 31.1% at 1000 °C to 1.2% at 1200 °C. This suggests that CaO fully reacts with other mineral components in this range,

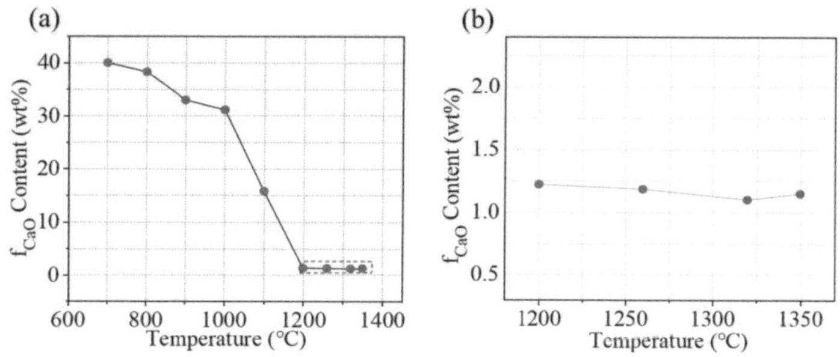

Fig. 3.3 The content of f-CaO in clinker at different calcination temperatures **a** 700–1400 °C, **b** 1200–1400 °C

accompanied by the formation of a large amount of minerals. As the calcination temperature further increases to the range of 1200–1350 °C, the content of f-CaO changes minimally. This indicates that the variation in mineral content is no longer the main feature in this stage. Instead, the optimization and development of mineral crystal structures and morphologies become dominant, and the clinker calcination tends to mature.

Notably, when the calcination temperature reaches 1350 °C, the content of f-CaO slightly increases. This may be due to the decomposition of $C_4A_3\check{S}$ into $CaSO_4$, which further decomposes into CaO at high temperatures (Eq. 3.15) (Guo & Xie, 2011).

$$CaSO_4 \rightarrow CaO + SO_2 + 0.5O_2 \tag{3.15}$$

3.3.3 *Mineral Phase Analysis*

The phase composition of the clinker is shown in Fig. 3.4. The mineralogical constituents remain relatively stable within the calcination temperature range of 700–900 °C, primarily consisting of CaO, $CaSO_4 \cdot 2H_2O$, and transitional minerals including gehlenite, anorthite, mayenite, and grossular. The relatively weak XRD peaks of these transition minerals can be attributed to their low content or underdeveloped crystal structure. During calcination, kaolinite from coal gangue reacts with $Ca(OH)_2$ from carbide slag to form substantial metakaolin and CaO, as described by Eqs. 3.16–3.17. Metakaolin exhibits high reactivity and participates in solid-state reactions to form gehlenite and anorthite, as described by Eqs. 3.18–3.19 (Ptáček et al., 2013; Hajjaji & Kacim, 2004). Mayenite is formed through the reaction between Al_2O_3 derived from coal gangue and f-CaO, as shown in Eq. 3.20 (Li et al., 2016). Grossular may form as a metastable mineral phase from the decomposition products of coal gangue and carbide slag (Hajjaji & Kacim, 2004).

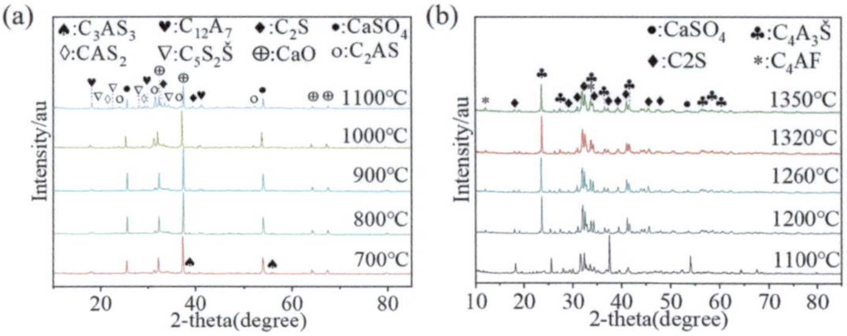

Fig. 3.4 XRD patterns of clinker at different calcination temperatures **a** 700–1100 °C, **b** 1100–1350 °C

When the calcination temperature reaches 1000 °C, new mineral phases appear in the clinker, including C_2S (Eq. 3.21) and $C_4A_3\check{S}$ (Eq. 3.22) (Bullerjahn et al., 2014). C_2S is formed through the reaction between amorphous SiO_2 (derived from calcined coal gangue) and CaO. Subsequently, $C_4A_3\check{S}$ are produced via solid-state reactions between C_2S and $CaSO_4$ (Li et al., 2016). As the calcination temperature rises to 1100 °C, the XRD diffraction peak intensities of C_2S, mayenite, gehlenite, and anorthite in the clinker increase significantly. This indicates that the mineral formation rate accelerates within the temperature range of 1000–1100 °C, which is consistent with the aforementioned analysis. The formation of these minerals consumes CaO and $CaSO_4$, leading to a noticeable decrease in their XRD diffraction peak intensities.

When the calcination temperature reaches 1200 °C, the mineral composition of the clinker undergoes significant changes (Fig. 3.4b)). The XRD diffraction peaks of CaO, $CaSO_4$, and transitional minerals such as gehlenite, anorthite, and calcium sulphate aluminate almost disappear, replaced by the diffraction peaks of $C_4A_3\check{S}$, C_2S, and C_4AF. This indicates that within the temperature range of 1100–1200 °C, the transitional minerals decompose or react with $CaSO_4$ and CaO, generating a large amount of C_2S and $C_4A_3\check{S}$. The reaction pathways are shown in Eqs. 3.21–3.27 (Bullerjahn et al., 2014; Ren, 2019).

As the calcination temperature further increases, the mineral phases in the clinker show no significant changes. However, the position, shape, and intensity of the main XRD diffraction peaks of C_2S and $C_4A_3\check{S}$ change (Fig. 3.5). This may be due to the substitution of Al^{3+} or Ca^{2+} in $C_4A_3\check{S}$ by metal ions such as Cr^{3+}, Fe^{3+}, and Mg^{2+} from the clinker, leading to changes in the crystal structure (Bullerjahn et al., 2014; Ren, 2019; Zhao, 2004). For C_2S, the distortion of its crystal structure may result from the solid solution of other oxides or the substitution of Si^{4+} by other ions during high-temperature calcination (Gazdič et al., 2019). Therefore, when the calcination temperature exceeds 1200 °C, the changes in the clinker minerals are mainly related to the development of crystal structures. When the calcination temperature reaches 1350 °C, the XRD diffraction peak intensity of $C_4A_3\check{S}$ decreases, which may be attributed to the decomposition of $C_4A_3\check{S}$.

$$Al_2Si_2O_7 \cdot 2H_2O \rightarrow Al_2Si_2O_7 + 2H_2O \quad (3.16)$$

$$Ca(HO)_2 \rightarrow CaO + H_2O \quad (3.17)$$

$$Al_2Si_2O_7 + 2\,CaO \rightarrow Ca_2Al_2SiO_{12} + SiO_2 \quad (3.18)$$

$$Ca_2Al_2SiO_{12} + Al_2Si_2O_7 + SiO_2 \rightarrow 2\,CaAl_2Si_2O_8 \quad (3.19)$$

$$7Al_2O_3 + 12CaO \rightarrow Ca_{12}Al_{14}O_{33} \quad (3.20)$$

$$SiO_2 + 2\,CaO \rightarrow Ca_2SiO_4 \quad (3.21)$$

$$2Ca_2SiO_4 + CaSO_4 \rightarrow Ca_5SiO_8(SO_4) \quad (3.22)$$

$$Ca_5SiO_8(SO_4) \rightarrow 2Ca_2SiO_4 + CaSO_4 \quad (3.23)$$

$$3Ca_2Al_2SiO_7 + CaSO_4 + 3CaO \rightarrow Ca_4Al_6SO_{16} + 6Ca_2SiO_4 \quad (3.24)$$

$$3CaAl_2Si_2O_8 + CaSO_4 + 8CaO \rightarrow Ca_4Al_6SO_{16} + 6Ca_2SiO_4 \quad (3.25)$$

$$3Ca_3Al_2Si_3O_{12} + CaSO_4 + 12CaO \rightarrow Ca_4Al_6SO_{16} + 9Ca_2SiO_4 \quad (3.26)$$

$$Ca_{12}Al_{14}O_{33} + 5Al_2O_3 + 4CaSO_4 \rightarrow 4Ca_4Al_6SO_{16} \quad (3.27)$$

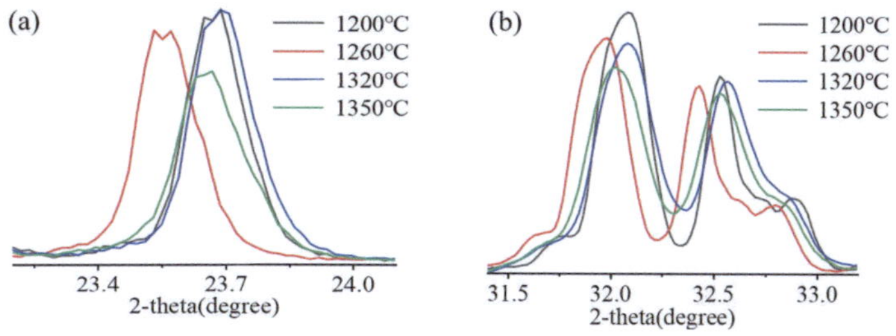

Fig. 3.5 Diffraction peaks at different calcination temperatures **a** $C_4A_3\check{S}$, **b** C_2S

3.3.4 Mineral Molecular Structure Analysis

The infrared spectra of clinker at different calcination temperatures are shown in Fig. 3.6a. Due to the complex mineral composition of the clinker, the infrared spectra of various minerals overlap, making direct analysis challenging. Therefore, the second derivative curves of the infrared spectra (Fig. 3.6b) were plotted to more clearly identify characteristic peaks. Table 3.3 lists the infrared absorption wavenumbers of different minerals in the clinker.

Analysis of Fig. 3.6a reveals that when the calcination temperature is between 700 and 1100 °C, a distinct absorption band appears at 3640 cm^{-1} in the clinker's infrared spectrum. This is caused by the asymmetric stretching vibration of hydroxyl (O-H) groups (Kalinkin et al., 2005), indicating the presence of $Ca(OH)_2$ in the clinker. The existence of $Ca(OH)_2$ may result from the absorption of moisture from the air by CaO in the clinker. Additionally, a broad absorption band is observed in the range of 1350–1500 cm^{-1}. This is attributed to the asymmetric stretching vibration of the C-O bond in $CaCO_3$ (Mutch et al., 2017). The $CaCO_3$ likely forms during raw material preparation through the reaction of $Ca(OH)_2$ with CO_2 in the air.

The infrared spectra of the clinker at different calcination temperatures are shown in Fig. 3.6a. Due to the multiple mineral components in the clinker, the infrared spectra of each mineral tend to overlap, which hinders the analysis. Therefore, the second derivative curve of the infrared spectra is plotted (Fig. 3.6b). Table 3.3 lists the infrared absorption wavenumbers of different minerals in the clinker. The absorption bands at 595, 613, 679, 1120, and 1154 cm^{-1} correspond to the characteristic absorption bands of $CaSO_4$, representing the bending and asymmetric stretching vibrations of the (SO_4) tetrahedral group (Yin et al., 2018; Wen, 1989). It is observed that the absorption bands of $CaSO_4$ change minimally when the calcination temperature is between 700 and 1000 °C. However, significant changes in the $CaSO_4$ absorption bands occur at 1100 and 1200 °C. This may be due to $CaSO_4$ participating in the formation reactions of other minerals, the specific mechanisms of which will be analyzed in detail in subsequent sections.

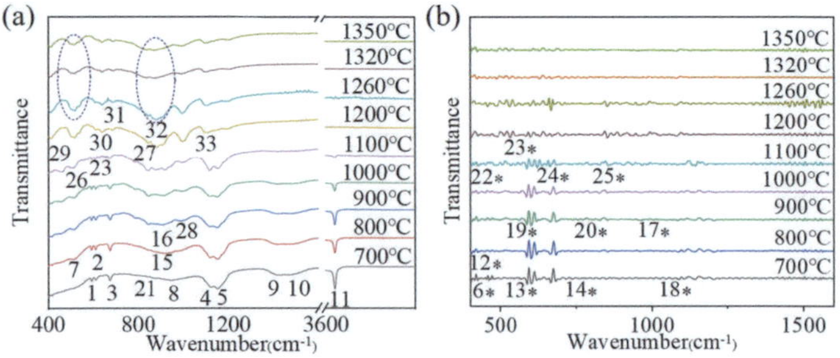

Fig. 3.6 **a** Infrared spectrum of the clinker at different calcination temperatures, **b** second derivative curve of infrared spectrum

Table 3.3 Infrared frequency characteristic table

No	Groups	Reference frequency (cm^{-1})	Compounds	References
1	$V_4(SO_4)$	595	$CaSO_4$	Yin et al. (2018) Wen (1989)
2	$V_4(SO_4)$	613	$CaSO_4$	Yin et al. (2018) Wen (1989)
3	$V_4(SO_4)$	679	$CaSO_4$	Yin et al. (2018) Wen (1989)
4	$V_3(SO_4)$	1120	$CaSO_4$	Yin et al. (2018) Wen (1989)
5	$V_3(SO_4)$	1154	$CaSO_4$	Yin et al. (2018) Wen (1989)
6*	V_4(Al-O)	435*	$Ca_3Al_2Si_3O_{12}$	Wen (1989) Peng (1982)
7	V_2(SiO4)	514	$Ca_3Al_2Si_3O_{12}$	Wen (1989) Peng (1982)
8	V_3(SiO4)	950	$Ca_3Al_2Si_3O_{12}$	Wen (1989) Peng (1982)
9	V_3(C-O)	1417	$CaCO_3$	Mutch et al. (2017)
10	V_3(C-O)	1472	$CaCO_3$	Mutch et al. (2017)
11	V_3(O-H)	3640	$Ca(OH)_2$	Kalinkin et al. (2005)
12*	V_2(Si-O-Si)	472*	$CaAl_2Si_2O_8$	Nirmala and Viruthagiri (2014)
13*	V_2(Si-O-Si)	558*	$CaAl_2Si_2O_8$	Nirmala and Viruthagiri (2014)
14*	V_1(Si-Si)	746*	$CaAl_2Si_2O_8$	Nirmala and Viruthagiri (2014)
15	V_3(Si-O)	920	$CaAl_2Si_2O_8$	Nirmala and Viruthagiri (2014)
16	V_3(Si(Al)-O)	908	$Ca_2Al_2SiO_7$	Perná et al. (2018)
17*	V_3(Si(Al)-O)	997*	$Ca_2Al_2SiO_7$	Perná et al. (2018)
18*	V_3(Si-O)	1059*	$Ca_2Al_2SiO_7$	Perná et al. (2018)
19*	V_4(AlO4)	573*	$Ca_{12}Al_{14}O_{33}$	Zapata et al. (2017)

(continued)

Table 3.3 (continued)

No	Groups	Reference frequency (cm^{-1})	Compounds	References
20*	V_3(Al-O)	783*	$Ca_{12}Al_{14}O_{33}$	Zapata et al. (2017)
21	V_3Al-O)	848	$Ca_{12}Al_{14}O_{33}$	Zapata et al. (2017)
22*	$V_2(SiO_4)$	443*	$Ca_5Si_2SO_{12}$	Liu et al. (2021)
23	$V_4(SO_4)$	630	$Ca_5Si_2SO_{12}$	Liu et al. (2021)
24*	$V_4(SO_4)$	661*	$Ca_5Si_2SO_{12}$	Liu et al. (2021)
25*	$V_1(SiO_4)$	836*	$Ca_5Si_2SO_{12}$	Liu et al. (2021)
26	$V_4(SiO_4)$	520	Ca_2SiO_4	Li (1984) Hughes et al. (1995)
27	$V_4(SiO_4)$	543*	Ca_2SiO_4	Li (1984) Hughes et al. (1995)
28	$V_3(SiO_4)$	995	Ca_2SiO_4	Li (1984) Hughes et al. (1995)
29	$V_4(AlO_4)$	413	$Ca_4Al_6SO_{16}$	Li (1984)
30	$V_3(AlO_4)/_4(SO_4)$	644	$Ca_4Al_6SO_{16}$	Li (1984)
31	$V_3(AlO_4)/_4(SO_4)$	692	$Ca_4Al_6SO_{16}$	Li (1984)
32	$V_3(AlO_4)$	884	$Ca_4Al_6SO_{16}$	Li (1984)
33	$V_3(SO_4)$	1103	$Ca_4Al_6SO_{16}$	Li (1984)
34	V_3(Si-O)	920	$CaAl_2Si_2O_8$	Wen (1989) Nirmala and Viruthagiri (2014)

In the temperature range of 700–1100 °C, the infrared spectra of the clinker exhibit two broad absorption regions, located at 420–550 and 900–1,250 cm^{-1} (Fig. 3.6a). These absorption regions are primarily caused by the bending vibrations of Si-O and Al-O groups and the asymmetric stretching vibrations of Si–O-Si and Si–O-Al groups in aluminosilicates. Specifically, these regions include the characteristic absorption bands of grossular (435, 514, 950 cm^{-1}) (Wen, 1989; Peng, 1982), gehlenite (472, 558, 746, 920 cm^{-1}) (Wen, 1989; Nirmala & Viruthagiri, 2014), and anorthite (908, 997, 1059 cm^{-1}) (Perná et al., 2018). These infrared absorption regions change with the calcination temperature, indicating continuous development of aluminosilicate minerals in the clinker within this temperature range. Additionally, the characteristic absorption bands of mayenite (573, 783, 848 cm^{-1}) (Zapata et al., 2017) are observed in the spectra between 700 and 1100 °C, and their intensity increases with rising calcination temperature. This suggests that mayenite is continuously formed within this temperature range. Notably, when the calcination temperature reaches 1100 °C, the characteristic absorption bands of $C_4A_3\check{S}$ (630, 661, and 836 cm^{-1}) appear in the spectra. These bands are attributed to the bending vibrations of (SO_4) groups and the stretching vibrations of (SiO_4) groups (Liu et al., 2021). The formation of $C_4A_3\check{S}$ consumes a certain amount of $CaSO_4$, which explains the weakening intensity of the $CaSO_4$ infrared absorption bands. Through the above analysis, it can be concluded that within the temperature range of 700–1100 °C, the minerals formed in the clinker are mainly transitional minerals such as grossular, gehlenite, and anorthite. This result is consistent with the conclusions drawn from previous analyses.

When the calcination temperature of clinker reaches 1200 °C, significant changes occur in the absorption bands of the infrared spectrum. The characteristic absorption bands of transitional minerals such as grossular, gehlenite, and anorthite, as well as $CaSO_4$, almost completely disappear, and are replaced by the infrared absorption bands of C_2S and $C_4A_3\check{S}$. The absorption bands of C_2S are mainly concentrated in two regions: 460–570 and 960–1070 cm^{-1}. Among them, the absorption bands at 520 and 543 cm^{-1} are caused by the out-of-plane bending vibrations of the (SiO_4) groups, while the absorption band at 995 cm^{-1} corresponds to the asymmetric stretching vibrations of the (SiO_4) groups (Li, 1984; Hughes et al., 1995). The absorption bands of $C_4A_3\check{S}$ are caused by the asymmetric contraction vibrations of the (AlO_4) groups (413 cm^{-1}), the bending vibrations of the (AlO_4) groups (884 cm^{-1}), the coupled stretching vibrations of the (SiO_4) and (AlO_4) groups (644, 692 cm^{-1}), and the asymmetric stretching vibrations of the (SO_4) groups (1103 cm^{-1}) (Li, 1984).

It is noteworthy that when the calcination temperature rises to 1320 and 1350 °C, the absorption bands of C_2S and $C_4A_3\check{S}$ become broader and their intensity weakens (Fig. 3.6a). According to the literature, when the crystal structure of a mineral is disrupted or its symmetry is reduced, the infrared absorption bands broaden and weaken (Mollah et al., 2000). This indicates that within the temperature range of 1320–1350 °C, the crystals of C_2S and $C_4A_3\check{S}$ may have partially melted, leading to the disruption of their crystal structure or a reduction in symmetry.

3.3.5 Thermogravimetric Comprehensive Analysis

Figure 3.7 shows the TG-DSC curve of the raw meal. Analysis reveals that the DSC curve exhibits distinct endothermic peaks at 156, 500, 540, and 710 °C. These peaks are attributed to the endothermic decomposition of $CaSO_4 \cdot 2H_2O$, $Ca(OH)_2$ in carbide slag, kaolinite in coal gangue, and $CaCO_3$, respectively, all accompanied by mass loss. The presence of $CaCO_3$ is likely due to the carbonation of $Ca(OH)_2$ in air. The decomposition of $Ca(OH)_2$, $CaCO_3$, and kaolinite generates a significant amount of reactive CaO, Al_2O_3, and SiO_2. Based on XRD and FTIR analysis, a small amount of intermediate minerals (such as grossular, gehlenite, anorthite, and mayenite) are formed in the temperature range of 540–710 °C. As the temperature increases, the DSC curve shows a broad endothermic region between 847 and 1013 °C, accompanied by mass loss, which is likely caused by the decomposition of $CaSO_4$. Combined with the analysis of f-CaO and XRD, it is inferred that transitional minerals continue to develop in this temperature range. The endothermic peak at 1004 °C may be attributed to the formation of C_2S (Mollah et al., 2000).

In the temperature range of 1013–1302 °C, the DSC curve exhibits two broad endothermic regions, while the TG curve shows no significant mass loss. This indicates the formation of a large number of minerals in this range. Based on the analysis

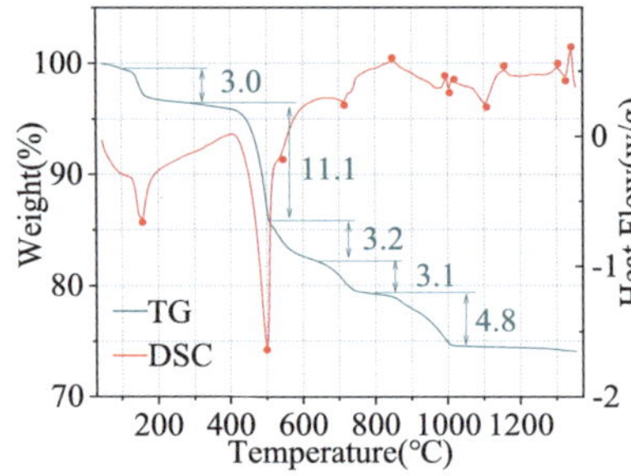

Fig. 3.7 TG-DSC curve

of f-CaO, XRD, and FTIR, a significant amount of transitional minerals forms in the range of 1013–1156 °C. In the range of 1156–1302 °C, these transitional minerals decompose or react to form final minerals ($C_4A_3\check{S}$, C_2S, and C_4AF). As the temperature further increases, the DSC curve shows an endothermic peak at 1324 °C, which may be due to the melting of solid phases in the system. When the temperature exceeds 1340 °C, the system exhibits continuous endothermic behavior accompanied by mass loss, likely caused by the decomposition of $C_4A_3\check{S}$.

3.3.6 Microstructural Analysis

The microstructures of clinker at different calcination temperatures are shown in Fig. 3.8. Analysis reveals that when the calcination temperature is 700 °C, the clinker primarily consists of spherical CaO, along with a small amount of rod-like $CaSO_4$, and transitional minerals (gehlenite, anorthite, mayenite, and grossular). The presence of trace impurity elements in the EDS analysis of the minerals may be due to the complex composition of the raw materials. When the calcination temperature increases to 900 °C, CaO remains the dominant phase in the clinker, but its particle size significantly increases. As the calcination temperature further rises, the size and proportion of CaO gradually decrease, while the size and proportion of transitional minerals increase significantly. At 1100 °C, the reduction in CaO becomes more pronounced, and the size and proportion of transitional minerals increase notably. This indicates that transitional minerals continue to develop within the temperature range of 700–1100 °C, with their rapid development mainly occurring between 1000 and 1100 °C. These results are consistent with the conclusions drawn from previous analyses.

When the calcination temperature reaches 1200 °C, the transitional minerals and CaO in the clinker almost completely disappear, replaced by hexagonal plate-like $C_4A_3\check{S}$ and elliptical C_2S. According to the EDS analysis, the $C_4A_3\check{S}$ and C_2S crystals at this stage contain a significant amount of impurity ions, and their crystal sizes are small, indicating incomplete maturation. As the calcination temperature increases to 1260 °C, the crystal sizes of $C_4A_3\check{S}$ and C_2S increase significantly, and their boundaries become more distinct, suggesting improved crystallinity.

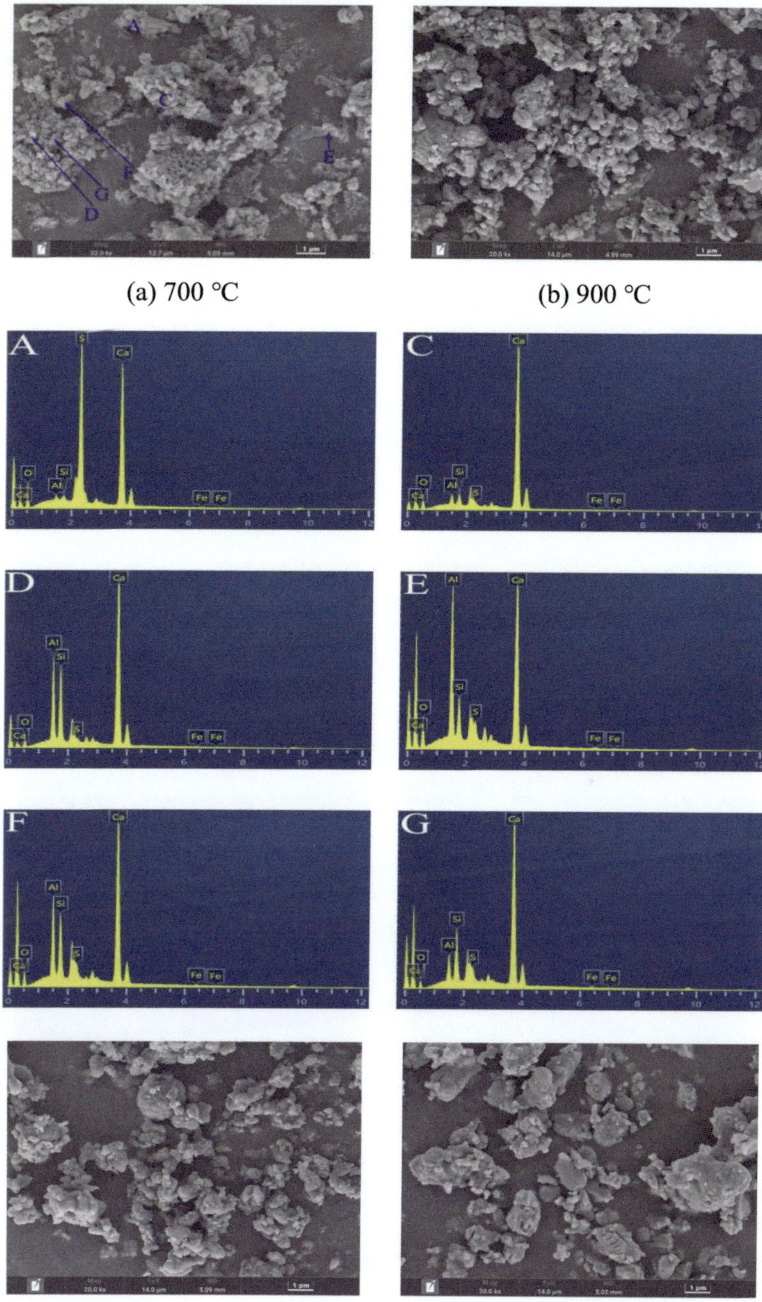

(a) 700 °C (b) 900 °C

(c) 1000 °C (d) 1100 °C

Fig. 3.8 SEM image and EDS spectrum of clinker at different calcination temperatures (A: $CaSO_4$, C: CaO, D: anorthite, E: $12CaO{\cdot}7Al_2O_3$, F: $2CaO{\cdot}Al_2O_3{\cdot}SiO_2$, G: garnet, B1: C_2S, Y1: $C_4A_3\hat{S}$, B2: C_2S, Y2: $C_4A_3\hat{S}$, I: C_4AF)

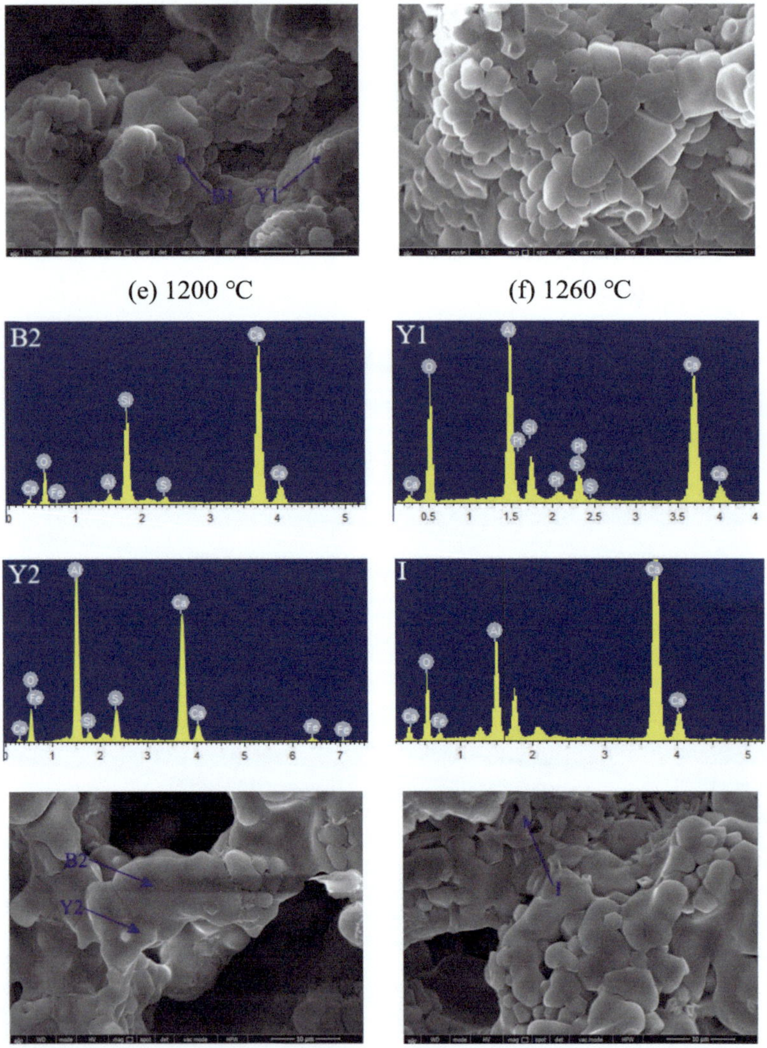

Fig. 3.8 (continued)

At a calcination temperature of 1320 °C, needle-like C_4AF appears in the clinker. Simultaneously, C_2S crystals exhibit melting phenomena, displaying irregular shapes and significantly increased sizes. Some $C_4A_3\check{S}$ crystals are enveloped by the molten phase. This phenomenon may explain the broadening of the infrared absorption bands of C_2S and $C_4A_3\check{S}$. Additionally, EDS analysis shows a reduction in impurity ion content within the crystals, indicating that C_2S and $C_4A_3\check{S}$ crystals continue to develop and mature within the temperature range of 1200–1320 °C. This also explains the changes in the main XRD diffraction peaks of C_2S and $C_4A_3\check{S}$ as the calcination temperature increases.

When the calcination temperature further rises to 1350 °C, C_4AF crystals become rod-like and increase in size. However, it is noteworthy that the size of $C_4A_3\check{S}$ crystals decreases. This may be due to the partial decomposition of $C_4A_3\check{S}$ crystals caused by excessively high calcination temperatures.

3.4 Optimal Calcination System

Based on the above analysis, the optimal clinkering temperature range for clinker is determined to be between 1200 °C and 1350 °C. This section further establishes the calcination regime for clinker. The workability of the prepared clinker is tested to evaluate its performance in practical applications.

3.4.1 Calcination Temperature

Determination of calcination temperature

The calcination temperature plays a decisive role in the formation of clinker minerals. If the temperature is too low, mineral reactions may be incomplete, while excessively high temperatures may cause mineral melting or decomposition. Therefore, this section aims to determine the optimal calcination temperature for the clinker. Since the addition of steel slag significantly improves the burnability of the clinker, the mix with the lowest steel slag content, LCCM-A, is selected as the research subject. Based on preliminary analysis, the suitable clinkering temperature range for LCCM is between 1200 and 1350 °C. Thus, calcination temperatures are set at 1200, 1230, 1260, 1290, 1320, and 1350 °C, with a preliminary calcination time of 40 min.

Figure 3.9 shows the XRD patterns of the clinker at different calcination temperatures. Analysis of the XRD patterns reveals that at 1200 °C, C_2S and $C_4A_3\check{S}$ are already formed in the clinker. However, a certain amount of CaO and mayenite still remain, indicating that the clinker has not yet reached optimal maturity. As the calcination temperature increases, the diffraction peak intensity of mayenite gradually weakens, while that of $C_4A_3\check{S}$ continues to strengthen. When the temperature reaches 1290 °C, the diffraction peak of mayenite almost disappears. At this stage, the main minerals in the clinker are C_2S, $C_4A_3\check{S}$, $CaSO_4$, and C_4AF, indicating that the clinker is approaching maturity. When the temperature rises to 1350 °C, the diffraction peak intensity of $C_4A_3\check{S}$ decreases, which may be due to the partial decomposition of $C_4A_3\check{S}$.

Figure 3.10 shows the SEM images of the minerals in the clinker. It is observed that as the calcination temperature increases, the size of the minerals in the clinker gradually increases, and the crystal edges become more distinct. Compared to 1290 °C, the

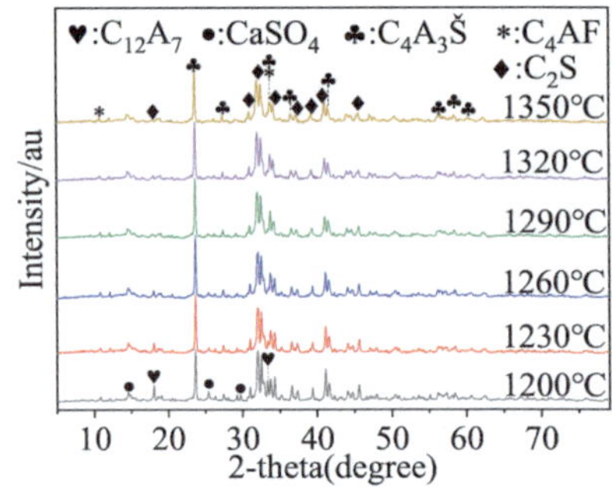

Fig. 3.9 XRD patterns of clinker at different calcination temperatures

mineral crystals at 1320 °C are larger and have more defined edges. However, when the calcination temperature rises to 1350 °C, the minerals begin to melt, indicating that the calcination temperature is too high.

The content of f-CaO in the clinker is an important indicator that directly reflects the maturity of the clinker. Figure 3.11 illustrates the variation in f-CaO content in the clinker at different calcination temperatures. Analysis reveals that within the temperature range of 1200–1320 °C, the f-CaO content gradually decreases as the calcination temperature increases. This indicates that the calcination process becomes more mature with rising temperature. However, when the calcination temperature reaches 1350 °C, the f-CaO content increases again. This may be due to the decomposition

(a) 1260 °C (b) 1290 °C

(c) 1320 °C (d) 1350 °C

Fig. 3.10 SEM image of clinker at different calcination temperatures

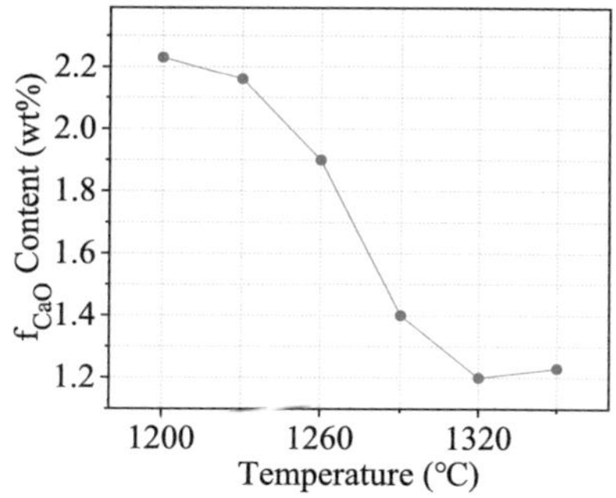

Fig. 3.11 The content of f-CaO in clinker at different calcination temperatures

of $C_4A_3\check{S}$, suggesting that the temperature has exceeded the optimal range, leading to over-calcination of the clinker.

Based on the above comprehensive analysis, we determine the calcination temperature for the clinker to be 1320 °C.

3.4.2 Calcination Time

The calcination time directly determines the formation and development of clinker minerals. Therefore, this section will focus on determining the holding time during the calcination of clinker. The calcination temperature for the clinker in this section is set at the previously determined temperature of 1320 °C. The calcination times for the clinker are set at 10, 20, 40, and 60 min. The analysis of clinker obtained at different calcination times is as follows.

XRD was used to perform phase analysis on clinker prepared at different holding times, and the XRD patterns are shown in Fig. 3.12. The analysis results indicate that the XRD patterns of clinker at different holding times are generally consistent, with the main mineral components being C_2S and $C_4A_3\check{S}$, along with small amounts of C_4AF and $CaSO_4$. Notably, within the holding time range of 10–40 min, the diffraction peak intensity of $C_4A_3\check{S}$ significantly increases with prolonged holding time. This suggests that sufficient solid-phase reaction time is beneficial for the formation of $C_4A_3\check{S}$ minerals. However, when the holding time is extended to 60 min, the change in the diffraction peak intensity of $C_4A_3\check{S}$ tends to stabilize. This indicates that the $C_4A_3\check{S}$ minerals have essentially matured by 40 min.

Figure 3.13 shows the SEM images of clinker at different holding times. The analysis reveals that within the holding time range of 10–40 min, the size of clinker minerals gradually increases with prolonged holding time. This suggests that the calcination process of the clinker is approaching maturity. However, when the holding time reaches 60 min, the change in mineral size is no longer significant.

The content of f-CaO in the clinker is shown in Fig. 3.14. It can be observed that the content of f-CaO in the clinker decreases with increasing calcination time. Particularly, within the time range of 10–40 min, the reduction in f-CaO content is

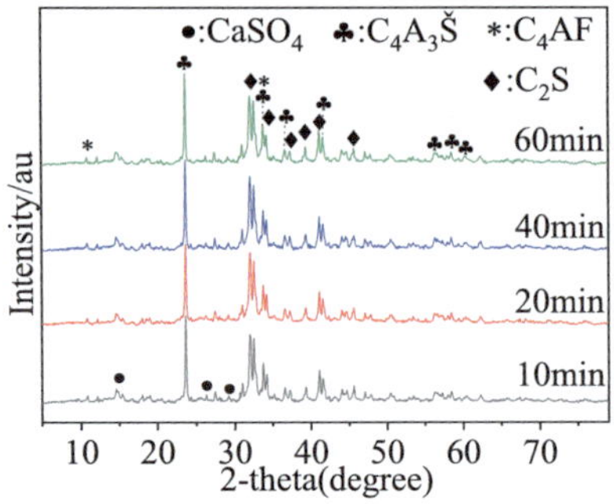

Fig. 3.12 XRD patterns of clinker with different calcination time

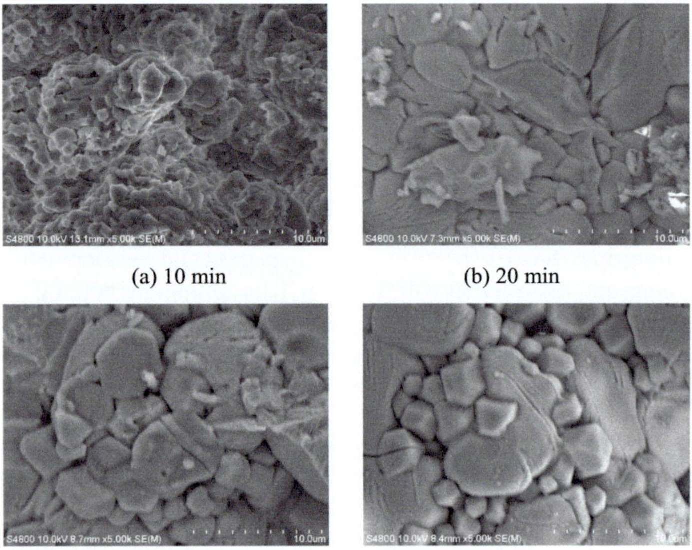

(a) 10 min (b) 20 min

(c) 40 min (d) 60 min

Fig. 3.13 SEM images of clinker with different calcination time

more pronounced. However, when the calcination time reaches 60 min, the change in f-CaO content becomes less significant. This indicates that the clinker has essentially matured after a calcination time of 40 min.

Based on the above comprehensive analysis, the calcination time for the clinker is determined to be 40 min.

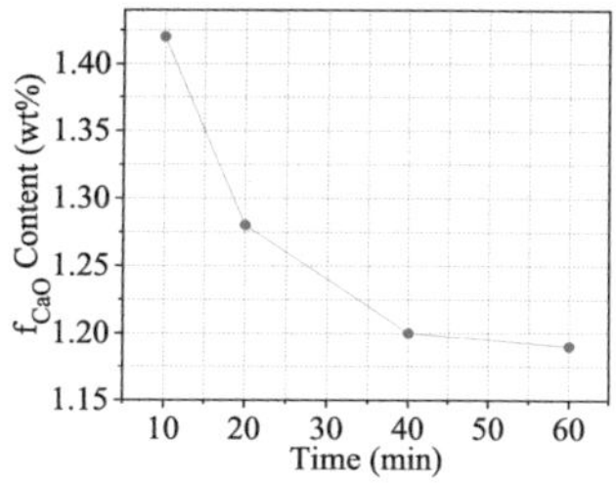

Fig. 3.14 The content of f-CaO in clinker with different calcination time

3.4.3 Mineral Composition Analysis

After establishing the calcination regime, to further investigate the mineral composition of LCCM clinker, an analysis was conducted on the LCCM clinker prepared by heating at a rate of 5 °C/min to 1320 °C and holding for 40 min. The results are as follows:

The content of f-CaO in the LCCM clinker is shown in Table 3.4. The data indicate that the f-CaO content in the clinker is approximately 1%, suggesting that the clinker has been adequately calcined.

The FTIR spectra of the clinker are presented in Fig. 3.15. The spectra reveal that the infrared spectra of the four types of clinker are largely consistent. According to literature, the absorption peak of $C_4A_3\check{S}$ at 882 cm^{-1} corresponds to the asymmetric stretching vibration of the (AlO_4) tetrahedron (Li, 1984). The absorption peak at 1098 cm^{-1} is attributed to the asymmetric stretching vibration of the (SO_4) tetrahedron (Li, 1984). Additionally, the absorption peaks at 616, 641, and 690 cm^{-1} result from the coupled vibrations of the (AlO_4) and (SO_4) tetrahedra (Li, 1984). For C_2S, the absorption peak at 520 cm^{-1} originates from the bending vibration of the (SiO_4) tetrahedron (Li, 1984), while the peaks at 847, 879, and 991 cm^{-1} correspond to the asymmetric stretching vibrations of the (SiO_4) tetrahedron (Li, 1984; Hughes et al., 1995). Comparative analysis reveals some differences in the wavenumbers corresponding to the absorption peaks of the clinker. Table 3.5 shows that during the clinker calcination process, C_2S and $C_4A_3\check{S}$ have partially dissolved metal ions, forming solid solutions. This solid solution phenomenon reduces the symmetry of the crystal structure, leading to a shift in the absorption peaks. Due to the low resolution of the infrared absorption peaks of C_4AF in the 400–750 cm^{-1} region, its characteristic absorption peaks are difficult to directly identify in the clinker's infrared spectra. Therefore, the second derivative curve of the infrared spectrum was further plotted (Fig. 3.4b). In the second derivative curve, the characteristic absorption peaks of C_4AF can be clearly observed at 443, 530, 610, 660, and 721 cm^{-1} (He et al., 2017).

Table 3.4 Content of f-CaO in clinker (wt.%)

LCCM-A	LCCM-B	LCCM-C	LCCM-D
1.19	0.96	1.1	0.7

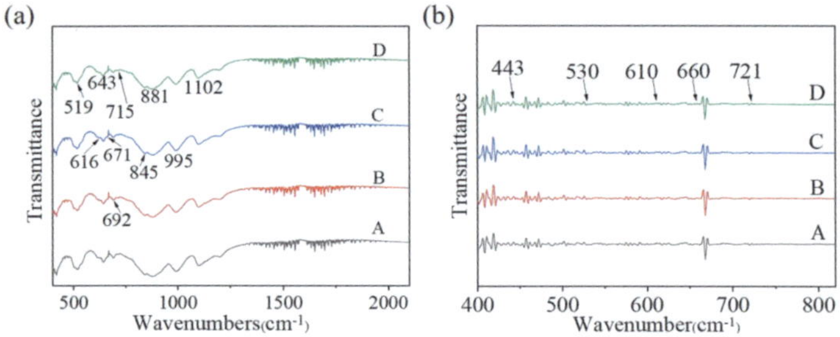

Fig. 3.15 **a** Infrared spectrum of the clinker, **b** second derivative curve of infrared spectrum

Table 3.5 EDS analysis (wt.%)

Number	Mineral	Ca	O	Si	Al	S	Fe	Mg	Na
1	C_2S	38.03	46.19	8.74	2.04	1.41	0.75	0.79	0.37
2	$C_4A_3\check{S}$	26.62	51.37	1.55	23.59	3.35	0.54	0.53	0.42

Phase analysis of the clinker was conducted using XRD, and the XRD pattern is shown in Fig. 3.17. The results indicate that the main mineral components of the clinker are $C_4A_3\check{S}$ and C_2S, with minor amounts of C_4AF and $CaSO_4$. This is consistent with the mineral design objectives. By comparing the XRD patterns of clinkers A-D, it is observed that as the steel slag content increases, the diffraction peak intensity of $C_4A_3\check{S}$ gradually decreases, while that of C_4AF gradually increases. This phenomenon indicates that with the increase in steel slag content, the amount of $C_4A_3\check{S}$ in the clinker gradually decreases, while the amount of C_4AF gradually increases, which aligns with the preliminary mineral composition design (Fig. 3.16).

Figure 3.17 shows the SEM images of LCCM-A and LCCM-D clinkers. Observations reveal that the mineral components in the clinker are well-crystallized. The $C_4A_3\check{S}$ crystals are small in size, hexagonal in shape, and mostly 2–5 μm in diameter. The C_2S crystals are slightly larger, spherical in shape, and range from 5 to 10 μm in diameter. Additionally, flaky-structured C_4AF minerals were observed.

Fig. 3.16 XRD patterns of the clinker

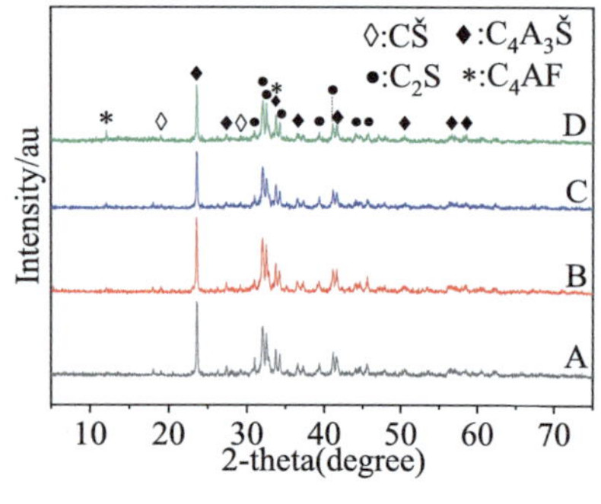

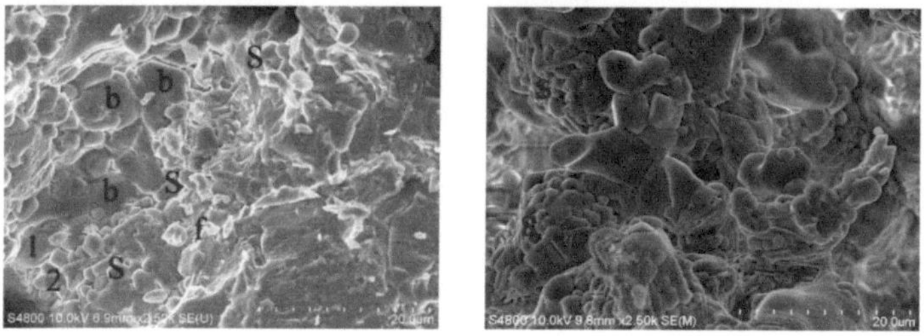

(a) Clinker A (b) Clinker D

Fig. 3.17 SEM of clinker A and D, b = C_2S, f = C_4AF, s = $C_4A_3\check{S}$

Table 3.5 presents the EDS analysis data of the clinker. The table shows that Fe, Mg, Na, and other metal ions are present in the $C_4A_3\check{S}$ and C_2S minerals. This indicates that some foreign ions were incorporated into the clinker minerals during their formation. This solid solution phenomenon contributes to enhancing the hydration activity of the minerals.

3.4.4 Workability Analysis

In this part of the experiment, the working performance of LCCM was tested, and P.O. 42.5 cement was used as a control group for synchronous testing.

According to the GB/T 1346–2011 standard, the working performance of LCCM was tested, and the results are shown in Table 3.6. It can be observed that the standard consistency water demand of LCCM is slightly lower than that of P.O. 42.5. This is primarily because, during the sieving process of LCCM using a 200-mesh sieve, approximately 3% of the residue was retained, resulting in a relatively larger average particle size of LCCM. This leads to a relatively smaller standard consistency water demand for LCCM. Additionally, Table 3.6 indicates that the setting time of LCCM is significantly shorter than that of P.O. 42.5. This is mainly attributed to the presence of a large amount of high-activity, fast-hydrating $C_4A_3\check{S}$ minerals in LCCM. Furthermore, by comparing the setting times of LCCM, it can be noted that as the content of steel slag increases, the setting time of LCCM slightly prolongs. This may be due to the gradual decrease in the content of $C_4A_3\check{S}$ minerals in the LCCM clinker as the steel slag content in the raw meal increases.

The compressive strength comparison between LCCM and P.O. 42.5 cement is shown in Fig. 3.18. It can be seen that the compressive strength of LCCM exhibits a significant growth trend with prolonged hydration time. Based on the analysis, its compressive strength development can be divided into three stages: 1–3 days, 3–28 days, and after 28 days.

In the early hydration stage of 1–3 days, the compressive strength of LCCM increases rapidly. The compressive strength at 3 days of LCCM-C can reach 62 MPa,

Table 3.6 LCCM water requirement of normal consistency and setting time

	Setting time (min)		Water requirement for normal consistency
	Initial setting time	Final setting time	Water/Cement
LCCM-A	19	29	0.27
LCCM-B	21	34	0.27
LCCM-C	23	34	0.26
LCCM-D	24	40	0.26
P.O. 42.5	130	190	0.29

Fig. 3.18 Compressive strength of LCCM

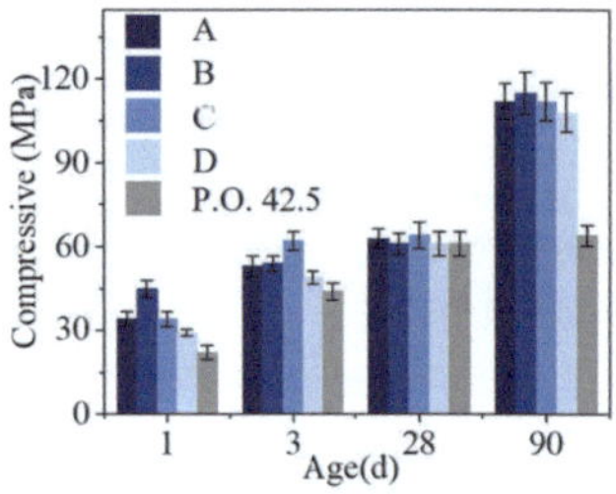

which is mainly attributed to the highly active $C_4A_3\check{S}$ minerals in LCCM, whose rapid hydration contributes significantly to the early strength (Ren et al., 2020). However, in the mid-term hydration stage of 3–28 days, the compressive strength growth of LCCM is relatively slow. This may be because the $C_4A_3\check{S}$ is almost entirely consumed during the early hydration stage (1–3 days), while C_2S begins to hydrate significantly only after 28 days, resulting in limited strength growth during the mid-term stage. When the hydration age enters the later stage of 28–90 days, the compressive strength of LCCM increases significantly again. The compressive strength of LCCM-C at 90 days can reach 112 MPa. The strength growth in this stage is mainly due to the continuous hydration of C_2S. Furthermore, by comparing Fig. 3.11, it can be observed that the compressive strength of LCCM after 3 days of hydration is close to that of P.O. 42.5 after 28 days of hydration. This indicates that LCCM exhibits significant early strength characteristics. At the same time, in the later hydration stage, the compressive strength of LCCM is also significantly higher than that of P.O. 42.5 cement. This further demonstrates the excellent long-term strength of LCCM.

3.5 Hydration Characteristics of Low Carbon Cementitious Materials

3.5.1 *Hydration Heat*

The hydration heat release curve of LCCM is shown in Fig. 3.19a. The heat release increases with the extension of curing time and is mainly concentrated in the first 3 days. The heat release on the first day reaches 261 J/g, accounting for approximately 73% of the total heat release over 90 days. This indicates that LCCM exhibits rapid and substantial heat release during the initial hydration process. This is primarily attributed to the presence of the $C_4A_3\check{S}$ mineral in the sample, which has a fast hydration rate and high heat release. When the hydration time exceeds 3 days, the rate of heat release significantly decreases but continues steadily. This phenomenon may be due to the fact that most of the $C_4A_3\check{S}$ has been consumed during the first 3 days of hydration. Simultaneously, the formation of ettringite (AFt) and aluminium hydroxide (AH_3) on the surface of unhydrated minerals creates a dense hydration layer, hindering further hydration. Additionally, the unhydrated minerals in the sample at this stage are mainly C_2S. However, C_2S has low hydration activity and releases less heat, typically starting significant hydration only after 28 days. Therefore, the heat release and its rate decrease significantly and stabilize after 3 days.

With the increase in steel slag content, the hydration heat release characteristics of the samples at different ages show significant differences, as shown in Fig. 3.19b. From the Figure, it can be observed that the heat release on the first day initially increases and then decreases with the increase in steel slag content. Studies have shown that during the formation of $C_4A_3\check{S}$, Fe^{3+} can be solid-solved into its crystal lattice, causing slight symmetry changes and structural distortions. This enhances the hydration activity and rate of $C_4A_3\check{S}$ (Bullerjahn et al., 2014; Idrissi et al., 2010). Therefore, a small amount of steel slag has little effect on the amount of $C_4A_3\check{S}$ generated but can effectively activate $C_4A_3\check{S}$. This leads to an increase in heat release with the increase in steel slag content. However, as the steel slag content further increases, iron oxide reacts with alumina and calcium oxide to form C_4AF, which

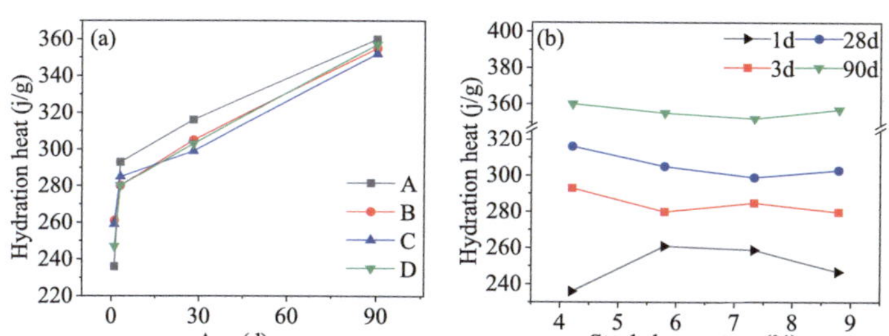

Fig. 3.19 Hydration heat **a** LCCM under different calcination systems, **b** LCCM under different curing time

has lower heat release and activity (Bullerjahn et al., 2014; Ectors et al., 2013; Idrissi et al., 2015). At the same time, the content of $C_4A_3\check{S}$ gradually decreases (Huang et al., 2007). This weakens the activation effect of Fe^{3+} on $C_4A_3\check{S}$, resulting in a decrease in heat release. Additionally, with the extension of hydration time, the heat release of the samples generally decreases with the increase in steel slag content. This may also be due to the reduction in $C_4A_3\check{S}$ content caused by the increase in steel slag content, thereby reducing the total heat release of the samples.

Through the analysis of hydration heat, it can be observed that LCCM exhibits high and concentrated heat release characteristics in the early stages of hydration. The heat release of LCCM at 3 days can reach 293 J/g. Under similar water-to-cement ratio (W/C) conditions, LCCM is about 90 J/g higher than P.O 42.5 cement (Dai et al., 2021). Combined with the mechanical performance test results, the compressive strength of LCCM at 3 days has already reached the level of P.O 42.5 cement after 28 days. This indicates that LCCM has high early strength. Therefore, LCCM is particularly suitable for engineering scenarios requiring high early strength, such as winter emergency repairs.

3.5.2 *Chemically Bound Water*

Chemically bound water is a product of the cement hydration process, and its content directly reflects the degree and rate of cement hydration (Darquennes et al., 2013). The chemically bound water content increases with age, showing a trend similar to that of hydration heat release. The growth of chemically bound water can be divided into two stages. The first stage is 1–3 days. During this stage, the chemically bound water increases rapidly, reaching 78% of the total bound water content (W_{90d}) at 3 days. This indicates that LCCM exhibits a high hydration rate in the early hydration stage. This is mainly attributed to the presence of $C_4A_3\check{S}$ mineral in LCCM, which has high activity and a fast hydration rate. The second stage is 3–90 days. During this stage, the growth rate of chemically bound water slows down but continues steadily. This suggests that the hydration rate of LCCM decreases after 3 days but still progresses. This may be due to the fact that most of the $C_4A_3\check{S}$ has been consumed during the early hydration stage (1–3 days), and the remaining unhydrated minerals are primarily C_2S, which has lower hydration activity.

From Fig. 3.20b, it can be observed that in the early hydration stage, the chemically bound water content is significantly influenced by the steel slag content. However, as the hydration age increases, the influence of steel slag content gradually weakens. Specifically, at 1 day of hydration, the chemically bound water content initially increases and then decreases with the increase in steel slag content. This trend is consistent with the change in hydration heat release on the first day. Since both hydration heat and chemically bound water are results of the hydration reaction, their trends should be influenced by the same factors. Therefore, the trend of chemically bound water on the first day may be related to the effect of steel slag on the activity of $C_4A_3\check{S}$. With the extension of hydration time, the chemically bound water content of

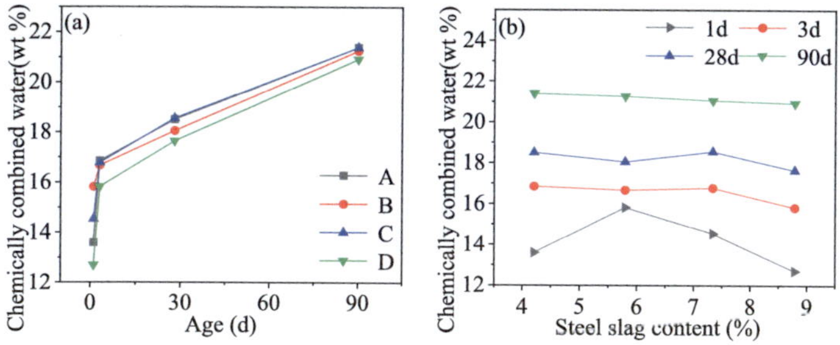

Fig. 3.20 Chemically combined water **a** LCCM under different calcination systems, **b** LCCM under different curing time

the samples generally decreases, and the differences between the samples gradually diminish. This may be due to the fact that the steel slag content is controlled within a specific range, resulting in minor changes in the content of $C_4A_3\check{S}$ and C_4AF. At the same time, in the hydration environment of gypsum and limestone, the chemically bound water content of the hydration products of $C_4A_3\check{S}$ and C_4AF is similar (Taylor, 1997; Radwan et al., 2011; Ma et al., 2019). Therefore, the final chemically bound water content of the samples tends to converge.

Through the analysis of chemically bound water, it can be concluded that LCCM exhibits a fast hydration rate during the early stage of hydration, reaching a higher degree of hydration in a short time. As the hydration age increases, the hydration rate significantly decreases, but the cement continues to undergo slow hydration during the later stages. Further analysis also reveals that within a certain range of steel slag content, the addition of steel slag has a relatively large impact on the early hydration of LCCM, but its effect on LCCM hydration becomes limited in the later hydration stage.

3.5.3 Hydration Products

The XRD patterns of LCCM at different ages are shown in Fig. 3.21a. By analyzing the XRD patterns, the changes in hydration products and unhydrated minerals during the hydration process can be understood. It can be observed that after 1 day of hydration, most of the $C_4A_3\check{S}$ has hydrated, generating a significant amount of AFt, as shown in Eq. 3.29 (Ma et al., 2014). When the hydration time is extended to 3 days, most of the $C_4A_3\check{S}$ in the sample has been nearly consumed. At this stage, the main mineral components in the sample are AFt, unhydrated C_2S, and $CaCO_3$. As the hydration age further increases, the diffraction peak intensity of AFt slightly strengthens, indicating that a small amount of $C_4A_3\check{S}$ continues to hydrate. When the hydration age reaches 90 days, the mineral composition of the sample shows no significant changes. However, the diffraction peak of C_2S exhibits slight changes, characterized by a narrower and sharper peak shape, along with a reduction in the

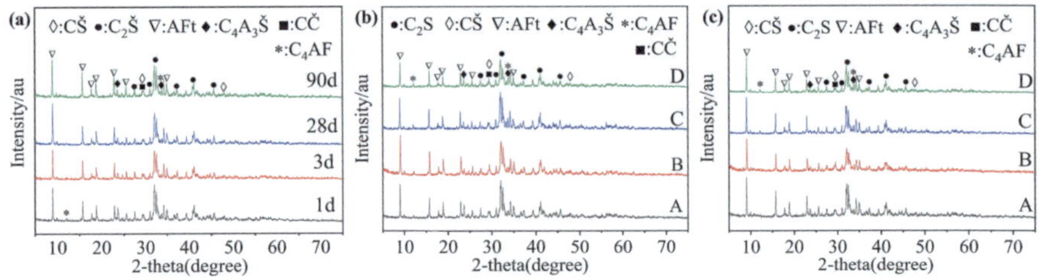

Fig. 3.21 XRD patterns **a** LCCM-A at different ages, **b** LCCM-A, LCCM-B, LCCM-C, LCCM-D at 1 d, **c** LCCM-A, LCCM-B, LCCM-C, LCCM-D at 90 d

intensity of the characteristic peak. This may be due to the onset of C_2S hydration as shown in Eqs. 3.30–3.31 (Ma et al., 2014; Kurdowski, 2014). Additionally, hydration products such as AH_3, $Ca(OH)_2$, and C-S-H may not exhibit distinct diffraction peaks in the XRD patterns due to their low crystallinity or low content.

Figure 3.21b shows the XRD patterns of each sample after 1 day of hydration. The analysis indicates that AFt is the primary hydration product at 1 day, and its diffraction peak intensity initially increases and then decreases with the increase in steel slag content. This suggests that the steel slag content influences the hydration process of $C_4A_3\check{S}$, thereby affecting the hydration behavior of the samples. This is consistent with the previous analysis results.

Figure 3.21c presents the XRD patterns of each sample after 90 days of hydration. The analysis reveals that the diffraction peak intensities of AFt in all samples are relatively similar, but the diffraction peaks of other hydration products are not observed. This may be due to the low content or poor crystallinity of other hydration products, causing them to remain undetected in the XRD patterns. Therefore, the influence of steel slag on the later-stage hydration of the samples cannot be directly analyzed through XRD. Further research will explore the mechanism of steel slag's impact on the hydration process

$$C_4A_3\check{S} + 2C\check{S} + 38H \rightarrow C_3A \cdot 3C\check{S} \cdot 32H + 2AH_3 \tag{3.28}$$

$$C_2S + AH_3 + 5H \rightarrow C_2ASH_8 \tag{3.29}$$

$$C_2S + H \rightarrow CH + \text{C-S-H} \tag{3.30}$$

Figure 3.22 shows the infrared spectra of LCCM after hydration. The infrared spectra can be used to analyze the hydration products of LCCM at the molecular structure level. Table 3.7 lists the infrared absorption wavenumbers corresponding to the minerals in the samples.

Figure 3.22a shows the infrared spectra of LCCM-A at different hydration ages (1 day, 3 days, 28 days, and 90 days). From the infrared spectra, it can be observed that the characteristic absorption peak of AFt appears at 1100 cm^{-1}, which is caused by

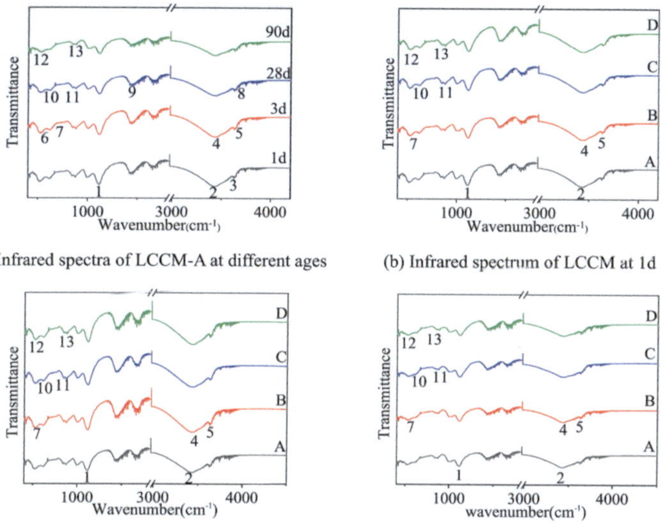

(a) Infrared spectra of LCCM-A at different ages

(b) Infrared spectrum of LCCM at 1d

(c) Infrared spectrum of LCCM at 28 d.

(d) Infrared spectrum of LCCM at 90d.

Fig. 3.22 Infrared spectrum

Table 3.7 Infrared frequency characteristic table

No	Group	Reference frequency (cm^{-1})	Mineral	References
1	$V_3(SO_4)$	1110	AFt	Fernández-Carrasco and Vázquez (1996)
2	$V_2(H_2O)$	3422	AFt	Fernández-Carrasco and Vázquez (1996)
3	OH	3635	AFt	Fernández-Carrasco and Vázquez (1996)
4	OH	3430	AH_3	Fernández-Carrasco and Vázquez (1996)
5	OH	3660	AH_3	Fernández-Carrasco and Vázquez (1996)
6	(AlO_6)	539	C_4ACH_{11}	Trezza and Lavat (2001)
7	$V_4(CO_3)$	716	C_4ACH_{11}	Trezza and Lavat (2001)
8	OH	3676	C_4ACH_{11}	Trezza and Lavat (2001)
9	$V_3(CO_3)$	1437	$CaCO_3$	Trezza and Lavat (2001)
10	(AlO_4)	616	C_3A_4S	Fernández-Carrasco and Vázquez (1996)
11	(AlO_4)	882	C_3A_4S	Fernández-Carrasco and Vázquez (1996)
12	(SiO_4)	520	C_2S	Fernández-Carrasco and Vázquez (1996)
13	(SiO_4)	847	C_2S	Fernández-Carrasco and Vázquez (1996)

the asymmetric stretching vibration of (SO_4). Additionally, the absorption peaks at 3422 and 3635 cm^{-1} correspond to the asymmetric stretching vibrations of (H_2O) and (OH), respectively (Fernández-Carrasco & Vázquez, 1996). The stretching vibrations of (OH) in AH_3 mainly appear at 3430 and 3660 cm^{-1} (Fernández-Carrasco & Vázquez, 1996). The asymmetric stretching vibration of the (AlO_4) functional group in $C_4A_3\check{S}$ is primarily observed at 882 cm^{-1}. The absorption peak at 616 cm^{-1} is attributed to the vibrational coupling between (AlO_4) and (SO_4)

When the sample was hydrated for 1 day, the characteristic absorption peak of $C_4A_3\check{S}$ could still be observed in the infrared spectrum, but its intensity was significantly lower than that of the AFt peak. This indicates that most of the $C_4A_3\check{S}$ had hydrated and generated a large amount of AFt during the first day of hydration. As the hydration age increased, the absorption peak of $C_4A_3\check{S}$ gradually weakened. This suggests that $C_4A_3\check{S}$ continued to hydrate throughout the hydration process. These results are consistent with the XRD analysis. Additionally, the absorption peak at 1437 cm^{-1} is caused by the stretching vibration of (CO_3) in calcium carbonate crystals. As the hydration reaction progressed, this absorption peak significantly weakened, especially at 90 days. This may be due to the participation of calcium carbonate in the hydration reaction.

At 3 days of hydration, monocarbonate-type hydrated calcium aluminate (AFm-CO_3) was detected in the sample, as shown in Eq. 3.32 (Ma et al., 2019). The absorption peaks of AFm-CO_3 at 539, 716, and 3676 cm^{-1} are attributed to the stretching vibrations of (AlO_6), (CO_3), and (OH), respectively (Trezza & Lavat, 2001). For C_2S, its main absorption peaks are located at 520 and 847 cm^{-1}. These correspond to the bending vibration and asymmetric stretching vibration of the (SiO_4) tetrahedron, respectively. Before 28 days, the absorption peaks of C_2S showed no significant changes, but they significantly weakened at 90 days. This indicates that the hydration of C_2S mainly occurs after 28 days. This phenomenon explains the continuous strength growth of the sample during the later hydration stage, which is consistent with the previous analysis results.

Through the analysis of the infrared spectra in Fig. 3.22b–d, it can be observed that the infrared spectra of the samples are generally consistent across different hydration ages. In the early stages of hydration, all samples generated a large amount of AFt and small amounts of AFm and AH_3. At this stage, a small amount of unhydrated $C_4A_3\check{S}$ and a large amount of C_2S remained in the samples. When the hydration age reached 90 days, the absorption peaks of $C_4A_3\check{S}$ in all samples almost disappeared, and the infrared absorption peaks of C_2S significantly weakened. This indicates that both $C_4A_3\check{S}$ and C_2S continued to hydrate in the later stages of hydration. Due to the minor differences in the infrared spectra of the samples, it is difficult to analysis the effect of steel slag content on hydration using infrared spectroscopy. In the following discussion, TG-DSC will be utilized to analysis the influence of steel slag content on hydration.

$$3C_4A_3\check{S} + 2C\check{C} + 72H \rightarrow 2C_4A\check{C}H_{11} + C_3A_3C\check{S} \times H_{32} + 6AH_3 \tag{3.31}$$

The results of the thermogravimetric comprehensive analysis of the samples are shown in Figs. 3.23, 3.24, 3.25 and 3.26. The temperatures corresponding to the endothermic decomposition of the hydration products are as follows:

(a) The dehydration endothermic characteristic peak of AFt occurs around 130–140 °C (Radwan & Heikal, 2005).

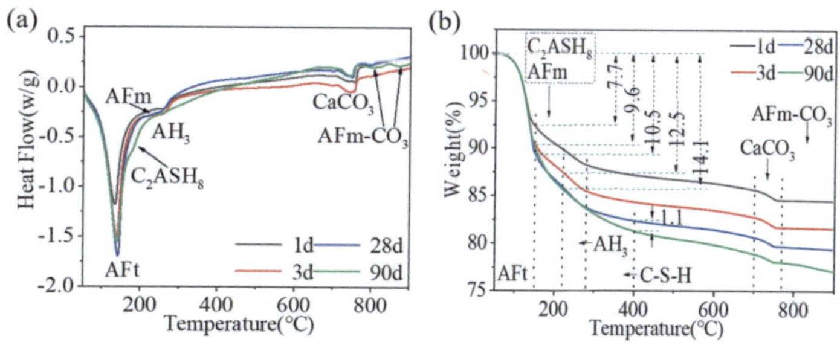

Fig. 3.23 **a** DSC curves of LCCM-A at different curing ages, **b** TG curves of LCCM-A at different curing ages

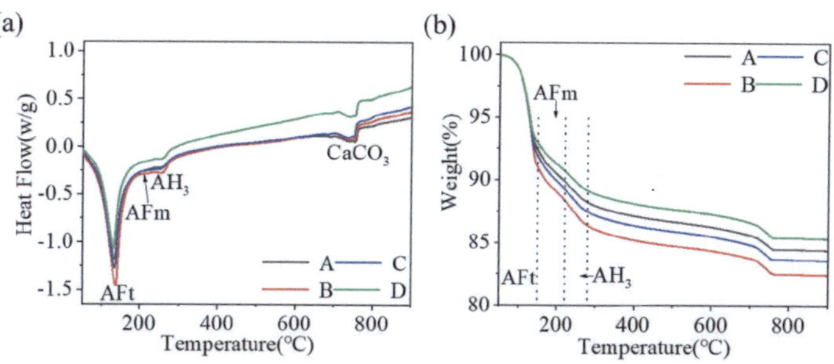

Fig. 3.24 **a** DSC curves of LCCM after hydration for 1 day, **b** TG curves of LCCM after hydration for 1 day

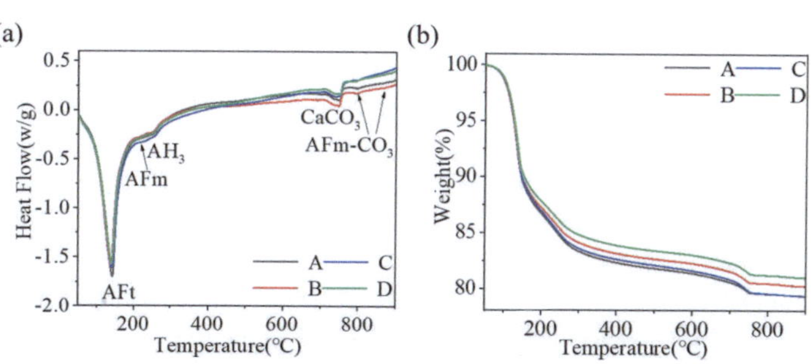

Fig. 3.25 **a** DSC curves of LCCM after hydration for 28 days, **b** TG curves of LCCM after hydration for 28 days

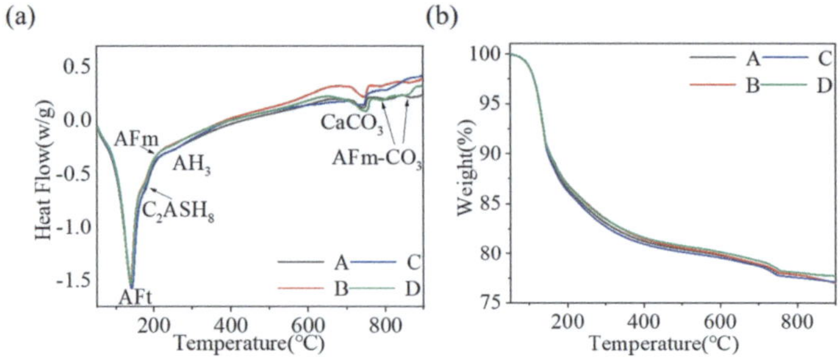

Fig. 3.26 **a** DSC curves of LCCM after hydration for 90 days, **b** TG curves of LCCM after hydration for 90 days

(b) The endothermic characteristic peak of AFm is around 22 5 °C (Radwan & Heikal, 2005).
(c) The weight loss temperature of the crystal phase C_2ASH_8 is about 160 °C (Winnefeld & Lothenbach, 2010).
(d) The dehydration endothermic peak of AH_3 is within the temperature range of 250–280 °C (Wang et al., 2017).
(e) The two endothermic peaks of monocarbonate-type sulphate aluminate hydrate ($AFm\text{-}CO_3$) are concentrated in the range of 800–900 °C (Hargis et al., 2014).

Figure 3.23 shows the thermogravimetric analysis (TGA) curves of sample LCCM-A at different hydration ages. By analyzing the heat flow and mass loss curves, it can be observed that the sample exhibits a rapid hydration rate in the early stages. At 1 day of hydration, a large amount of AFt and small amounts of AFm, and AH_3 were generated in the sample, as shown in Eq. 3.33 (Wang et al., 2017). This is the reason for the high compressive strength of LCCM in the early hydration stage. When the hydration age reached 3 days, the types of hydration products in the sample did not change, but the intensity of the AFt endothermic peak significantly increased, with a mass loss increase of approximately 25%. This is the result of the continued hydration of $C_4A_3\check{S}$. After 3 days, the formation rate of $C_4A_3\check{S}$ hydration products noticeably slowed, leading to a significant reduction in the hydration rate of the sample. This is likely the reason for the slow strength growth of LCCM during the 3–28 days hydration period. These findings are consistent with previous discussions.

From the mass loss curves, it can be observed that when the hydration age reached 28 days, the content of AFt increased only slightly, but the mass loss of AFm increased by approximately 24%. This indicates that a small amount of $C_4A_3\check{S}$ continued to hydrate slowly. However, due to the depletion of gypsum, the hydration products of $C_4A_3\check{S}$ were mainly AFm.

Further analysis of the heat flow curves revealed that when the hydration age reached 90 days, endothermic peaks corresponding to the decomposition of C_2ASH_8 and $AFm\text{-}CO_3$ appeared on the heat flow curve, as shown in Eqs. 3.32–3.34 (Hargis et al., 2014). In the corresponding temperature range, the mass loss curve also showed significant mass loss. This indicates that the formation of $AFm\text{-}CO_3$ and C_2ASH_8

mainly occurred in the later hydration stage. Additionally, compared to the 28-day mass loss curve, the 90-day mass loss curve exhibited more pronounced mass loss in the 280–400 °C temperature range. This is attributed to the mass loss of C-S-H with a mass loss of approximately 1.1% (Stepkowska et al., 2004). The reasons for these phenomena may be related to the low solubility of calcium carbonate and the low reactivity of C_2S. Due to the relatively low solubility of $CaCO_3$, its hydration reactions with $C_4A_3\check{S}$ and AFm mainly occurred in the later hydration stage. Similarly, due to the low reactivity of C_2S, its reactions with AH_3 and its own hydration became significant only after 28 days. From Eqs. 3.30, 3.32, to 3.34, it can be observed that the formation of AFm-CO_3 and C_2ASH_8 consumes some calcium carbonate and AH_3. This is also the reason for the reduction in the endothermic peaks of calcium carbonate and AH_3 in the sample. At this stage, the endothermic peak of AFt also decreased, as gypsum was nearly depleted in the later hydration stage, and some AFt transformed into AFm (Jiahui et al., 2006). The continued hydration of LCCM in the later stage is the main reason for its high long-term strength.

The heat flow and mass loss curves of LCCM-A to LCCM-D after 1 day of hydration are shown in Fig. 3.24. From the heat flow curves, it can be observed that the main hydration products of LCCM at 1 day are AFt, AFm, and AH_3. However, by comparing the mass loss curves, significant differences in the content of hydration products among different samples can be identified. Analysis reveals that the content of hydration products initially increases and then decreases with the increase in steel slag content. This indicates that the hydration rate of LCCM in the early stage first increases and then decreases with the increase in steel slag content, which is consistent with previous analysis results. When the hydration age reaches 28 days, the heat flow and mass loss curves of all samples gradually converge, as shown in Fig. 3.25. This suggests that the influence of steel slag content on the hydration process diminishes as hydration time increases. Further observation of the heat flow and mass loss curves at 90 days of hydration (Fig. 3.26) reveals that the curves of all samples almost completely overlap. This indicates that the composition and content of hydration products in all samples are nearly identical at this stage, and the influence of steel slag content on the later hydration stage is significantly reduced.

$$C_4A_3\check{S} + 18H \rightarrow C_3A \cdot C\,\check{S} \cdot H_{12} + 2AH_3 \tag{3.32}$$

$$3C_3A \cdot C\check{S} \cdot H_{12} + 2C\check{C} + 18H \rightarrow 2C3A \cdot 3C\check{C} \cdot 11H + C_3A_3C\check{S} \cdot H_{32} \tag{3.33}$$

By integrating the thermal analysis results at 1 day, 28 days, and 90 days, it can be concluded that in the early hydration stage, steel slag primarily regulates the hydration process of LCCM by influencing the hydration of $C_4A_3\check{S}$. However, as the hydration reaction progresses, the influence of steel slag on LCCM hydration gradually weakens. This conclusion is consistent with previous analysis results.

3.5.4 Micro-morphology of Hydration Products

The microstructures of LCCM-A at hydration ages of 1 day, 3 days, and 90 days are shown in Fig. 3.27.

(a) Microstructure of LCCM-A after 1 day hydration

(b) Microstructure of LCCM-A after hydration for 3 days

(c) Microstructure of LCCM-A after hydration for 90 days

Fig. 3.27 Micromorphology of the LCCM-A at 1 day, 3 days, 90 days. a = AH_3, e = AFt, m = AFm, s = $C_4A_3\check{S}$, f = AFm-CO_3, g = C_2ASH_8

From Fig. 3.27a, it can be observed that after 1 day of hydration, a small amount of columnar and tubular AFt, as well as flaky AH_3, are present in the LCCM-A sample. At this stage, unhydrated $C_4A_3\check{S}$ is also observed. Although previous analyses indicate that a large amount of AFt should have formed at 1 day of hydration, its crystals may not have fully developed and were still in the early stages of nucleation and growth. Therefore, AFt is not fully visible in the microstructure.

Figure 3.27b shows the microstructure of the sample after 3 days of hydration. Compared to 1 day, the amount of unhydrated $C_4A_3\check{S}$ significantly decreased, while a large amount of columnar AFt and flaky AH_3 became clearly visible. This indicates that the sample continued to hydrate between 1 and 3 days, with hydration products such as AFt and AH_3 continuously developing. This is consistent with previous analysis results. The interweaving of AFt and AH_3 and their embedding in the matrix significantly enhanced the internal structural density of the hydrated sample. This is also the main reason for the high compressive strength of the sample in the early hydration stage

As shown in Fig. 3.27c, at 90 days of hydration, the sample's texture became further densified. This change is mainly attributed to the continued reactions of minerals such as C_2S and $C_4A_3\check{S}$ in the later hydration stage. The microstructure shows a reduction in AH_3 content, indicating its participation in hydration reactions and transformation into other hydration products. Meanwhile, the columnar AFt significantly thickened and formed tightly arranged bundles. These phenomena indicate that the sample continued to hydrate in the later stage. The interweaving of AFt formed a spatial skeleton structure, while the cementing and filling effects of hydration products such as AH_3 and AFm further enhanced the sample's density. The combined effects of these factors ensured that the LCCM-A sample maintained high strength and good mechanical properties in the later hydration stage.

3.6 Conclusions

This section successfully prepared LCCM based on the synergistic effects of various solid wastes, including coal gangue, carbide slag, steel slag, and desulfurization gypsum. Through systematic research, the following main conclusions were drawn:

(1) The mineral composition and structural evolution of LCCM clinker are closely related to the calcination temperature. When the calcination temperature is below 1100 °C, the main mineral components of the clinker are f-CaO, grossular, gehlenite, anorthite, and $C_4A_3\check{S}$, accompanied by a small amount of C_2S. When the temperature rises to 1100–1200 °C, intermediate minerals decompose or react with $CaSO_4$ and CaO in solid phases, forming $C_4A_3\check{S}$ and C_2S. When the temperature exceeds 1200 °C, the mineral formation process weakens, primarily manifesting as the evolution and development of crystal structures. The crystal

sizes of $C_4A_3\check{S}$ and C_2S increase with rising temperature. When the calcination temperature reaches 1350 °C, the clinker minerals undergo melting and decomposition.

(2) The LCCM clinker prepared by maintaining a temperature of 1320 °C for 40 min consists mainly of $C_4A_3\check{S}$ and C_2S as the primary mineral phases, along with small amounts of C_4AF and $CaSO_4$. This cementitious material exhibits excellent mechanical properties. Its 3-day hydration compressive strength reaches the level of P.O. 42.5 cement at 28 days. After 90 days of hydration, the compressive strength further increases to 112 MPa.

(3) The hydration characteristics of LCCM show distinct stage-specific features. In the early hydration stage, due to the high hydration activity of $C_4A_3\check{S}$, the material exhibits a rapid hydration rate, which is the main reason for its early high strength development. In the later hydration stage, hydration products such as C-S-H, C_2ASH_8, and AFm-CO_3 are generated in the paste. The formation of these products mainly results from the continuous hydration of low-activity C_2S and low-solubility $CaCO_3$, ensuring sustained strength growth in the later stages.

(4) The content of steel slag has a significant impact on the hydration behavior of LCCM. Research shows that within a specific range of steel slag content, the hydration activity of LCCM initially increases and then decreases with increasing steel slag content. This effect is mainly observed in the early hydration stage, while its influence on the later hydration process is relatively limited.

References

Bullerjahn, F., Schmitt, D., & Haha, M. B. (2014). Effect of raw mix design and of clinkering process on the formation and mineralogical composition of (ternesite) belite calcium sulphoaluminate ferrite clinker. *Cement and Concrete Research, 59*, 87–95.

Dai, J., Wang, Q., Lou, X., Bao, X., Zhang, B., Wang, J., & Zhang, X. (2021). Solution calorimetry to assess effects of water-cement ratio and low temperature on hydration heat of cement. *Construction and Building Materials, 269*, 121222.

Darquennes, A., Espion, B., & Staquet, S. (2013). How to assess the hydration of slag cement concretes? *Construction and Building Materials, 40*, 1012–1020.

Ectors, D., Neubauer, J., & Goetz-Neunhoeffer, F. (2013). The hydration of synthetic brownmillerite in presence of low Ca-sulfate content and calcite monitored by quantitative in-situ-XRD and heat flow calorimetry. *Cement and Concrete Research, 54*, 61–68.

Fernández-Carrasco, L., & Vázquez, T. (1996). Aplicación de la espectroscopia infrarroja al estudio de cemento aluminoso. *Materiales De Construcción, 46*(241), 39–51.

Gazdič, D., Fridrichová, M., Dvořák, K., Halešová, A., & Všianský, D. (2019). Influence of firing regime and potassium ions on synthetic preparation of Belite clinker. *Solid State Phenomena, 296*, 64–69.

Guo, H., & Xie, J. (2011). Thermodynamics and kinetics of calcium sulphoaluminate. *Journal of Wuhan University of Technology-Materials Science Edition, 26*(4), 719–722.

Hajjaji, M., & Kacim, S. (2004). Clay-calcite mixes: Sintering and phase formation. *British Ceramic Transactions, 103*(1), 29–32.

Hargis, C. W., Telesca, A., & Monteiro, P. J. (2014). Calcium sulphate aluminate (Ye'elimite) hydration in the presence of gypsum, calcite, and vaterite. *Cement and Concrete Research, 65*, 15–20.

He, Z., Li, Y., & Shao, Y. (2017). Low-temperature synthesis of C4AF by self-propagating combustion reaction. *Advances in Cement Research, 29*(1), 33–43.

Huang, Y., Shen, X., Ma, S., Chen, L., & Zhong, B. (2007). Effect of Fe2O3 on the formation of calcium sulphoaluminate mineral. *Kuei Suan Jen Hsueh Pao/journal of the Chinese Ceramic Society, 35*(4), 485–488.

Hughes, T. L., Methven, C. M., Jones, T. G., Pelham, S. E., Fletcher, P., & Hall, C. (1995). Determining cement composition by Fourier transform infrared spectroscopy. *Advanced Cement Based Materials, 2*(3), 91–104.

Idrissi, M., Diouri, A., Damidot, D., Greneche, J. M., Talbi, M. A., & Taibi, M. (2010). Characterisation of iron inclusion during the formation of calcium sulphate aluminate phase. *Cement and Concrete Research, 40*(8), 1314–1319.

Idrissi, M., Diouri, A., Talbi, M. A., Sassi, O., & Damidot, D. (2015). Hydration behavior of iron doped calcium sulphate aluminate phase at room temperature (N: 1P-368). In *INVACO2: International Seminar, Innovation & Valorization in Civil Engineering & Construction Materials, Morocco, Rabat*.

Jiahui, P., Jianxin, Z., & Jindong, Q. (2006). The mechanism of the formation and transformation of ettringite. *Journal of Wuhan University of Technology-Materials Science Edition, 21*, 158–161.

Kalinkin, A. M., Kalinkina, E. V., Zalkind, O. A., & Makarova, T. I. (2005). Chemical interaction of calcium oxide and calcium hydroxide with CO 2 during mechanical activation. *Inorganic Materials, 41*, 1073–1079.

Koumpouri, D., & Angelopoulos, G. N. (2016). Effect of boron waste and boric acid addition on the production of low energy belite cement. *Cement and Concrete Composites, 68*, 1–8.

Kurdowski, W. (2014). *Cement and Concrete Chemistry*. Springer Science & Business.

Li, D. (1984). Fourier transform infrared spectrum study of sulphoaluminate cement. *Journal of the Chinese Ceramic Society, 01*, 121–127.

Li, J. (2013). *Study on high belite-sulphoaluminate cement* [Doctoral dissertation, Wuhan University of Technology, Wuhan].

Li, L., Zhang, Y., Zhang, Y., Sun, J., & Hao, Z. (2016). The thermal activation process of coal gangue selected from Zhungeer in China. *Journal of Thermal Analysis and Calorimetry, 126*, 1559–1566.

Liu, L., Zhang, W., Ren, X., Ye, J., Zhang, J., & Qian, J. (2021). Formation, structure, and thermal stability evolution of ternesite based on a single-stage sintering process. *Cement and Concrete Research, 147*, 106519.

Ma, B., Ma, M., Shen, X., Li, X., & Wu, X. (2014). Compatibility between a polycarboxylate superplasticizer and the belite-rich sulphate aluminate cement: Setting time and the hydration properties. *Construction and Building Materials, 51*, 47–54.

Ma, J., Yu, Z., Ni, C., Shi, H., & Shen, X. (2019). Effects of limestone powder on the hydration and microstructure development of calcium sulphoaluminate cement under long-term curing. *Construction and Building Materials, 199*, 688–695.

Mollah, M. Y. A., Yu, W., Schennach, R., & Cocke, D. L. (2000). A Fourier transform infrared spectroscopic investigation of the early hydration of Portland cement and the influence of sodium lignosulfonate. *Cement and Concrete Research, 30*(2), 267–273.

Mutch, G. A., Anderson, J. A., & Vega-Maza, D. (2017). Surface and bulk carbonate formation in calcium oxide during CO2 capture. *Applied Energy, 202*, 365–376.

Nirmala, G., & Viruthagiri, G. (2014). FT-IR characterization of articulated ceramic bricks with wastes from ceramic industries. *Spectrochimica Acta Part a: Molecular and Biomolecular Spectroscopy, 126*, 129–134.

Peng, W. (1982). *Infrared Spectra of Minerals* (p. 303). Science Press.

Perná, I., Šupová, M., & Hanzlíček, T. (2018). Gehlenite and anorthite formation from fluid fly ash. *Journal of Molecular Structure, 1157*, 476–481.

Ptáček, P., Opravil, T., Šoukal, F., Havlica, J., & Holešinský, R. (2013). Kinetics and mechanism of formation of gehlenite, Al–Si spinel and anorthite from the mixture of kaolinite and calcite. *Solid State Sciences, 26*, 53–58.

Radwan, M. M., & El, H. S. (2011). Hydration characteristics of tetracalcium alumino-ferrite phase in the presence calcium carbonate. *Ceramics-Silikaty, 55*(4), 337–342.

Radwan, M. M., & Heikal, M. (2005). Hydration characteristics of tricalcium aluminate phase in mixes containing β-hemihydate and phosphogypsum. *Cement and Concrete Research, 35*(8), 1601–1608.

Ren, C. (2019). *Experimental study on preparation and application of sulphoaluminate phosphate cement composites using industrial solid waste* [Doctoral dissertation, Shandong University].

Ren, C., Wang, W., Yao, Y., Wu, S., & Yao, X. (2020). Complementary use of industrial solid wastes to produce green materials and their role in CO2 reduction. *Journal of Cleaner Production, 252*, 119840.

Stepkowska, E. T., Blanes, J. M., Franco, F., Real, C., & Pérez-Rodrıguez, J. L. (2004). Phase transformation on heating of an aged cement paste. *Thermochimica Acta, 420*(1–2), 79–87.

Taylor, H. F. (1997). *Cement Chemistry* (Vol. 2, p. 459). Thomas Telford.

Test methods for heat of hydration of cement: GB/T 12959-2008 [S]. (2008). General Administration of Quality Supervision, Inspection and Quarantine of the People's Republic of China.

Test methods for water requirement of normal consistency, setting time and soundness of the Portland cement: GB/T1346-2011[S]. (2012). Beijing: General Administration of Quality Supervision, Inspection and Quarantine of the People's Republic of China.

Trezza, M. A., & Lavat, A. E. (2001). Analysis of the system 3CaO· Al2O3–CaSO4· 2H2O–CaCO3–H2O by FT-IR spectroscopy. *Cement and Concrete Research, 31*(6), 869–872.

Wang, P., Li, N., & Xu, L. (2017). Hydration evolution and compressive strength of calcium sulphoaluminate cement constantly cured over the temperature range of 0 to 80 C. *Cement and Concrete Research, 100*, 203–213.

Wen, L. (1989). *Infrared Spectroscopy of Minerals* (pp. 62–65). Chongqing University Press.

Winnefeld, F., & Lothenbach, B. (2010). Hydration of calcium sulphate aluminate cements—Experimental findings and thermodynamic modelling. *Cement and Concrete Research, 40*(8), 1239–1247.

Yin, Y., Yin, J., Zhang, W., Tian, H., Hu, Z., Ruan, M., ... & Chen, D. (2018). FT-IR and micro-Raman spectroscopic characterization of minerals in high-calcium coal ashes. *Journal of the Energy Institute, 91*(3), 389–396.

Zapata, J. F., Gomez, M., & Colorado, H. A. (2017). Structure-property relation and Weibull analysis of calcium aluminate cement pastes. *Materials Characterization, 134*, 9–17.

Zhang, S., & Guolong, S. (2011). Batching calculation of Belite Sulphoaluminate cement clinker. *Journal of Nanchang University (Engineering & Technology), 33*(1), 30–32.

Zhao, G. (2004). *The research of desulfurization behavior of cement raw and the formation mechanism of sulphoaluminate* [Doctoral dissertation, Master's thesis. Wuhan University of Technology].

Chapter 4
AC Impedance Spectrum of Low-Carbon Cementitious Materials During the Hydration Process

4.1 Introduction

The AC impedance method, also known as electrochemical impedance spectroscopy (EIS), serves as an analytical technique for investigating hydration reaction mechanisms in electrochemical systems. This methodology involves applying small-amplitude sinusoidal AC voltage perturbations at varying frequencies to an electrochemical system while measuring the corresponding alternating current response. Alternatively, direct measurement of system impedance (or admittance) can be conducted to analyze hydration mechanisms. By establishing the correlation between perturbation signals and electrode system responses, the AC impedance characteristics of electrochemical systems can be determined. Based on these characteristics, an equivalent circuit model can be proposed according to the system's intrinsic properties, thereby enabling the exploration of kinetic processes within the electrochemical system. Through analysis of AC impedance responses—particularly variations in equivalent circuit component parameters (including charge transfer resistance, double-layer capacitance, and diffusion coefficients during charge transfer)—kinetic parameters of the electrochemical system can be effectively estimated.

The hydration process of cementitious materials involves ion transfer, rearrangement, and redistribution phenomena. When mixed with water, these materials can be conceptualized as specialized electrochemical systems. Notably, AC impedance measurements are influenced not only by the solid matrix and pore structure of hardened pastes but also by the conductivity of pore solutions. Compared with DC measurement techniques, AC impedance testing demonstrates enhanced accuracy due to reduced directional ion migration in pore solutions under sinusoidal current excitation. The parameters derived from equivalent circuit modeling of AC impedance data effectively reflect internal pore structure evolution and mechanical property development in cementitious materials. While conventional microstructural characterization methods typically involve destructive testing, AC impedance

C. Yan et al., *Low-Carbon Cementitious Materials with 100% Solid Wastes*,
https://doi.org/10.1007/978-981-96-8742-8_4

technology offers distinct advantages of simplicity, rapid implementation, and non-destructive nature. These merits have driven its rapid development in applications related to cement-based materials.

This study systematically investigates the hydration process and microstructural evolution of low-carbon cementitious materials prepared from coal gangue, carbide slag, and desulfurized gypsum using alternating current (AC) impedance testing. An equivalent circuit model was developed to quantify the degree of hydration by analyzing impedance spectra under varying water-to-binder ratios and hydration ages. The evolution of impedance parameters was correlated with hydration kinetics through complementary techniques including calorimetry, chemically bound water measurement, and pore structure analysis. Final, an autoregressive integrated moving average (ARIMA) model was successfully implemented to predict and validate key performance parameters (impedance characteristics, bound water content, and porosity) across different mix formulations. The integrated experimental and modeling approach provides new insights into the structure–property relationships of sustainable cementitious materials.

4.2 Test Survey

For low-carbon cementitious materials with varying water-to-binder ratios, mortar specimens were subjected to alternating current (AC) impedance testing. The specimens were prepared with a binder-to-sand ratio of 1:3 and water-to-binder ratios of 0.42, 0.47, and 0.52. The mixed paste was cast into specially designed molds (40 mm × 40 mm × 160 mm) for impedance measurements. Two stainless steel mesh electrodes (3 mm aperture, 42 mm × 55 mm × 1 mm) were fixed 20 mm from the inner walls on both longitudinal sides of the molds. After filling, the molds were consolidated on a vibrating table to minimize air entrainment. The cast specimens were sealed with plastic membranes to reduce moisture evaporation from the upper surface and interior, then cured under standard conditions in a controlled environment chamber. Specimens remained undemolded throughout the testing period and were extracted for measurements at designated hydration ages: 1 h, 2 h, 4 h, 6 h, 12 h, 1 d, 3 d, 7 d, 14 d, and 28 d.

AC impedance measurements were conducted using a frequency sweep from 0.01 to 1 MHz with a minimum high-frequency current range of 2 mA and an AC amplitude of 10 mV. Ten data points were recorded per frequency decade under impedance-frequency sweep mode. Electrical connections were established using alligator-clipped cables securely attached to the stainless steel meshes, ensuring stable contact. A schematic diagram of the testing configuration is illustrated in Fig. 4.1.

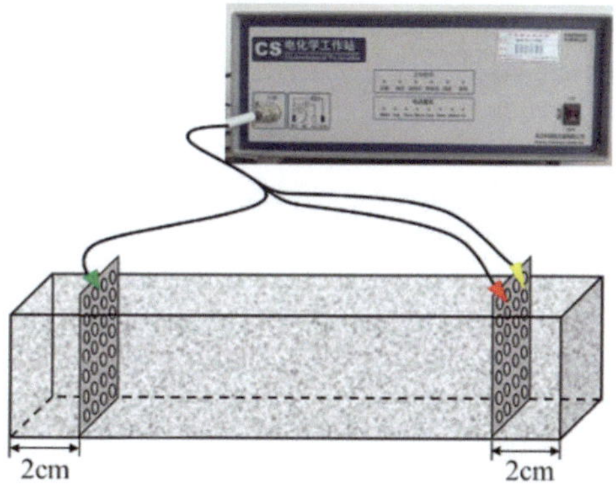

Fig. 4.1 AC impedance measurement

4.3 Hydration and AC Impedance Characteristics of LCCM

4.3.1 Hydration Characteristics Analysis

Hydration product test results and analysis

Figure 4.2 shows the XRD testing results for the low-carbon cementitious materials. By analyzing the XRD patterns, it can be observed that the main mineral components in the prepared low-carbon cementitious materials include $C_4A_3\overline{S}$, C_2S, and $CaSO_4$. Among these, $C_4A_3\overline{S}$ primarily provides the early strength of the cementitious materials, while the increase in later strength is supplemented by the hydration of C_2S. This indicates that the low-carbon cementitious materials prepared using coal gangue, calcium carbide slag, and desulfurization gypsum as the main raw materials are feasible.

Figure 4.3 shows the XRD patterns of low-carbon cementitious materials with a water-to-binder ratio of 0.47 at different hydration ages. By analyzing the patterns,

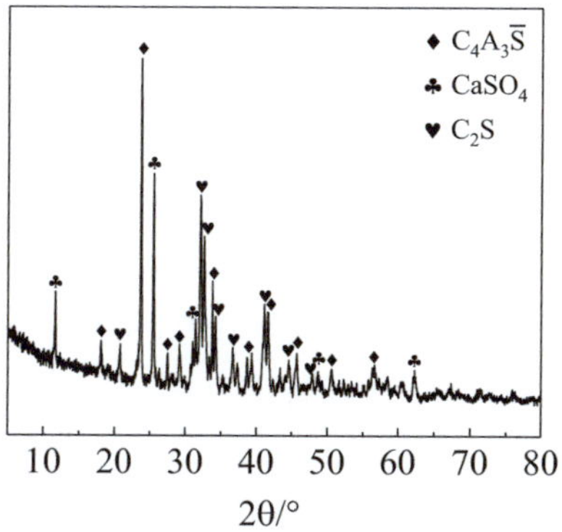

Fig. 4.2 XRD pattern of low-carbon cementitious materials

we can gain insights into the changes in the mineral composition of the cementitious materials during the hydration process. It can be seen from the Fig. 4.3 that after 6 h of hydration, the diffraction peak for $CaSO_4$ is no longer observable in the cementitious material system. At this time, most of the $C_4A_3\overline{S}$ has reacted with all the $CaSO_4$ and water in the hydration reaction (Eq. 4.1), resulting in a significant amount of ettringite (AFt) formation. A weak diffraction peak for $C_4A_3\overline{S}$ can still be observed at a diffraction angle of 23.78°, but it becomes weaker as the hydration time increases. Concurrently, the diffraction peaks for AFt gradually strengthen due to the continued hydration of a small amount of $C_4A_3\overline{S}$, leading to an increase in the hydration product AFt. Regarding the other mineral component, C_2S, its diffraction peaks are mainly concentrated between diffraction angles of 31° and 33°. By the time of 28 days of hydration, there are no significant changes in the primary mineral components of the cementitious materials, although the diffraction peak for C_2S slightly decreases. This indicates that the hydration rate of C_2S is slower than that of $C_4A_3\overline{S}$. At this stage, C_2S undergoes hydration reactions as described in Eqs. 4.2–4.3, which, to some extent, supplements the development of late strength in low-carbon cementitious materials.

$$C_4A_3\overline{S} + 2C\overline{S} + 38H \rightarrow C_3A \cdot 3C\overline{S} \cdot 32H + 2AH_3 \tag{4.1}$$

$$C_2S + H \rightarrow CH + C\text{-}S\text{-}H \tag{4.2}$$

$$C_2S + AH_3 + 5H_2O \rightarrow C_2ASH_8 \tag{4.3}$$

In the XRD patterns, no diffraction peaks for aluminum gel (AH_3) or hydrated silicate gel (C-S-H) are observed among the hydration products. This may be attributed to their poor crystallinity or low quantities, making them undetectable in the XRD spectra. Additionally, no diffraction peaks for $Ca(OH)_2$ are found throughout the patterns, which is due to the reaction between $Ca(OH)_2$ and C-S-H to form AFt, resulting in the consumption of $Ca(OH)_2$ within the system.

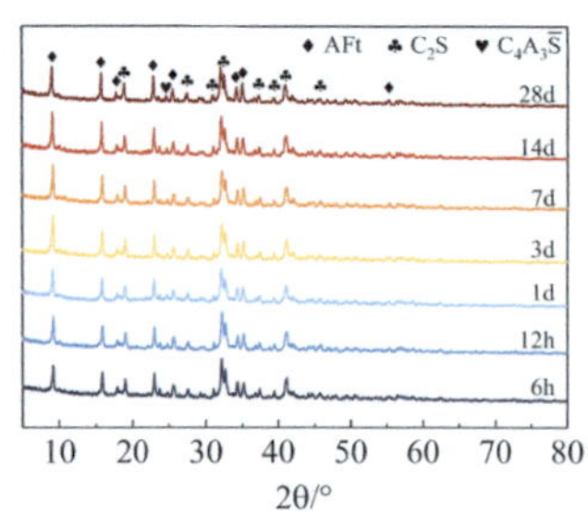

Fig. 4.3 XRD patterns of low-carbon cementitious materials at different hydration ages

Hydration heat test results and analysis

The heat released during the hydration of cementitious materials results from chemical reactions between multiple components and water. The hydration heat release directly reflects the reaction rate and degree of hydration in cementitious systems. Investigating the hydration progression of low-carbon cementitious materials enables a clearer understanding of their internal reaction mechanisms, thereby facilitating the establishment of corresponding hydration models. To examine the heat evolution characteristics of low-carbon cementitious materials, isothermal calorimetry was employed to measure both the hydration heat release rate and cumulative heat release over a 3-day hydration period, as shown in Figs. 4.4 and 4.5, respectively.

Taking the test results of low-carbon cementitious materials with a water-to-binder ratio of 0.42 as an example, it can be seen from Fig. 4.4 that there are three main exothermic peaks during the hydration of low-carbon cementitious materials. The first exothermic peak appears at approximately 15 min of hydration, corresponding to the dissolution heat when the low-carbon cementitious materials first come into contact with water (Sargam & Wang, 2021). This peak is the highest among the three exothermic peaks, with the maximum dissolution heat rate reaching 0.0458 W/g. In the initial stage of hydration, the exothermic heat rate increases linearly because there is a certain amount of free gypsum and free calcium in the low-carbon cementitious materials, which react upon contact with water, releasing a substantial amount of

Fig. 4.4 Hydration heat release curve

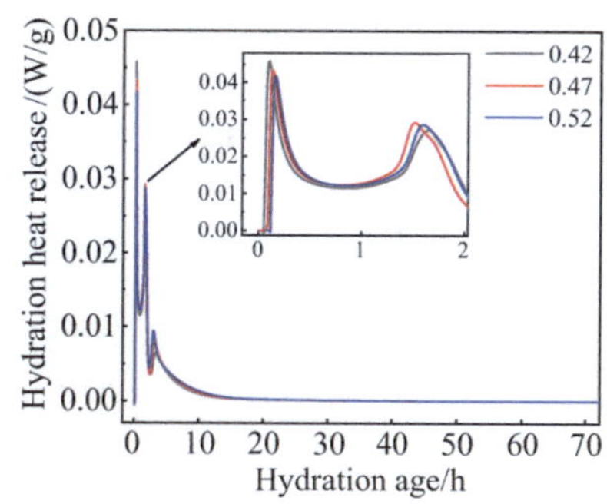

Fig. 4.5 Cumulative hydration heat curve

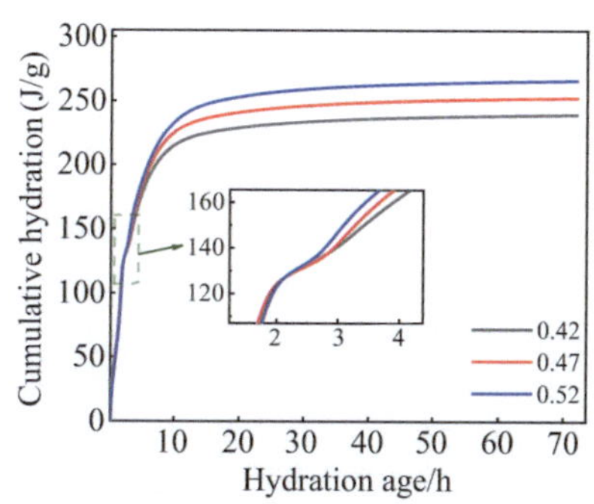

dissolution heat and reaction heat, thereby promoting the dissolution and hydration of other minerals within the cementitious material. As hydration time extends, a second exothermic peak occurs approximately 1 h after mixing. The release of hydration heat enters an acceleration phase during this period, where $CaSO_4$ continuously dissolves into ions, and $C_4A_3\overline{S}$ reacts with $CaSO_4$ and water to form AFt precipitation, releasing a large amount of heat. The heat release rate during this stage is relatively high, around 0.027 W/g, which, although lower than that of the first exothermic peak, shows an extended duration of heat release. This indicates a higher content of $C_4A_3\overline{S}$ in the low-carbon cementitious material, resulting in a longer hydration time, with the acceleration phase continuing until the $SO4^{2-}$ ions in the solution are consumed (Chi et al., 2022; Tambara et al., 2020). The third exothermic peak appears around 3 h after mixing with water and corresponds to the reaction heat generated from the localized formation of AFt and monosulfate sulfoaluminate (AFm) around the particles of the cementitious material. The corresponding heat release rate during this phase is approximately 0.0064 W/g. This reaction heat occurs after the formation of a framework that connects solid particles, mainly influenced by ion diffusion and migration. After the third exothermic peak, the hydration heat rate of the cementitious material gradually slows down, indicating that the hydration process has entered a deceleration phase.

Figure 4.4 shows that the difference in dissolution exothermic peak values among cementitious materials with different water-to-binder ratios is not significant. The peak dissolution heat rates for cementitious materials with water-to-binder ratios of 0.42, 0.47, and 0.52 are 0.0458, 0.0433, and 0.0417 W/g, respectively. For a given quantity of cementitious material particles, the added water is adsorbed onto the surface of the particles, and the dissolution heat should remain constant. The reaction systems of cementitious materials with different water-to-binder exhibit notable differences at the third exothermic peak. The exothermic peak value gradually decreases with a lower water-to-binder ratio. The time at which the third hydration exothermic peak appears advances with an increase in the water-to-binder ratio. This is because, in reaction systems with higher water-to-binder ratios, cementitious material particles are more likely to undergo hydration reactions, resulting in faster reaction rates.

From the cumulative hydration heat release curve of the low-carbon cementitious materials, a distinct inflection point is observed at around 3 h of hydration. This characteristic point corresponds to the time when the third hydration heat peak occurs, indicating a transition of hydration products from AFt to AFm. After 3 h of hydration, the hydration reaction systems of the cementitious materials with the three different water-to-binder ratios begin to develop at different rates. As seen in Fig. 4.5, for the same hydration age, a higher water-to-binder ratio corresponds to a greater cumulative heat release. This is because a lower water-to-binder ratio reduces the amount of water available for hydration within the cementitious materials, leading to a decreased degree of hydration. Approximately 20 h after mixing with water, the cumulative heat release curves of low-carbon cementitious materials with different water-to-binder ratios all enter a slow growth phase.

At 72 h of hydration, the cumulative heat release values for low-carbon cementitious materials with water-to-binder ratios of 0.42, 0.47, and 0.52 are 239.18, 252.78, and 266.17 J/g, respectively. The cumulative heat release values of the hydration reaction systems for cementitious materials with different water-to-binder ratios at various time intervals are shown in Table 4.1. It is important to note that although a higher heat of hydration indicates a higher degree of hydration, the strength of the hardened paste of the cementitious materials does not solely depend on the degree of hydration. The development trend of compressive strength in relation to the water-to-binder ratio is opposite to that of the heat of hydration.

Chemically bound water test results and analysis

The chemically bound water in the hardened cementitious material paste refers to the moisture lost due to evaporation or internal cohesion. The degree of hydration of the cementitious material paste is primarily related to the chemically bound water, which continuously increases with the accumulation of hydration products (Darquennes et al., 2013). Chemically bound water can be used to assess the long-term hydration kinetics of low-carbon cementitious materials and is significant for exploring the hydration rate and the amount of hydration products generated.

The amounts of chemically bound water in low-carbon cementitious materials with different water-to-binder ratios at various hydration ages are shown in Fig. 4.6. It can be seen that the amount of chemically bound water increases significantly during the first 3 days of hydration, reaching approximately 70% of $W_{28\text{d}}$ (the amount of bound water after 28 days of hydration) at 3 days. This indicates that low-carbon cementitious materials have a rapid hydration rate in the early stages, primarily because they contain a relatively high amount of reactive $C_4A_3\overline{S}$, allowing the cementitious materials to acquire a significant amount of bound water through hydration reactions in a short time. From 7 to 28 days of hydration, the amount of chemically bound water gradually stabilizes, as most of the $C_4A_3\overline{S}$ has been consumed during the earlier hydration reactions. During this period, the unhydrated minerals primarily consist of the less reactive C_2S. Furthermore, a higher water-to-binder ratio results in a relatively higher amount of chemically bound water, indicating that the degree of hydration increases with the water-to-binder ratio. This observation is consistent with the results of the hydration heat tests.

Through the analysis of the chemically bound water content, it is found that low-carbon cementitious materials exhibit a rapid reaction rate in the early stages of hydration, allowing for a high degree of hydration to be achieved in a short period. This is also the reason for the high early strength of low-carbon cementitious

Table 4.1 Cumulative hydration heat values of low carbon cementitious materials (J/g)

w/b	6 h	12 h	24 h	48 h	72 h
0.42	191.32	220.63	231.03	237.07	239.18
0.47	198.69	232.11	243.79	250.46	252.78
0.52	203.50	241.85	256.13	263.63	266.17

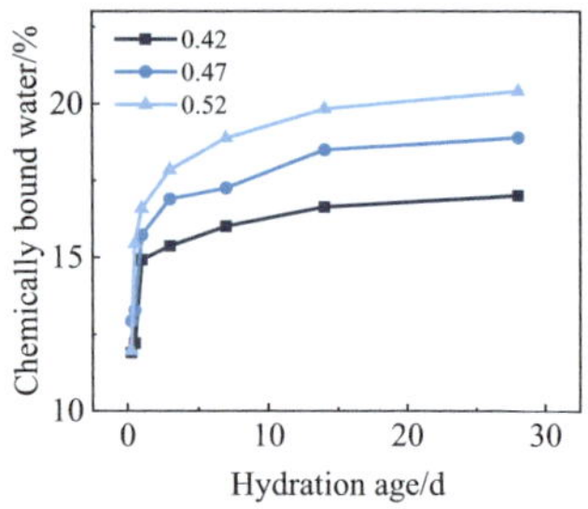

Fig. 4.6 Chemically bound water content of low-carbon cementitious materials with different water-to-binder ratios

materials. As the hydration age increases, the amount of chemically bound water in low-carbon cementitious materials with different water-to-binder ratios continues to increase, but the rate of increase gradually decreases. This indicates that sustained hydration reactions continue to occur within the cementitious materials during the mid to late stages of hydration.

4.3.2 AC Impedance Spectrum Analysis

For the analysis of AC impedance testing results, the most commonly used method is to establish corresponding equivalent circuit models to fit the reactions occurring within the cementitious materials. The equivalent circuit model is composed of various circuit elements arranged in series and parallel. In studying the hydration of cementitious materials, the essence of AC impedance technology is to investigate the variations of the parameters of each element in the equivalent circuit under the influence of a sinusoidal alternating current.

The equivalent circuit model is a primary method for the analysis and application of AC impedance. In the study of the electrical properties of cementitious materials, many researchers have proposed equivalent circuit models suitable for cementitious materials, such as the "unit model" (McCarter, 1994), "brick model" (Gu et al., 1994), and T and I models (Xie et al., 1993). These models can help explain the impact of solid, liquid, and solid–liquid interfaces on the AC impedance behavior of cementitious materials. The equivalent circuit model adopted in this study is shown in Fig. 4.7. This model considers the Faradaic processes between the paste and the electrode interface, while also accounting for the dispersion effects arising from the rough solid surfaces during charge transfer reactions. The code for the equivalent circuit model is $R_s(CPE_1(R_{ct1}W_1))(CPE_2(R_{ct2}W_2))$. Here, R_s represents the pore solution resistance of the low-carbon cementitious materials, R_{ct1} indicates the resistance generated within the cementitious material due to the charge transfer process, W_1 is the Warburg impedance resulting from ionic diffusion within the cementitious material, and CPE_1 represents the constant phase element describing the solid/liquid interface of the cementitious matrix. R_{ct2} is the resistance generated during the charge transfer process between the low-carbon cementitious material and

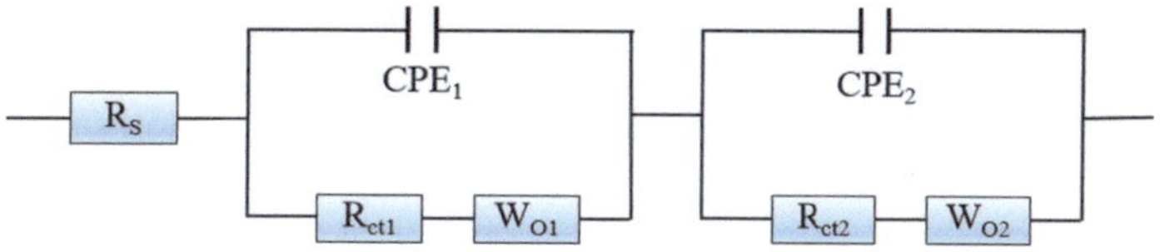

Fig. 4.7 Equivalent circuit model

the electrode interface, W_2 is the Warburg impedance due to ionic diffusion at the electrode surface, and CPE_2 indicates the constant phase element characterizing the double layer properties at the cementitious materials-electrode interface. This study primarily investigates the changes in impedance parameters within the cementitious matrix.

Nyquist diagram analysis

Figure 4.8 shows a representative Nyquist plot for the hardened paste samples of low-carbon cementitious materials, which consists of a semicircle in the high-frequency range and a straight line in the low-frequency range. It can be observed that the AC impedance spectrum presents a quasi-Randles type. In the Nyquist plot, the vertical axis represents the imaginary part of the AC impedance (resistive value), while the horizontal axis indicates the real part of the resistive value, with frequency gradually decreasing as the horizontal coordinate increases. At the intersection of the high-frequency semicircle and the low-frequency straight line, the horizontal coordinate corresponds to the charge transfer resistance R_{ct1} in the cementitious material, at which point the absolute value of the imaginary part of the resistance is minimized. In general, a complete Nyquist plot for cementitious materials should feature three capacitive arcs; however, in actual testing, due to instrument limitations, the first ultra-high frequency capacitive arc may often disappear. Therefore, in most cases, the Nyquist plot essentially starts from the second capacitive arc. The second capacitive arc corresponds to the semicircle in the high-frequency region, and the high-frequency semicircle tested in this experiment is within the range of $10–10^6$ Hz, where the electrochemical reactions occur at the solid/liquid interface within the cementitious material matrix. The third capacitive arc is represented by a large-diameter semicircle in the low-frequency range. Due to the instrumental limitations on the maximum testing frequency, the low-frequency portion of the Nyquist plot can only be represented as a straight line. The low-frequency range tested in this experiment is between 0.01 and 10 Hz, and the electrochemical reactions occurring in this low-frequency region take place between the cementitious material matrix and the stainless steel mesh interface. During this stage, the actual AC impedance curve appears as a slanted line that deviates at an angle of 45° from the real axis. This deviation is caused by the dispersion effect between the hardened paste of low-carbon cementitious material and the electrode interface, resulting in a deflection of the semicircle in the high-frequency region and an angle deviation of the low-frequency straight line from the horizontal axis. In the equivalent circuit model, this

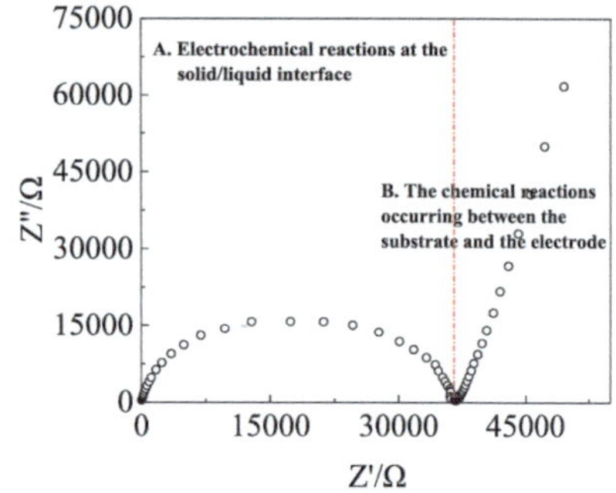

Fig. 4.8 Representative Nyquist diagram and frequency distribution

is represented as the double layer capacitance being replaced by a constant phase element (*CPE*).

Figures 4.9, 4.10 and 4.11 present Nyquist plots of low-carbon cementitious materials with water-to-binder ratios of 0.42, 0.47, and 0.52, respectively, at varying hydration ages. Figure 4.9a displays Nyquist plots from initial mixing to hydration durations of 1, 2, 4, and 6 h, while Fig. 4.9b illustrates plots corresponding to hydration periods of 12 h, 1 d, 3 d, 7 d, 14 d, and 28 d. As shown in Fig. 4.9, the impedance value at the interface between the high-frequency semicircular arc and the low-frequency straight line progressively increases with hydration time. The real impedance components measured at 1, 2, 4, 6, 12 h, 1 d, 3 d, 7 d, 14 d, and 28 d are 7,140 Ω, 11,600 Ω, 18,776 Ω, 23,881 Ω, 26,106 Ω, 31,550 Ω, 32,184 Ω, 33,665 Ω, 34,644 Ω, and 36,097 Ω, respectively. This trend is visually manifested as a continuous enlargement of the high-frequency semicircular arc diameter and a progressive rightward shift in the arc position within the plots.

It can be observed that in the early stages of hydration, the diameter of the high-frequency semicircle changes significantly. As the hydration of the cementitious material progresses, the rate of increase in the diameter of the high-frequency semicircle becomes smaller. This is because, in the early stages of hydration, the generation rate of AFt is relatively fast, while the hydration reaction rate slows down as

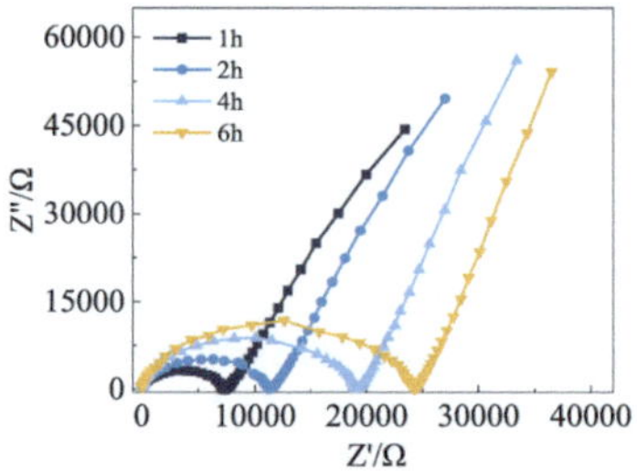

(a) Early hydration

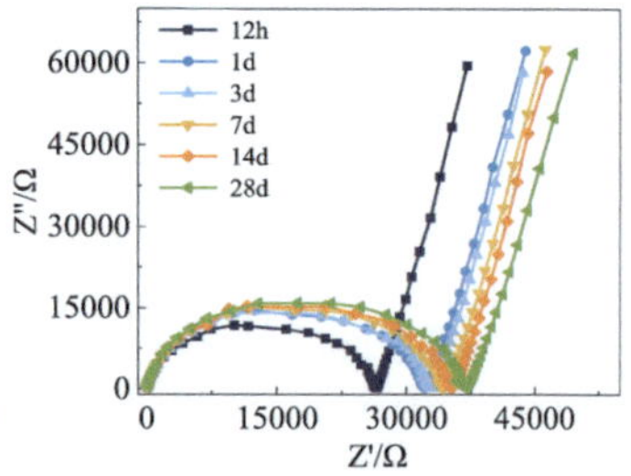

(b) Mid-late hydration

Fig. 4.9 Nyquist curve of low-carbon cementitious materials with water-binder ratio of 0.42

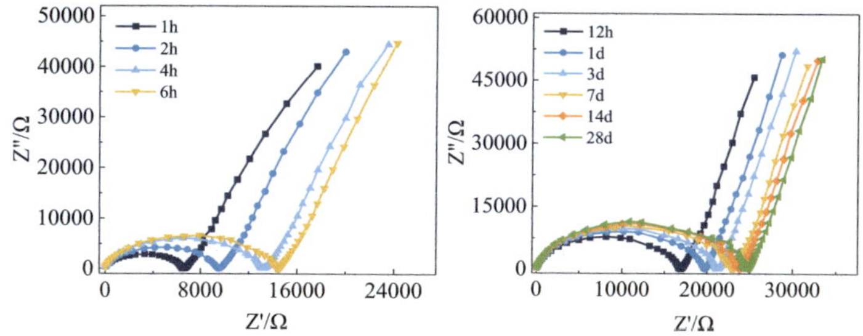

(a) Early hydration (b) Mid-late hydration

Fig. 4.10 Nyquist curve of low-carbon cementitious materials with water-binder ratio of 0.47

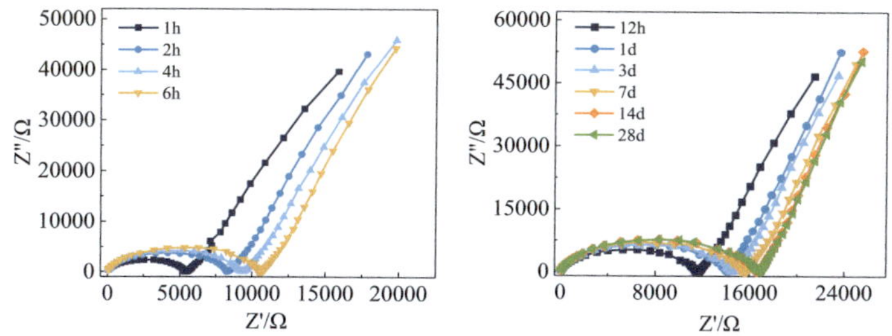

(a) Early hydration (b) Mid-late hydration

Fig. 4.11 Nyquist curve of low-carbon cementitious materials with water-binder ratio of 0.52

hydration time increases. Among the ions in the pore solution of low-carbon cementitious materials, OH^- ions have a higher conductivity. After a period of hydration, the SO_4^{2-} ions in the pore solution of the paste react with Ca^{2+} ions from the hydration products to form calcium sulfate precipitate, resulting in a decrease in the concentration of OH^- ions in the pore solution of the cementitious material system, which subsequently leads to a deterioration in the conductivity of the cementitious materials. At the same time, as hydration continues within the cementitious material, it continuously dissolves into ions and generates various hydration products. The formation of these products reduces the amount of free ions within the pores of the cementitious materials, which leads to a decrease in ion concentration. As the hydration process progresses, the conductivity of the internal system of the cementitious materials deteriorates, making the transport of free ions within increasingly difficult. Therefore, the resistance value R_{ct1} of the charge transfer process continues to increase, which is reflected in the Nyquist curve as a continual increase in the diameter of the semicircle in the high-frequency region.

From Figs. 4.9, 4.10 to 4.11, it can be seen that the diameter of the high-frequency semicircle in the Nyquist plot of the cementitious material paste with a water-to-binder ratio of 0.42 is significantly larger than that of the semicircles for other water-to-binder ratios. This means that as the water-to-binder ratio decreases, the diameter of the high-frequency semicircle in the Nyquist plot increases. This is because an increase in the water-to-binder ratio leads to an increase in the porosity of the paste at the same hydration age. As the water-to-binder ratio increases, the contact area between water and unhydrated cementitious particles also increases, resulting in more cementitious particles being dissolved and an increase in the concentration of ions within the paste. Therefore, as the water-to-binder ratio increases, the movement of free ions within the cementitious material system becomes easier, and the resistance value generated from the ion transfer process continues to decrease, which is reflected in the Nyquist plot as a gradual decrease in the diameter of the high-frequency semicircle. In the low-frequency phase of the AC impedance, the slope of the straight line decreases with a reduction in the water-to-binder ratio, indicating that as the water-to-binder ratio decreases, the structure of the hardened paste becomes denser, making ion diffusion more difficult.

Bode diagram analysis

The impedance modulus |Z| represents the hindrance effect of current conduction in the hardened paste of cementitious materials. In the Bode plot, the presence of a peak or a trough indicates that the system has a time constant within the tested frequency range. The Bode plots for low-carbon cementitious materials with different water-to-binder ratios are shown in Fig. 4.12.

It can be observed that in the impedance-frequency plots of low-carbon cementitious materials with different water-to-binder ratios, the low-frequency section appears as a straight line at approximately 135°, while the mid-high frequency section is a line parallel to the horizontal axis. As the hydration time increases, the slope of the low-frequency straight line gradually increases, and the impedance modulus value in the mid-high frequency section shows an increasing trend. For the phase angle curve, in the high-frequency range of 10^3–10^5 Hz, the phase angle generally trends upward with increasing hydration time. In the medium frequency range of 1–10^3 Hz, changes are not significant, while in the low-frequency range of 0.01–1 Hz, the phase angle shows the opposite trend of that observed in the high-frequency range, gradually decreasing with increasing hydration time.

4.3.3 AC Impedance Parameters Analysis

Pore solution resistance R_s

Under the same conditions, the magnitude of the pore solution resistance R_s of low-carbon cementitious materials is inversely related to the ion concentration in the

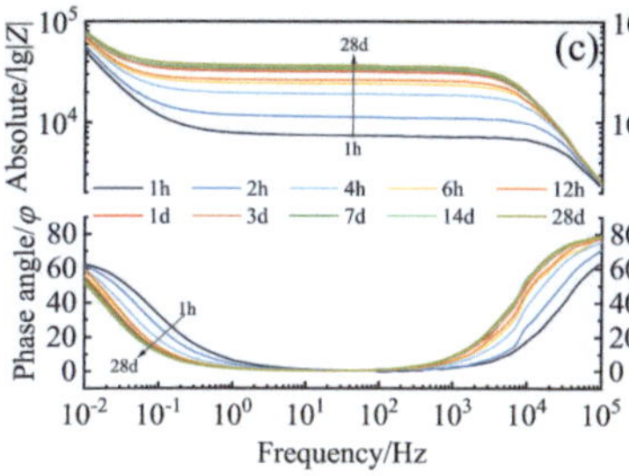

(a) Water-to-binder ratio of 0.42

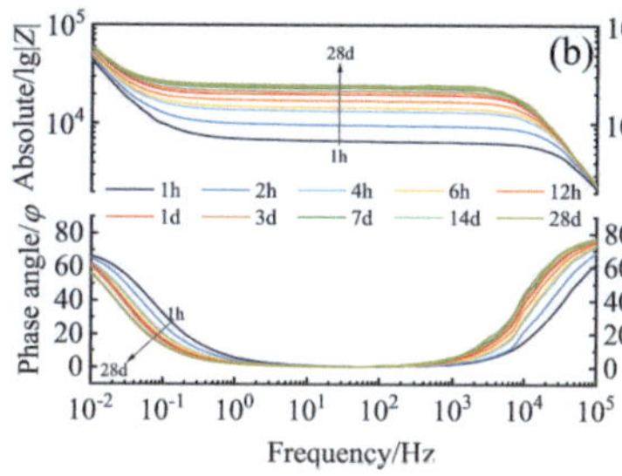

(b) Water-to-binder ratio of 0.47

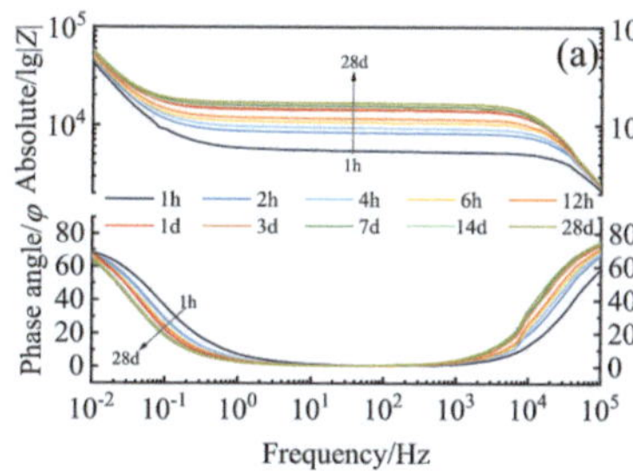

(c) Water-to-binder ratio of 0.52

Fig. 4.12 Bode diagram of low-carbon cementitious materials with different water-to-binder ratio

pore solution and the porosity of the cementitious materials. Therefore, its value is determined by both the total porosity and the ion concentration in the pore solution.

Figure 4.13 shows the variation curves of *R*s for low-carbon cementitious materials with different water-to-binder ratios at various hydration ages. Theoretically, as the hydration process progresses, the concentration of ions dissolved in the cementitious paste rapidly increases and quickly reaches saturation within a short time, including anions (SO_4^{2-}, OH^-) and cations (Ca^{2+}, Al^{3+}). The presence of charged ions leads to a decrease in the pore solution resistance *R*s of the cementitious paste. However, the data show that as hydration time increases, *R*s generally presents an increasing trend. This is because the hydration products generated by the hydration process continuously fill the pore structure, and the decrease in porosity is greater than the increase in solution ion concentration. As the hydration process continues, the rate of increase in *R*s gradually diminishes, because once hydration reaches a certain level, the ion concentration in the pore solution tends to stabilize, and there is less space available for hydration products to occupy, resulting in a gradual decrease in the extent of porosity reduction. Among the two factors determining *R*s, the reduction in total porosity dictates the trend of *R*s. For low-carbon cementitious materials with different water-to-binder ratios, the variation curves of the *R*s resistance parameter values exhibit a consistent trend. At the same hydration age, as the water-to-binder ratio increases, the *R*s value gradually decreases. This is because, under the same

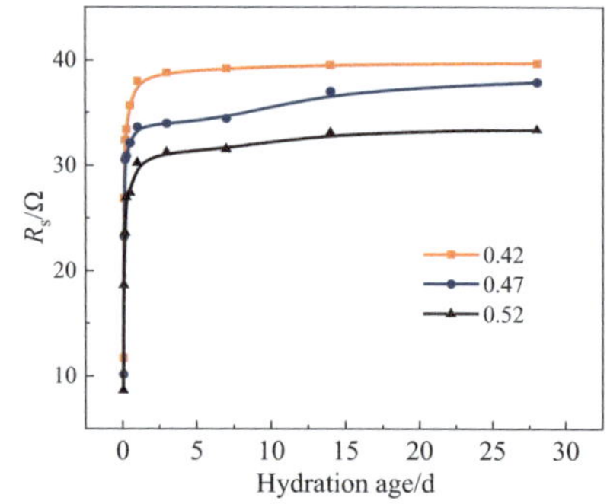

Fig. 4.13 Change curves of pore solution resistance R_s of low-carbon cementitious materials

conditions, an increase in the water-to-binder ratio leads to a lower ion concentration in the pore solution within the cementitious material system, while the porosity continues to increase. Thus, porosity remains the primary factor affecting *R*s, leading to its gradual decrease.

Charge transfer resistance R_{ct1}

R_{ct1} represents the resistance generated by charge transfer of hydration electrons within low-carbon cementitious materials, and its magnitude is closely related to the degree of hydration of the cementitious material and the quantity of hydration products. Factors influencing the value of R_{ct1} include porosity, average pore size, and the concentration of ions in the pore solution. Due to the rapid early hydration rate characteristic of low-carbon cementitious materials, the ion concentration in the pore solution tends to stabilize early in the hydration process, meaning that changes in the impedance parameter R_{ct1} effectively reflect variations in the degree of hydration of the cementitious material.

For low-carbon cementitious materials with different water-to-binder ratios, the variation curves of the impedance parameter R_{ct1} are shown in Fig. 4.14. It is evident that as the hydration process continues, the resistance R_{ct1} associated with the charge transfer reactions of hydration electrons exhibits an increasing trend, reaching over 90% of the 28-days value by the 3-days hydration stage. This is attributed to the characteristic rapid early hydration of low-carbon cementitious materials. As hydration progresses, the structure of the material gradually becomes denser, and the solid phase volume of the cementitious material increases, which makes ionic transfer more difficult. The growth rate of R_{ct1} stabilizes during the hydration period from 3 to 28 days, where the C_2S in the cementitious material begins to consume a significant amount of the hydration product AH_3, with the consumption rate of AH_3 exceeding its generation rate. This also results in a more rapid reduction of total porosity than the increase in ion concentrations (such as Al^{3+}, Ca^{2+}, SO_4^{2-}, etc.).

In addition, during the mid to late stages of hydration, C_2S and C_2ASH_8 in the cementitious material system undergo reactions as described in Eq. 4.4. At this stage, C_2ASH_8 cannot stably exist, and the reduction of Al^{3+} cannot suppress the increase of Ca^{2+} concentration in the pore solution of the cementitious material. The hydration

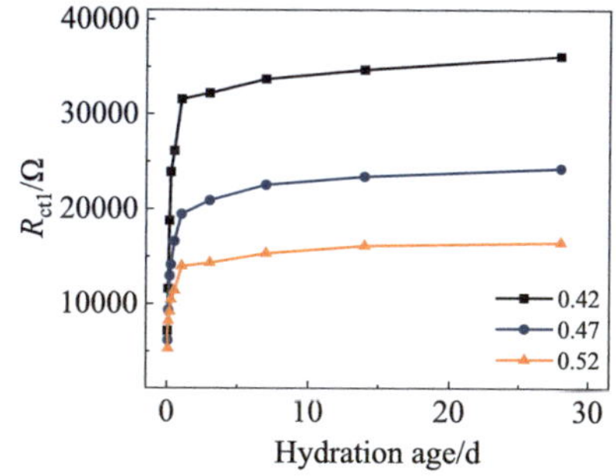

Fig. 4.14 Pore solution resistance R_{ct1} under different hydration ages

products generated continue to fill the pores, and the reduction in total porosity leads to a continuous increase in R_{ct1} from 3 to 28 days of hydration.

$$C_2S + C_2ASH_8 \rightarrow C_3ASH_4 + \text{C-S-H} \quad (4.4)$$

As shown in Fig. 4.15, at the same hydration age, the impedance parameter value R_{ct1} gradually decreases with the increase in the water-to-binder ratio. This is because a smaller water-to-binder ratio results in less solution within the low-carbon cementitious materials, leading to a reduction in the ion content within the structure of the paste. As hydration time increases, the trend of the impedance parameter value R_{ct1} decreasing with an increasing water-to-binder ratio becomes more pronounced. This is because a smaller water-to-binder ratio leads to a denser internal structure of the cementitious materials, making it more difficult for ions to migrate within the pore solution. Conversely, when the water-to-binder ratio is higher, the cementitious particles can be highly dispersed, increasing the contact area between water and the cementitious materials, which accelerates the hydration rate. Therefore, as the hydration process continues, the trend of decreasing R_{ct1} values becomes more significant.

Electric double layer capacitance *CPE*

The *CPE* is the electric double layer capacitance of the solid/liquid interface in the hardened paste system of low-carbon cementitious materials (Jain & Neithalath,

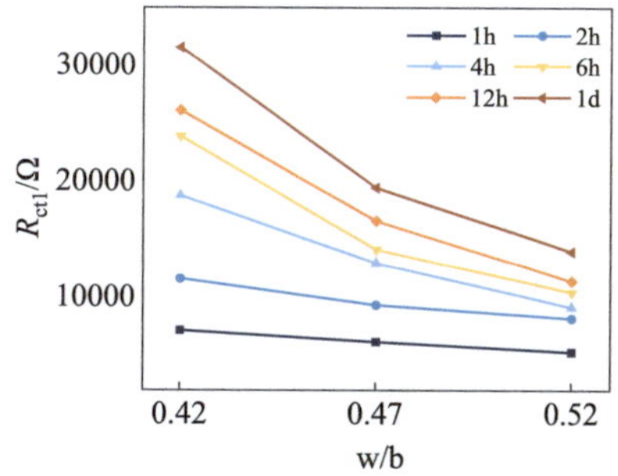

Fig. 4.15 Pore solution resistance R_{ct1} under different water-to-binder ratios

2011). The calculation formula of the electric double layer capacitance CPE is as follows:

$$CPE = \frac{1}{\omega^{*} R_{ct1}} \quad (4.5)$$

where: ω^{*}-the frequency value corresponding to the highest point of the high frequency semicircle. R_{ct1}-charge transfer resistance.

Many factors influence the value of the constant phase element (*CPE*) capacitance, such as changes in the porosity of the material, variations in average pore size, and changes in the ion concentration of the pore solution. Table 4.2 lists the *CPE* values for low-carbon cementitious materials with different water-to-binder ratios at various hydration ages. From the table, it can be observed that as the hydration process progresses, the capacitance value *CPE* shows a decreasing trend, and the rate of decrease is gradually diminishing. The Nyquist plot of the cementitious materials indicates that as hydration time increases, the diameter of the high-frequency semicircle also gradually increases, leading to a corresponding increase in the frequency value at the highest point. As previously analyzed, with the increase in hydration time, the paste structure becomes denser, resulting in a continuous increase in the resistance R_{ct1} during the charge transfer process. Therefore, the reduction in the capacitance value CPE is evident. The capacitance value *CPE* characterizes the electrical properties of the gel formed during hydration; as the hydration process continues, more gel products are generated, gradually filling the pores, which makes the decrease in capacitance values become slower over time.

Resistivity analysis

The variation in resistivity can reflect the hydration process of the cementitious material paste, with resistivity development over hydration time indicating a reduction in porosity and an increase in the solid phase. The main factors influencing resistivity

Table 4.2 *CPE* values of low carbon cementitious materials (pF)

Hydration time	w/b = 0.42	w/b = 0.47	w/b = 0.52
1h	1694.4	1734.2	1706.5
2h	1397.0	1439.5	1514.2
4h	1287.6	1326.1	1409.6
6h	1288.0	1310.7	1347.7
12h	1268.9	1295.9	1352.4
1d	1207.7	1284.2	1312.7
3d	1200.2	1257.7	1261.3
7d	1188.0	1241.7	1251.4
14d	1181.9	1204.5	1249.8
28d	1188.0	1200.4	1235.0

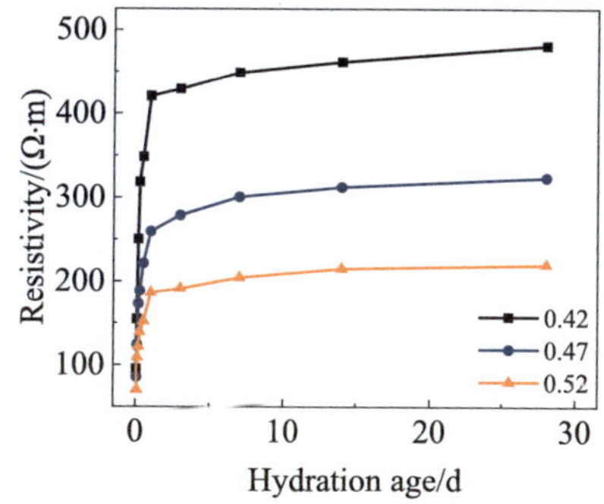

Fig. 4.16 Resistivity curves of low-carbon cementitious materials

are pore structure and pore solution. Equation 4.6 provides the calculation formula for resistivity ρ.

$$\rho = R_{\mathrm{ct1}} \frac{A}{L} \tag{4.6}$$

where: R_{ct1} is the charge transfer resistance obtained from the AC impedance test. A is the area of contact between the stainless steel mesh and the cementitious material paste, taken as $\mathrm{A} = 0.0016\mathrm{m}^2$. L is the effective distance between the two stainless steel meshes, taken as $L = 0.12$ m.

Figure 4.16 shows the variation curves of resistivity for low-carbon cementitious materials with different water-to-binder ratios. Due to the rapid hardening characteristics of the prepared low-carbon cementitious materials, the resistivity increases rapidly during the first 3 days of hydration. During this stage, hydration products are generated quickly, and the increase in resistivity is evident. As the hydration age progresses, the rate of increase in resistivity slows down after the 3-day mark. During this phase, the pores are continuously filled with hydration products, resulting in fewer interconnected pores. Therefore, as the curing age increases, the extent of growth in resistivity decreases. The resistivity of low-carbon cementitious materials decreases with an increasing water-to-binder ratio because, as the water-to-binder ratio decreases, the porosity and connectivity of the pores within the cementitious materials diminish, leading to a denser structure (Wu et al., 2018). This denser micropore structure hinders the transmission of electrical signals (Farzanian et al., 2016). AC impedance testing is a simple and effective method for characterizing changes in the microstructure.

4.3.4 Relationship Between AC Impedance Characteristics and Hydration Degree

1. **Relationship between resistivity and hydration heat**

Figure 4.17 shows the curves of cumulative hydration heat release and resistivity for low-carbon cementitious materials with different water-to-binder ratios over a hydration period of 3 days. It can be observed that during the first 2 h of hydration, the curves of cumulative hydration heat release and resistivity for the hardened paste of low-carbon cementitious materials with different water-to-binder ratios are nearly coincident. After 2 h of hydration, the impact of water-to-binder ratio on resistivity becomes more pronounced compared to its effect on hydration heat release. The variation curve of resistivity indicates that, within the cementitious material systems with different water-to-binder ratios, the slope of the resistivity decreases with an increasing water-to-binder ratio. This is because, within the first 2 h of hydration, the hydration reactions within the entire cementitious material system are mainly controlled by the kinetics of hydration. The charge transfer process within the cementitious material relies on the electrons in the pore solution, and the reaction has not yet entered the charge diffusion control stage. As the hydration time increases, the microstructure of the hardened paste gradually becomes denser, obstructing the interconnected pathways within the system. The hydration reaction transitions from one that is controlled by hydration kinetics to one that is controlled by charge diffusion. During this period, the influence of water-to-binder ratio on the changes in the microstructure of the cementitious material becomes increasingly significant, which is reflected in the marked differences in resistivity at different water-to-binder ratios.

Figure 4.18 shows the relationship curves between cumulative hydration heat release and resistivity for low-carbon cementitious materials with different water-to-binder ratios at hydration ages of 2, 4, 6, 12 h, 1 day, and 3 days. As hydration products continue to form, the cumulative heat release increases, and the resistivity of the paste gradually increases. It can be observed that the resistivity of the hardened

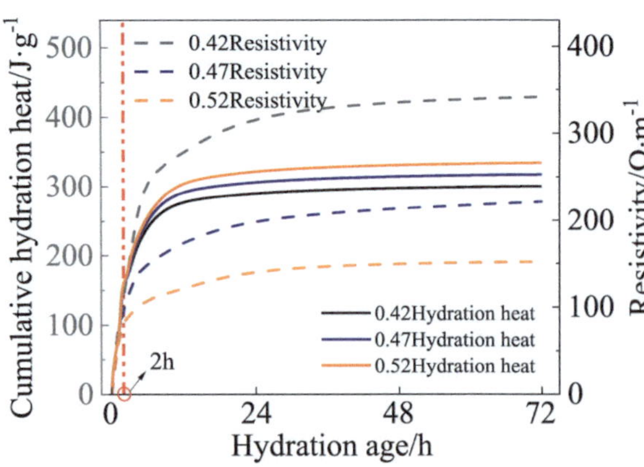

Fig. 4.17 Cumulative hydration heat and resistivity curves of low-carbon cementitious materials

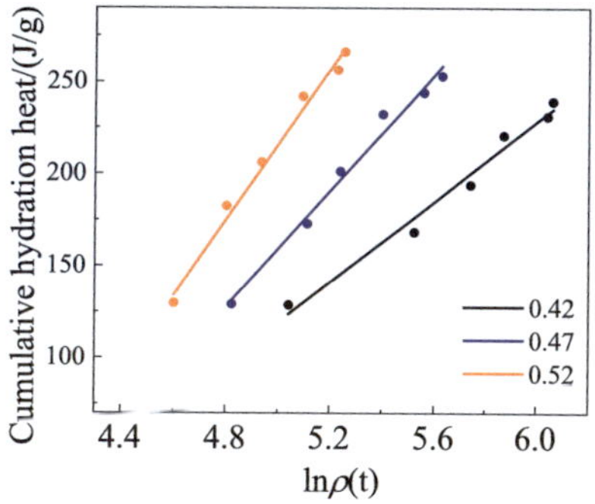

Fig. 4.18 Fitting curves of cumulative hydration heat and resistivity

paste for low-carbon cementitious materials with different water-to-binder ratios steadily increases with the accumulation of hydration heat release, indicating a good correlation between the two.

The fitting Equations for the relationships between cumulative hydration heat release and resistivity at different water-to-binder ratios are represented by Eqs. 4.7–4.9, with correlation coefficients close to 1. Furthermore, the relationship between the cumulative hydration heat release and resistivity of low-carbon cementitious materials can be summarized by Eq. 4.10. This result indicates that the resistivity obtained from AC impedance testing can serve as a real-time monitoring parameter and may be used in the future for predicting the hydration heat of cementitious materials.

$$Q(t)_{0.42} - Q(t_0) = 108.318 \ln(\rho_{0.42}(t)) - 422.713 \; R^2 = 0.976 \tag{4.7}$$

$$Q(t)_{0.47} - Q(t_0) = 157.208 \ln(\rho_{0.47}(t)) - 626.947 \; R^2 = 0.984 \tag{4.8}$$

$$Q(t)_{0.52} - Q(t_0) = 204.702 \ln(\rho_{0.52}(t)) - 607.526 \; R^2 = 0.968 \tag{4.9}$$

$$Q(t) - Q(t_0) = a \ln \rho(t) + b = a\prime \ln(b\prime \rho(t)) \tag{4.10}$$

where: $Q(t)$ represents the cumulative hydration heat release of low-carbon cementitious materials at time t, J/g. $Q(t_0)$ represents the cumulative hydration heat release of low-carbon cementitious materials at time t_0 (when ($t_0 = 2$ h)), J/g. $\rho(t)$ represents the resistivity of low-carbon cementitious materials at time t, Ω·m. a' and b' are the fitting coefficients.

2. **Relationship between resistivity and chemically bound water**

The degree of hydration of cementitious materials in the early stages can be characterized by the heat of hydration, while the degree of hydration in the later stages can be represented by the amount of chemically bound water. The linear relationship curve between the chemically bound water content and the logarithm of resistivity

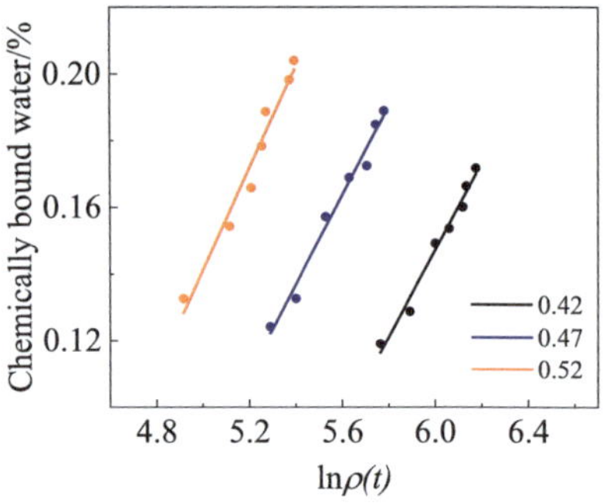

Fig. 4.19 Fitting curves of chemical combined water content and resistivity

for low-carbon cementitious materials with different water-to-binder ratios is shown in Fig. 4.19.

It can be observed that there is a strong correlation between the logarithm of the resistivity of hardened paste from low-carbon cementitious materials and the amount of chemically bound water, with the correlation coefficient close to 1. The fitting Equations for the relationship between the chemically bound water content and the logarithm of resistivity for cementitious materials of different water-to-binder ratios are presented as Eqs. 4.11–4.13. Further summarization leads to the relationship Equation between the chemically bound water content and the logarithm of resistivity for the low-carbon cementitious paste, as shown in Eq. 4.14.

$$W_{0.42}(t) = -0.632 \ln \rho_{0.42}(t) + 0.130\ R^2 = 0.976 \tag{4.11}$$

$$W_{0.47}(t) = -0.589 \ln \rho_{0.47}(t) + 0.134\ R^2 = 0.978 \tag{4.12}$$

$$W_{0.52}(t) = -0.635 \ln \rho_{0.52}(t) + 0.155\ R^2 = 0.957 \tag{4.13}$$

$$W(t) = m \ln \rho(t) + n = m\prime \ln(n\prime \rho(t)) \tag{4.14}$$

where: $W(t)$ represents the amount of chemically bound water in low-carbon cementitious materials at time t, expressed as a percentage (%). $\rho(t)$ represents the resistivity of low-carbon cementitious materials at time t, Ω·m. m' and n' are the fitting coefficients.

3. **Relationship between resistivity and hydration degree**

Generally, the degree of hydration of cementitious materials can be calculated using the ratio of the chemically bound water amount at a specific age to the total bound water amount, or the ratio of the cumulative hydration heat release at a specific age to the total cumulative hydration heat release, as shown in Eq. 4.15.

$$\alpha = \frac{W(t)}{W_0} = \frac{Q(t)}{Q_0} \tag{4.15}$$

where: α represents the degree of hydration of the cementitious material. $W(t)$ is the amount of chemically bound water in 1 g of low-carbon cementitious material at time t, expressed as a percentage (%). W_0 is the amount of chemically bound water in 1 g of low-carbon cementitious material when fully hydrated, expressed as a percentage (%). $Q(t)$ is the cumulative hydration heat release for 1 g of low-carbon cementitious material at time t, J/g. Q_0 is the cumulative hydration heat release for 1 g of low-carbon cementitious material when fully hydrated, J/g.

Based on the existing equations for calculating the hydration degree of cementitious materials, we analyze the hydration reaction parameters of hardened paste of low-carbon cementitious materials by incorporating the discussion on the correlation between the resistivity of cementitious materials, cumulative hydration heat, and chemically bound water. By introducing the resistivity, an alternating current impedance parameter, into the original theoretical Equations, we can extend the theoretical equations for the hydration degree of cementitious materials, as shown in Eq. 4.16.

$$\alpha = \frac{W(t)}{W_0} = \frac{Q(t)}{Q_0} = K\frac{\ln(k_1\rho(t))}{\ln(k_2\rho_s(t))} \tag{4.16}$$

where: $\rho(t)$ represents the resistivity of the hardened paste of 1 g of cementitious material at time t, Ω·m. $\rho_s(t)$ represents the resistivity of the hardened paste of 1 g of cementitious material when fully hydrated, Ω·m. k_1, k_2, and K are fitting parameters.

By establishing the relationship between resistivity and key physical and chemical property parameters, the hydration characteristics of the cementitious materials can be monitored in real time. The resistivity obtained from AC impedance testing can serve as an additional computational parameter to be incorporated into the theoretical equation for hydration degree. By using electrical signals to represent the hydration degree of the cementitious material system, this method overcomes the errors of traditional hydration testing methods. It provides a new approach for assessing the hydration degree of cementitious materials and deriving their theoretical equations for hydration degree.

4.4 Microstructure Evolution and AC Impedance Characteristics of LCCM

4.4.1 Equivalent Circuit Model Based on Microstructure

This study references the microscopic structural model proposed by Song (2000), which identifies three primary conductive pathways in low-carbon cementitious materials: 1. Continuous Conductive Pathway (CCP): Composed of continuously connected conductive pores. 2. Discontinuous Conductive Pathway (DCP): Formed by the alternation of the cementitious material system and conductive pores, where

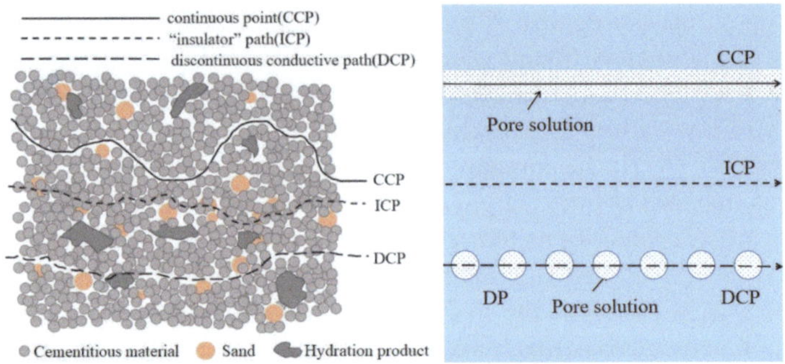

(a) Conductive path (b) Simplified conductive path

Fig. 4.20 Conductive path of cementitious material microstructure

continuity is obstructed by hydration products. This pathway includes discontinuity points (DP). 3. Insulating Pathway (ICP): This refers to other pathways aside from the DCP and CCP, which act as "insulators" within the entire system. The three conductive pathways for low-carbon cementitious materials are illustrated in Fig. 4.20.

As shown in Fig. 4.20b, the diagram illustrates the simplified conductive pathways of the microstructure of low-carbon cementitious materials. The white areas represent the porosity within the low-carbon cementitious material (filled with pore solution), which includes the continuous conductive pathway (CCP) and the discontinuous points (DP) within the discontinuous conductive pathway (DCP). The blue area represents the matrix of the waste-based cementitious materials. The porosity of the low-carbon cementitious material system can be expressed as the ratio of the white area to the total volume, which includes both the light blue area and the white area. The conductivity of low-carbon cementitious materials depends on the impedance of all three conductive pathways. The conductive model formed by these pathways is shown in Eq. 4.17. Consequently, the total impedance of the cementitious material can be expressed as:

$$\frac{1}{Z} = \frac{1}{Z_{\mathrm{CCP}}} + \frac{1}{Z_{\mathrm{ICP}}} + \frac{1}{Z_{\mathrm{DCP}}} \tag{4.17}$$

where: Z_{CCP} represents the impedance of the continuous conductive pathway. Z_{ICP} represents the impedance of the insulating pathway. Z_{DCP} represents the impedance of the discontinuous conductive pathway (Fig. 4.21).

Theoretically, the current conduction through the Continuous Conductive Pathway (CCP) is caused by the migration of ions in the pore solution, thus Z_{CCP} can be described as R_{CCP}. In contrast, the impedance of the Discontinuous Conductive Pathway (DCP) is more complex. The impedance of the DCP (Z_{DCP}) consists of two parts: the continuous section of the DCP (Z_{CP}) and the discontinuity points (Z_{DP}). The expression for the impedance of the DCP is given by Eq. 4.18.

Fig. 4.21 Conductive model of low-carbon cementitious material

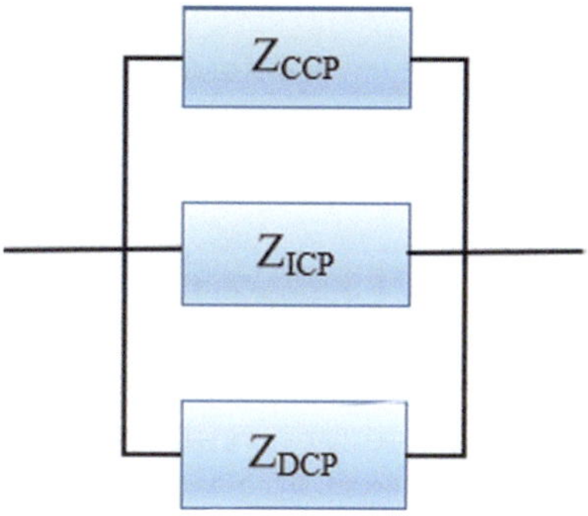

$$Z_{DCP} = Z_{CP} + Z_{DP} \tag{4.18}$$

The impedance Z_{DP} of the DP discontinuity point in the low-carbon cementitious material can be written as Eq. 4.19. The CP part of the DCP path has an impedance Z_{CP} similar to the CCP path, so it can be expressed by R_{CP}.

$$Z_{DP} = \frac{1}{j\omega C_{DP}} \tag{4.19}$$

where: j is the imaginary unit, $j = \sqrt{-1}$. ω is the alternating current frequency, $\omega = 2\pi f$; C_{DP} is the capacitance of the parallel plates at the discontinuity points.

The impedance of Z_{ICP} can be expressed by the following equation:

$$Z_{ICP} = \frac{1}{j\omega C_{mat}} \tag{4.20}$$

In the matrix of low-carbon cementitious materials, C_{mat} represents the capacitance when its two end faces are used as a double parallel plate capacitor. Based on the analysis of the three conductive pathways mentioned above, corresponding equivalent circuit elements are used to represent their internal electrochemical reaction characteristics, allowing for the derivation of the equivalent circuit model for low-carbon cementitious materials.

Strictly speaking, under alternating current, the matrix of low-carbon cementitious materials is not a true insulator. This is because the hydration product gel within the cementitious matrix has limited conductivity, which can be represented as a resistance R_{mat}. Since the resistivity of the gel phase hydration products in the cementitious matrix is significantly high compared to the resistivity of the pore solution within the cementitious material (by several orders of magnitude), its conductivity can be considered negligible compared to other conductive pathways. Furthermore, the pathway characterizing R_{mat} in the cementitious material can be effectively equated to an open circuit. In most cases, R_{mat} in the matrix of low-carbon cementitious

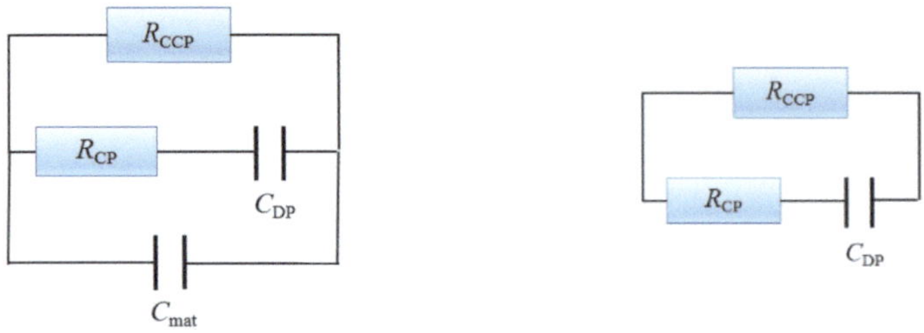

(a) Equivalent circuit model (b) Simplified equivalent circuit model

Fig. 4.22 Equivalent circuit model based on microstructure

materials can be disregarded, and the corresponding equivalent circuit model can be further simplified. The simplified equivalent circuit model is illustrated in Fig. 4.22a.

Based on Fig. 4.22a, the total impedance Z of the low-carbon cementitious materials can be expressed as:

$$Z = \frac{1}{\left\{ j\omega C_{\text{mat}} + \frac{1}{R_{\text{CCP}}} + \frac{1}{\left(R_{\text{CP}} + \frac{1}{j\omega C_{\text{DP}}} \right)} \right\}} \tag{4.21}$$

In actual AC impedance testing, the Nyquist plots of most cementitious materials typically begin from the second capacitive arc. In this case, the equivalent circuit can be further simplified; compared to C_{DP}, C_{mat} is very small ($C_{\text{mat}} \to 0$). As a result, the equivalent circuit is transformed into the model shown in Fig. 4.22b. By comparing the traditional circuit with the equivalent circuit model based on the microstructure, it becomes easy to establish the relationships between the parameters of different equivalent circuit elements.

$$R_{\text{CCP}} = R_{\text{S}} + R_{\text{ct1}} \tag{4.22}$$

$$R_{\text{CP}} = \frac{(R_{\text{S}} + R_{\text{ct1}})R_{\text{S}}}{R_{\text{ct1}}} \tag{4.23}$$

$$C_{DP} = CPE_1 \left(\frac{R_{ct1}}{R_{ct1} + R_S} \right)^2 \tag{4.24}$$

$$\frac{R_{\text{CP}}}{R_{\text{CCP}}} = \frac{R_{\text{S}}}{R_{\text{ct1}}} \tag{4.25}$$

where: R_{CCP} represents the resistance in the CCP pathway of the cementitious material, and this value increases with the reduction of interconnected pores within the cementitious material. R_{CP} represents the resistance of the pore solution in the DCP pathway, and its value is determined by both ($R_{\text{s}} + R_{\text{ct1}}$) and ($R_{\text{s}}/R_{\text{ct1}}$). As hydration products continue to form, the interconnected pores in the low-carbon cementitious matrix gradually convert to non-interconnected pores. The impedance parameter value R_{CP}/R_{CCP} can be used to evaluate the extent to which interconnected pores in low-carbon cementitious materials transition to non-interconnected pores at different hydration ages. C_{DP} is the capacitance of the discontinuity point DP in the DCP pathway, which is inversely proportional to the thickness of the non-interconnected pores; its value is determined by CPE_1.

4.4.2 Analysis of AC Impedance Parameters Based on Microstructure

Based on the aforementioned microstructural model of the cementitious materials, a comparison is made with the equivalent circuit model. This analysis focuses on the variations in AC impedance parameters of low-carbon cementitious materials with different water-to-binder ratios across various hydration ages. Using Eqs. 4.22–4.25, the corresponding impedance parameter values R_{CCP}, R_{CP}, and C_{DP} are calculated. The variations in these parameter values are then used to study the evolution of the microstructure of low-carbon cementitious materials.

The values of R_{CP} and R_{CP}/R_{CCP}

Figure 4.23a illustrates the variation of R_{CP} values for low-carbon cementitious materials with different water-to-binder ratios over hydration ages, while Fig. 4.23b shows the change in the R_{CP}/R_{CCP} values over hydration ages. The R_{CP} value reflects the state of non-interconnected pores within the cementitious materials. It can be seen that the R_{CP} value is continuously increasing, while the R_{CP}/R_{CCP} value initially decreases rapidly and then levels off. This indicates that, within the low-carbon cementitious materials, the interconnected pores responsible for conductivity are gradually being clogged. In the early stages of hydration, the R_{CP} value increases rapidly, and the growth rate slows down afterward. Taking the cementitious material with a water-to-binder ratio of 0.47 as an example, the R_{CP} value increases by 18 Ω during the hydration period of 1 to 12 h, while only increasing by 5.77 Ω during the period from 12 h to 28 days. This is because, as the hydration process continues, the interconnected pores within the cementitious material matrix gradually convert to non-interconnected pores. In the initial hours of hydration, a significant number of interconnected pores transition to non-interconnected pores; in the mid to late hydration stage, the number of pores that can be filled by hydration products decreases, leading to a reduced quantity of interconnected pores converting to non-interconnected pores. Consequently, the R_{CP}/R_{CCP} ratio reflects this trend as it initially declines quickly and then levels off.

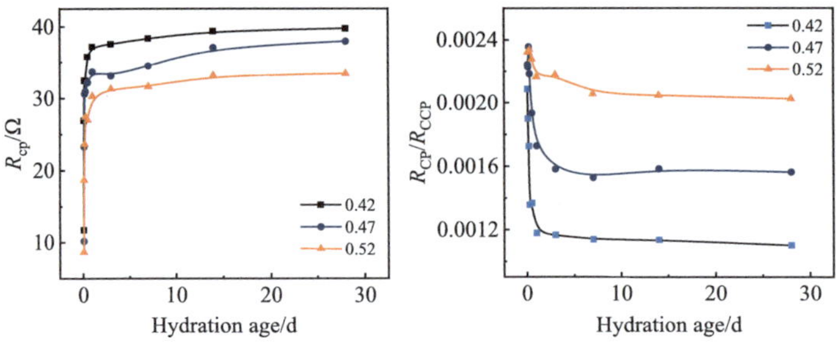

(a) R_{CP} (b) R_{CP}/R_{CCP}

Fig. 4.23 Variation curves of R_{CP} value and R_{CP} / R_{CCP} value

The values of R_{CCP}

As shown in Fig. 4.24, the development curves of R_{CCP} values for low-carbon cementitious materials with different water-to-binder ratios over hydration ages are presented. R_{CCP} effectively evaluates the degree of hydration of the cementitious materials. It can be observed that as the hydration process progresses, the R_{CCP} values for all three groups with different water-to-binder ratios increase. After 1 h of hydration, the R_{CCP} values for the cementitious materials with water-to-binder ratios of 0.42, 0.47, and 0.52 are 7054.91, 6327.75, and 5590.54 Ω, respectively. After 28 days of hydration, these values increase to 36,136.69, 24,306.89, and 16,510.39 Ω, respectively.

It has been found that the impact of hydration time on the R_{CCP} value is significantly greater than that on R_{CP}. Taking the low-carbon cementitious material with a water-to-binder ratio of 0.47 as an example, its R_{CCP} value after 28 days increases by 17,979.14 Ω compared to the initial value after 1 h of hydration, while the R_{CP} value only increases by 23.77 Ω. This is because the formation of hydration products not only narrows the pathway of non-interconnected pores but also increases the thickness of the discontinuous points, causing more interconnected pores to transform into non-interconnected pores, which to some extent offsets the increase in the R_{CP} value.

Fig. 4.24 Variation curve of R_{CCP} value

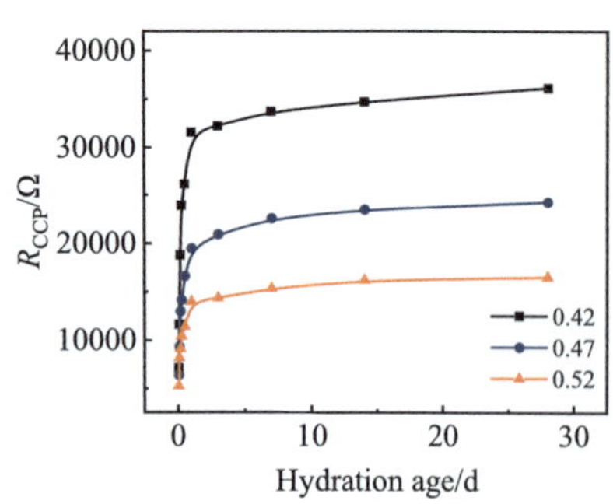

From Fig. 4.24, it can also be observed that the water-to-binder ratio has a significant influence on the R_{CCP} values. At 1 h of hydration, the R_{CCP} values for the cementitious materials with different water-to-binder ratios are quite similar, indicating that the rates of hydration reactions within the cementitious materials are relatively consistent at this stage. As hydration time increases, a larger water-to-binder ratio gradually leads to a decrease in the R_{CCP} value. At 3 days of hydration, the R_{CCP} value for the cementitious material with a water-to-binder ratio of 0.42 is 11286.5 and 17,853.33 Ω higher than those for the water-to-binder ratios of 0.47 and 0.52, respectively. This occurs because a smaller water-to-binder ratio leads to a greater reduction in the number of pore sizes within the system and a larger degree of narrowing of the interconnected pores. As the hydration process continues, this difference becomes more pronounced. From 3 to 28 days of hydration, the R_{CCP} value for the cementitious material with a water-to-binder ratio of 0.42 is significantly greater than that for the other two water-to-binder ratios. At 28 days of hydration, the R_{CCP} value for low-carbon cementitious materials with a water-to-binder ratio of 0.42 is 1.49 times and 2.19 times greater than those for cementitious materials with water-to-binder ratios of 0.47 and 0.52, respectively, indicating that the overall microstructural development within the lower water-to-binder ratio low-carbon cementitious material system is better, with a lower porosity.

The values of C_{DP}

Figure 4.25 shows the development trend curve of the C_{DP} values for low-carbon cementitious materials with different water-to-binder ratios. It can be observed that the C_{DP} values continuously decrease with increasing hydration time. There is a rapid decline in C_{DP} values within the first 3 days of hydration, followed by a gradual leveling off after 3 days. The decrease in C_{DP} values is mainly attributed to the increase in the thickness of discontinuous points within the cementitious materials. Hydration products continuously fill the pores, causing larger diameter pores to gradually transform into smaller diameter pores. The hydration products play a critical role in increasing the thickness of the non-interconnected pores, and this effect is particularly significant in the early stages of hydration, becoming less pronounced in the mid to late stages of hydration.

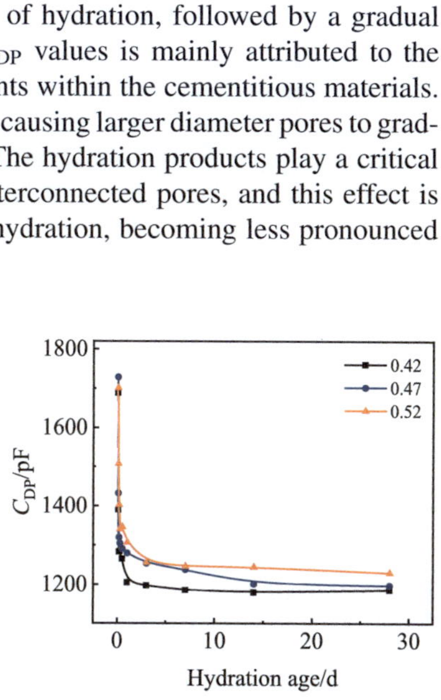

Fig. 4.25 Variation curve of C_{DP} value

The water-to-binder ratio also has a certain impact on the magnitude of the C_{DP} values. This effect is more pronounced in the early stages of hydration. At 1 day of hydration, the C_{DP} value of the cementitious material with a water-to-binder ratio of 0.42 is reduced by 74.91 pF and 102.17 pF compared to the C_{DP} values of cementitious materials with water-to-binder ratios of 0.47 and 0.52, respectively. This is because the water-to-binder ratio plays a significant role in the increase in the thickness of discontinuous points within the cementitious matrix. As the hydration process continues, the influence of the water-to-binder ratio on the C_{DP} values becomes less significant, indicating that this increasing effect is no longer obvious in the mid to late stages of hydration.

Fractal dimension

The fractal dimension is an important parameter that characterizes the microstructure of cementitious materials and can be divided into the fractal dimension of pore surface properties d_s and the fractal dimension of pore volume properties d. The fractal dimension of pore surface properties d_s is closely related to the pore structure parameters of the cementitious materials. A smaller fractal dimension d_s indicates that the average pore size and the median pore size within the cementitious materials are smaller, resulting in a smoother surface and a denser structure. In this study, the constant phase element (CPE) is used to replace the double layer capacitance C_d, which can be expressed as Eq. 4.26. The degree of flattening of the high-frequency semicircle can be represented by the constant phase exponent q. The quantitative relationship between the fractal dimension d_s and q is shown in Eq. 4.27. The values of q and the fractal dimension d_s for low-carbon cementitious materials with different water-to-binder ratios at various hydration ages are presented in Table 4.2.

$$CPE = K(j\omega)^{-q} \tag{4.26}$$

where: K represents the constant of pure capacitance. q is the constant phase exponent. ω is the angular frequency, $\omega = 2\pi f$. j is the imaginary unit, $j^2 = -1$.

$$d_s = 3 - q \tag{4.27}$$

where: d_s represents the fractal dimension of the pore surface properties. q is the constant phase exponent.

From Table 4.3, it can be seen that the range of fractal dimension d_s values is between 2.046 and 2.075, which is largely influenced by factors such as the type of cementitious materials and supplementary materials. As the hydration time increases, the q values continuously increase, and the d_s values correspondingly decrease. This indicates that during the hydration process, the number of small pores increases while the number of large pores decreases, leading to a more uniform pore size distribution. Generally, the q value ranges from 0 to 1, and in this experiment, all q values are greater than 0.9, suggesting that there is a dispersion effect within the cementitious material system, resulting in the deflection of the high-frequency semicircle.

The fractal dimension d for pore volume can characterize the complexity of the pore structure. The quantitative relationship between the dispersion index p and the fractal dimension d is illustrated in Eq. 4.28, while the magnitude of the dispersion index p can be obtained from the angle between the low-frequency linear part and the horizontal axis, as shown in Eq. 4.29. The p and d values for low-carbon cementitious materials with different water-to-binder ratios are presented in Table 4.4. Theoretically, the p value should be around 0.5; however, in this experiment, all p values are slightly less than 0.5, which is attributed to the existence of the dispersion angle causing the low-frequency line to deviate from 45°. From the analysis of the table, the range of the pore volume fractal dimension d values is between 3.546 and 3.595. As hydration time increases, the fractal dimension d gradually increases, indicating that the pore structure within the cementitious material system is continuously changing, and the pore size distribution is becoming increasingly uniform.

$$d = 4 - p \tag{4.28}$$

where: d represents the fractal dimension of the pore volume. p is the dispersion index that characterizes the degree of flattening of the high-frequency semicircle.

$$p=1 - \frac{2\theta}{\pi} \tag{4.29}$$

where: θ represents the dispersion angle.

Table 4.3 The q values and d_s values

Hydration time	w/b = 0.42		w/b = 0.47		w/b = 0.52	
	q	d_s	q	d_s	q	d_s
1h	0.92756	2.07244	0.9254	2.0746	0.92566	2.07434
2h	0.94163	2.05837	0.93902	2.06098	0.93503	2.06497
4h	0.94681	2.05319	0.94489	2.05511	0.94029	2.05971
6h	0.94693	2.05307	0.94653	2.05347	0.94329	2.05671
12h	0.94856	2.05144	0.94604	2.05396	0.94319	2.05681
1d	0.95222	2.04778	0.94945	2.05055	0.94573	2.05427
3d	0.95231	2.04769	0.94745	2.05255	0.94782	2.05218
7d	0.95132	2.04868	0.94889	2.05111	0.94813	2.05187
14d	0.95248	2.04752	0.95158	2.04842	0.94829	2.05171
28d	0.95358	2.04642	0.95123	2.04787	0.94955	2.05045

Table 4.4 p values and d values of low carbon cementitious materials with different water-to-binder ratios

Hydration time	w/b = 0.42		w/b = 0.47		w/b = 0.52	
	p	d	p	d	p	d
1h	0.40488	3.59512	0.42771	3.57229	0.42480	3.5752
2h	0.41405	3.58595	0.43248	3.56752	0.43334	3.56666
4h	0.42639	3.57361	0.43480	3.5652	0.43417	3.56583
6h	0.43314	3.56686	0.43546	3.56454	0.43702	3.56298
12h	0.44045	3.55955	0.43803	3.56197	0.43871	3.56129
1d	0.44362	3.55638	0.44197	3.55803	0.44310	3.5569
3d	0.44577	3.55423	0.44639	3.55361	0.44318	3.55682
7d	0.44867	3.55133	0.44704	3.55296	0.45002	3.54998
14d	0.44771	3.55229	0.45143	3.54857	0.45072	3.54928
28d	0.45045	3.54955	0.45157	3.54843	0.45382	3.54618

4.4.3 Pore Structure Analysis

Basic principles of nuclear magnetic resonance

Nuclear Magnetic Resonance (NMR) is a physical process in which atomic nuclei with non-zero magnetic moments resonate under the influence of an external magnetic field, thereby absorbing radiofrequency radiation of certain frequencies (Mlynárik, 2017). In the presence of an external magnetic field, the atoms undergo regular rotation, and this interaction generates signals. During the application of radiofrequency pulses, the atomic nuclei resonate and reach a high-energy state, causing the nuclear magnetization vector to deviate from the direction of the magnetic field. The process in which the atomic nuclei return to their pre-resonance equilibrium state after the external radiofrequency field disappears is called relaxation. NMR technology converts the fluid signals detected in the pores of cementitious materials into T_2 relaxation time spectra through mathematical inversion, allowing for the analysis of the state of fluids within the specimen.

The NMR testing utilizes the Carr-Purcell-Meiboom-Gill (CPMG) pulse sequence, detecting the relaxation responses of protons ^{1}H in the saturated low-carbon cementitious material paste. This provides the corresponding decay signals of the echo train, and the collection of decay amplitudes forms the T_2distribution (Zou et al., 2022). Based on the principles of NMR, the NMR transverse relaxation rate $1/T_2$ is given by:

$$\frac{1}{T_2}=\frac{1}{T_{2\text{free}}}+\rho_2\left(\frac{S}{V}\right)_{\text{Porosity}}+\frac{D(\gamma G T_{\text{E}})^2}{12} \tag{4.30}$$

where: T_{2free} represents the free relaxation time of the pore fluid within the cementitious material, ms. ρ_2 is the transverse surface relaxation intensity of T_2, which is related to the surface properties of the pore in the sample. S is the surface area of the pores within the cementitious material, cm^2. V is the volume of the pore fluid within the cementitious material, cm^3. D is the diffusion coefficient. γ is the gyromagnetic ratio, rad/(s·T). G is the magnetic gradient, Gs/cm. T_E is the echo interval of the measurement sequence, ms.

During the nuclear magnetic resonance (NMR) testing process, it can be assumed that there is only one type of fluid present within the pores of the cementitious material. Consequently, a difference exists between the volume relaxation time and the size relaxation time, and the parameter value representing the fluid's free relaxation time T_{2free} can be considered negligible. In the NMR testing process, assuming a uniform magnetic field (with a very small magnetic gradient G and a relatively small T_E value, the diffusion relaxation time can also be neglected (Liang et al., 2019; Li et al., 2016). Therefore, Eq. 4.31 can be simplified to Eq. 4.32:

$$\frac{1}{T_2} = \rho_2 \left(\frac{S}{V}\right)_{\text{Porosity}} \tag{4.31}$$

Assuming that the pores in the low-carbon cementitious material are spherical, Eq. 4.33 can be further transformed into the relationship between the T_2 relaxation time and pore diameter, expressed as:

$$T_2 = \frac{r_c}{3\rho_2} \tag{4.32}$$

where: r_c represents the pore radius, μm. ρ_2 is the surface relaxation rate of T_2, μm/ms. ρ_2 is a constant value related to the relevant physical properties of the sample. Let the constant $\mathrm{C} = 3\rho_2$, thus:

$$r = \mathrm{C}T_2 \tag{4.33}$$

From Eqs. 4.30 to 4.33, it can be understood that the pore radius is directly proportional to the relaxation time T_2. The distribution of T_2 can reflect the size information of the pores; that is, a smaller T_2 indicates smaller pores, while conversely, larger pores correspond to larger T_2.

Porosity analysis

Figure 4.26 shows the changes in porosity of low-carbon cementitious materials with different water-to-binder ratios over hydration age. It can be observed that as the hydration process progresses, the porosity of the low-carbon cementitious materials gradually decreases, with the extent of reduction becoming smaller over time. This indicates that the volume of pores changes significantly in the early stages of hydration. This is because, during the early hydration of the cementitious materials, hydration products are generated rapidly, continuously occupying the interconnected pores in the CCP pathway and the discontinuous points in the DCP pathway, leading to noticeable changes in porosity during the early hydration phase. As the water-to-binder ratio increases, the porosity of the low-carbon cementitious materials gradually increases. This is due to the increase in water, which enlarges the contact area between water and the cementitious particles, resulting in a higher number of small-pore pores and leading to an increase in the overall porosity.

Nuclear magnetic resonance T_2 spectrum analysis

In the analysis of Nuclear Magnetic Resonance (NMR) test results, the amount of liquid in the pores of cementitious materials can be characterized by the variation in the T_2 spectrum area. In the T_2 spectrum distribution of cementitious materials, the horizontal x-axis represents the relaxation time, which is related to the pore size of the cementitious material. The longer the relaxation time, the larger the pore diameter in the cementitious material. The vertical y-axis represents the pore signal intensity; the stronger the signal, the greater the number of pores within that pore size range in the cementitious material.

As shown in Figs. 4.27a–c, the Nuclear Magnetic Resonance (NMR) T_2 spectrum distribution curves for low-carbon cementitious materials with water-to-binder ratios of 0.42, 0.47, and 0.52, respectively, over the hydration period, exhibit similar trends. The T_2 spectrum curve for a water-to-binder ratio of 0.42 displays a three-peak distribution, while the T_2 spectrum curves for water-to-binder ratios of 0.47 and 0.52 primarily display a two-peak distribution, with the third peak (Peak3) being almost indistinguishable. Among the T_2 spectrum curves for the three water-to-binder ratios, the first peak (Peak1) appears at a short relaxation time, with signal intensity significantly higher than that of the second peak (Peak2) and the third peak (Peak3). This indicates that the proportion of small-diameter pores is the largest, while the proportion of large-diameter pores is smaller within the matrix of the low-carbon cementitious materials.

Analysis of the T_2 spectrum curves of cementitious materials with different water-to-binder ratios reveals that their relaxation times predominantly range from 0.1 to 3000 ms. Specifically, Peak1 appears near 1 ms, Peak2 appears near 45 ms, and Peak3 appears near 1000 ms. As the hydration period increases, the height of the main peak

Fig. 4.26 Porosity change curves

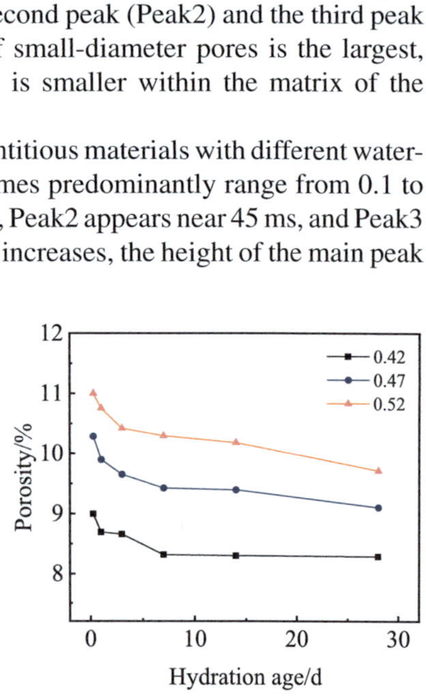

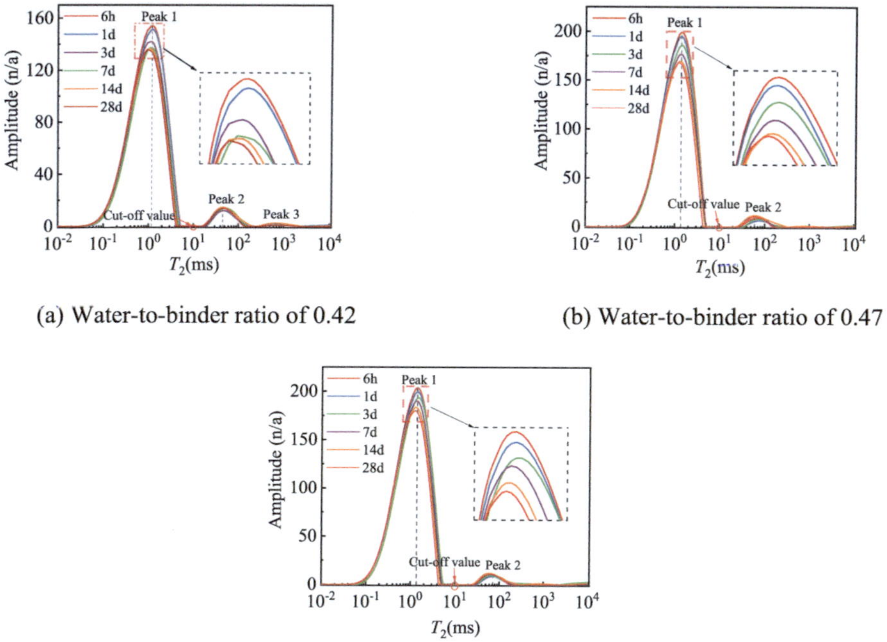

(a) Water-to-binder ratio of 0.42

(b) Water-to-binder ratio of 0.47

(c) Water-to-binder ratio of 0.52

Fig. 4.27 T_2 spectrum of cementitious materials with different water-to-binder ratio

gradually decreases, and the peak value shifts to the left. This indicates that, with the accumulation of hydration products, the number of small-diameter pores in the cementitious material system decreases, as these pores are progressively filled with hydration products, forming smaller pores.

As the hydration process progresses, the area of the first peak remains relatively unchanged, but the second peak shifts overall to the left. This suggests that large-diameter pores are gradually transforming into small-diameter pores, likely due to the expansion of cementitious particles blocking *CCP* pathways and the generation of hydration products narrowing D_{CP} pathways, thereby increasing the thickness of discontinuities. Furthermore, comparative analysis of the T_2 spectrum curves for different water-to-binder ratios shows that with an increase in the water-to-binder ratio, the amplitudes of Peak1 and Peak2 both increase. This is because the number of small and medium pores in the cementitious material increases with a higher water-to-binder ratio. The Peak3 proportion is more pronounced in cementitious materials with a water-to-binder ratio of 0.42, due to the wider pore radius distribution and the corresponding increase in the number of large pores when the water-to-binder ratio is lower.

Nuclear magnetic resonance T_2 spectrum area change analysis

Based on the T_2 spectrum distribution curves of the cementitious materials, the total area enclosed by the entire T_2 curve and the horizontal axis can be calculated, as well as the areas enclosed by each peak. The changes in the T_2 spectrum areas for different water-to-binder ratios are presented in Table 4.5.

From the analysis of Table 4.5, it can be seen that as the hydration process progresses, the total peak area and the area of Peak 1 for the three groups of water-to-binder ratio specimens gradually decrease. The area of Peak 2 shows overall fluctuating changes; during the early stage of hydration, the area of Peak 2 gradually decreases, while in the later stages of hydration, the area of Peak 2 slightly increases again. This is due to the rapid hydration of a significant amount of $C_4A_3\overline{S}$ in the early hydration phase, generating hydration products that occupy a large number of pores, resulting in a gradual reduction of larger pores. In the mid to late hydration stages,

Table 4.5 Changes of T_2 spectrum are of low-carbon cementitious materials

Hydration time	w/b = 0.42		w/b = 0.47		w/b = 0.52	
6 h	Peak1	4957.421	Peak1	5586.15	Peak1	6178.421
	Peak2	252.317	Peak2	99.289	Peak2	148.109
	Peak3	23.128	Peak3	3.955	Peak3	6.326
	Total	5232.866	Total	5689.395	Total	6332.856
1d	Peak1	4762.994	Peak1	5410.115	Peak1	6019.553
	Peak2	250.057	Peak2	125.285	Peak2	132.966
	Peak3	20.221	Peak3	5.707	Peak3	6.424
	Total	5033.273	Total	5541.106	Total	6158.943
3d	Peak1	4646.771	Peak1	5374.727	Peak1	5743.128
	Peak2	227.875	Peak2	125.315	Peak2	191.535
	Peak3	65.031	Peak3	24.012	Peak3	42.027
	Total	4939.677	Total	5508.016	Total	5976.69
7d	Peak1	4598.355	Peak1	5222.054	Peak1	5699.475
	Peak2	279.378	Peak2	194.591	Peak2	20.367
	Peak3	37.691	Peak3	7.778	Peak3	11.349
	Total	4915.424	Total	5424.423	Total	5945.408
14d	Peak1	4459.208	Peak1	5187.421	Peak1	5574.979
	Peak2	327.303	Peak2	204.57	Peak2	228.253
	Peak3	62.624	Peak3	14.391	Peak3	25.601
	Total	4849.135	Total	5406.382	Total	5828.832
28d	Peak1	4450.654	Peak1	5097.822	Peak1	5498.143
	Peak2	285.999	Peak2	247.307	Peak2	239.356
	Peak3	58.661	Peak3	12.507	Peak3	12.549
	Total	4795.314	Total	5357.637	Total	5750.093

due to the expansion characteristics and looser structure of needle-like ettringite, discontinuity points (DP) formed near the micropores cause a small portion of the pores to gradually transition from discontinuous pathways to continuous pathways. At the same hydration age, as the water-to-binder ratio increases, the total area of the T_2 spectrum also increases, indicating a continuous rise in the proportion of pores.

From the analysis of Table 4.5, it can be observed that with the extension of hydration age, during the hydration period from 6 h to 3 days, the total peak areas of the cementitious materials with water-to-binder ratios of 0.42, 0.47, and 0.52 decreased by 5.6, 3.2, and 5.6%, respectively. During the hydration period from 3 to 14 days, the reductions were 1.8, 1.8, and 2.5%, respectively. During the hydration period from 14 to 28 days, the reductions were 1.1, 0.9, and 1.3%, respectively. By the 28-days hydration mark, the T_2 spectrum areas for the three groups with different water-to-binder ratios decreased to 4735.314, 5257.637, and 5590.093, respectively. This is consistent with the changes in the porosity of the cementitious materials, indicating that during the hydration process, the continuous generation of hydration products leads to a reduction in the total pore volume, a gradual decrease in the pathways of interconnected pore sizes (CCP), an increase in structural density, and a decrease in porosity.

From Table 4.5, it can be analyzed that the proportion of Peak 1 is the largest, with the Peak 1 values for the three groups of water-to-binder ratios all exceeding 90% of the total peak area. At the same hydration age, the T_2 spectrum area gradually increases with a higher water-to-binder ratio. This is because, as the water-to-binder ratio increases, the contact area between the cementitious material particles and water expands, leading to an increase in the internal pore volume of the cementitious material. Additionally, at the initial formation of the low-carbon cementitious material samples, the bubbles that were not expelled during the mixing process lead to the formation of internal pores during the setting process. The number of closed pores internally is at its highest at the initial setting time, which contributes to the increase in the T_2 spectrum area.

Pore size distribution analysis

According to the conversion formula of Eq. 4.32, the pore size distribution curves for low-carbon cementitious materials can be obtained, as shown in Fig. 4.28.

The pore size distribution curves over time for low-carbon cementitious materials with water-to-binder ratios of 0.42, 0.47, and 0.52 are represented in Figs. 4.28a–c, respectively. It can be observed that the changes in the pore size distribution curves are consistent with those of the T_2 spectra. At the 6-h hydration age, the proportion of pore sizes in the cementitious materials is at its highest value. As the hydration age increases, the pore size distribution curves for the cementitious materials with different water-to-binder ratios exhibit similar trends of decrease as the age increases. Additionally, at the same hydration age, as the water-to-binder ratio increases, the corresponding proportion of the pore size distribution also gradually increases.

Based on the impact of pore size on strength, the pores within the cementitious material system can be categorized as follows: harmless pores (<20 nm), slightly harmful pores (20–50 nm), harmful pores (50–200 nm), and extremely harmful pores

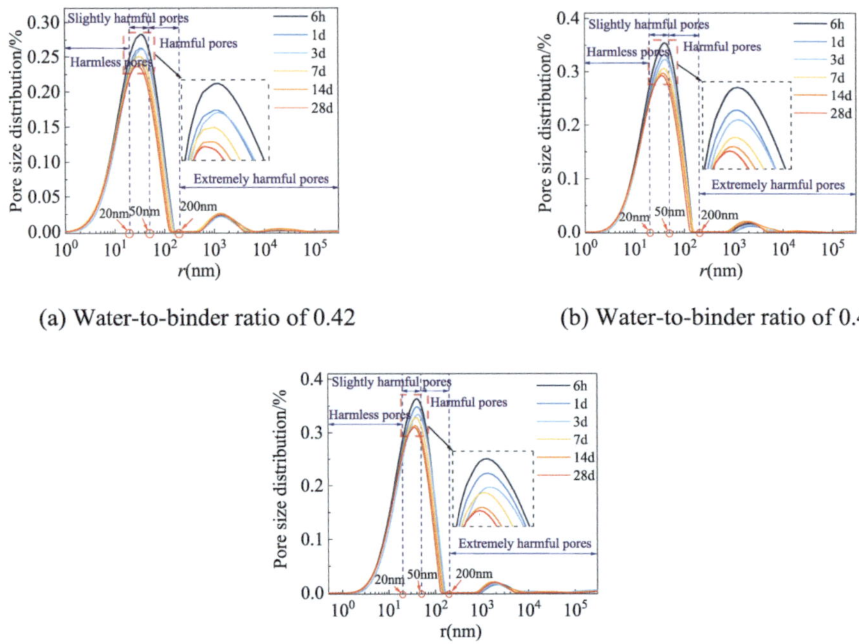

(a) Water-to-binder ratio of 0.42

(b) Water-to-binder ratio of 0.47

(c) Water-to-binder ratio of 0.52

Fig. 4.28 Pore size distribution curves

(>200 nm). Using the pore size distribution curves, the evolutionary characteristics of the proportions of different ranges of pore sizes within the low-carbon cementitious material system can be further analyzed. Figure 4.29 presents the proportion of various levels of pore sizes for low-carbon cementitious materials with different water-to-binder ratios.

From the analysis of the Fig. 4.29, it can be observed that among the four categories of pore sizes, extremely harmful pores have the smallest proportion, while slightly harmful pores have the largest proportion. As the hydration process progresses, the proportion of harmless pores gradually increases, the proportion of slightly harmful pores rises overall, and the proportion of extremely harmful pores significantly decreases, while the proportion of harmful pores remains relatively stable. For the cementitious materials with water-to-binder ratios of 0.42, 0.47, and 0.52, the proportions of harmless pores increase by 20.95, 22.41, and 23.47% respectively from hydration ages of 6 h to 28 days, while the proportions of harmful pores decrease by 28.86, 24.86, and 27.16%, respectively. This indicates that with the increase in hydration time, due to the continuous generation of hydration products, the harmful pores within the cementitious materials gradually transition into slightly harmful pores, and some slightly harmful pores are gradually transforming into harmless pores. This is because the number of discontinuous points within the cementitious material system increases, the width of the interconnected pores continues to narrow,

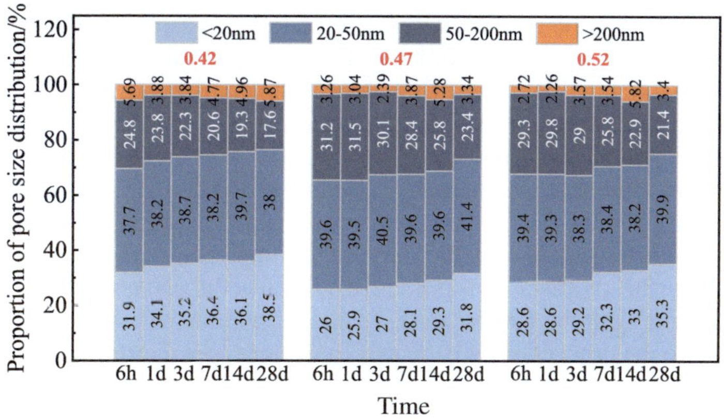

Fig. 4.29 Pore size ratio diagram of cementitious materials with different water-to-binder ratio

and these interconnected pores gradually transition into non-interconnected pores, resulting in an increase in the number of small pores and a gradual reduction in the number of large pores, thereby making the pore size distribution more uniform.

4.4.4 Relationship Between AC Impedance Parameters and Pore Structure

The pore structure is a key factor affecting the durability performance of structures, as it is closely related to phenomena such as chloride ion ingress, carbonation, sulfate corrosion, and the material's freeze–thaw resistance. Research indicates that AC impedance parameters are primarily associated with the ion concentration in the pore solution, pore size distribution, and total porosity. This section explores the relationship between the impedance parameter value R_{CCP} of low-carbon cementitious materials and the product of the most probable pore diameter and porosity, as well as the relationship between porosity and the R_{CP}/R_{CCP} ratio. From the perspective of pore structure, AC impedance technology is used to explain the evolutionary patterns of the microstructure of the cementitious materials.

Relationship between R_{CCP} value and the product of the most probable pore size and porosity

The most probable pore diameter characterizes the size of the pores where the pore density is highest within the cementitious material; it corresponds to the pore diameter at the highest point of the pore distribution. The most probable pore diameter of low-carbon cementitious materials with different water-to-binder ratios changes continuously over hydration time. As the hydration age increases, the main peak of the pore size distribution gradually shifts to the left, indicating that the most probable pore diameter is decreasing. Conversely, as the water-to-binder ratio increases, the most probable pore diameter of the cementitious material gradually increases, although the extent of this increase is quite small.

Gu et al. studied the relationship between the pore interfacial resistance of cementitious materials and pore structure. The research found that the product of porosity and equivalent pore diameter in porous materials can serve as a basis for the appearance of the high-frequency semicircle in the AC impedance spectrum. The magnitude of the interfacial resistance is primarily influenced by the total porosity and pore size distribution, with the interfacial resistance of porous materials being inversely proportional to the product of porosity and equivalent pore diameter. The pore interfacial resistance of low-carbon cementitious materials can be characterized by the resistance of the continuous path R_{CCP}. Since the proportion of the main peak area in the T_2 spectrum of the cementitious materials is over 90%, the most probable pore diameter of the cementitious material samples is used as the equivalent pore diameter. The relationship between the R_{CCP} value of low-carbon cementitious materials and the product of porosity and the most probable pore diameter is illustrated in Fig. 4.30a–c.

The relationship curves for different water-to-binder ratios exhibit an exponential correlation, and the relationship Eqs. can be categorized as Eqs. 4.34–4.35. This

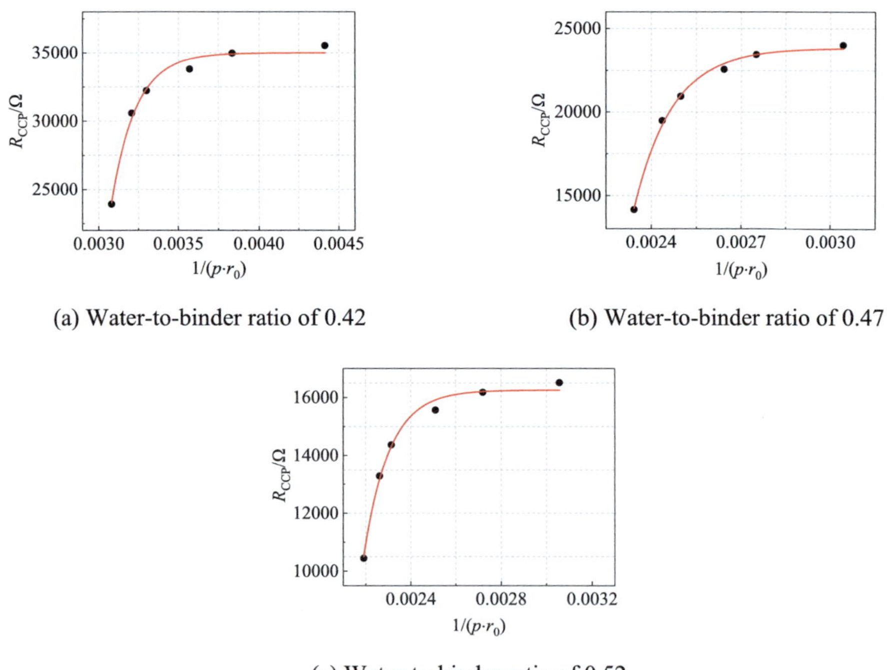

(a) Water-to-binder ratio of 0.42

(b) Water-to-binder ratio of 0.47

(c) Water-to-binder ratio of 0.52

Fig. 4.30 Relation curve between R_{CCP} and $1/(p{\cdot}r_0)$

indicates that AC impedance technology is one of the effective methods for studying the pore structure of cementitious materials and can serve as a basis for investigating the evolution of their microstructure.

$$R_{CCP} = Ae^{-\frac{k}{t}} + B \tag{4.34}$$

$$k = \frac{1}{p \cdot r_0} \tag{4.35}$$

where: A, B, and t are fitting parameters. P is the porosity of the low-carbon cementitious material. r_0 is the most probable pore diameter of the low-carbon cementitious material.

Relationship between R_{CP}/R_{CCP} value and porosity

During the development process of the pore structure in cementitious materials, hydration products continuously fill the internal pores, causing interconnected pathways to transform into non-interconnected pore sizes. The ratio of the AC impedance parameters R_{CP}/R_{CCP} can characterize the extent to which interconnected pores are converted into non-interconnected pores. The variation in porosity over time reflects the evolutionary patterns of the various pore sizes within the cementitious material.

Research has found a certain correlation between the impedance parameter ratio R_{CP}/R_{CCP} and porosity, as shown in Fig. 4.31. It indicates a good correlation between the two, and the fitting equation. between R_{CP}/R_{CCP} and porosity p is presented as Eq. 4.36, with the fitting correlation coefficient close to 1. Therefore, the impedance parameter ratio R_{CP}/R_{CCP} can effectively evaluate the changes in the microstructure during the hydration process of the cementitious materials.

$$p = 1991.672R_{CP}/R_{CCP} + 6.225 \quad R^2 = 0.965 \tag{4.36}$$

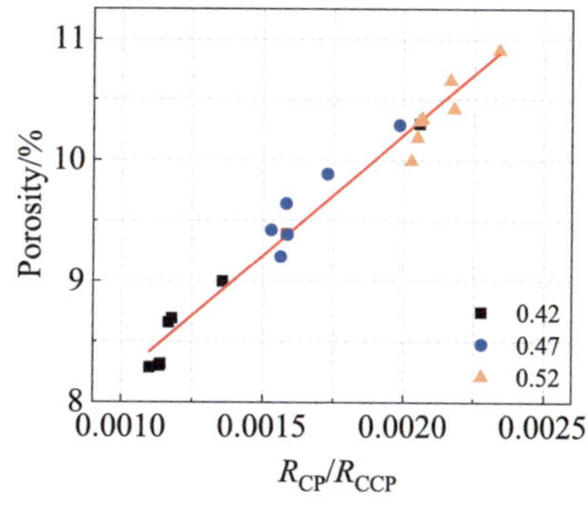

Fig. 4.31 Relation curve between R_{CP}/R_{CCP} and porosity

4.5 Prediction of AC Impedance Parameters of LCCM Based on ARIMA Model

4.5.1 Basic Principle of ARIMA Model

The ARIMA model was proposed by American scholar Box and British statistician Jenkins in the 1970s, and is also known as the Box-Jenkins model. The basic idea of the ARIMA model analysis method is to treat the time-varying data series as a random sequence. By constructing a model to describe it, the method uses weighted modeling of historical values, current values, and lagged random disturbance terms from the data series to achieve predictions of future values. Typically, the time series selected for the ARIMA model is a weakly stationary series. By applying differencing to the weakly stationary series, a stationary time series can be obtained, which can then be modeled using the ARMA(p, q) model. The parameters of the constructed model are determined based on the statistical properties of the autocorrelation function and the partial autocorrelation function.

The ARIMA(p, d, q) model includes the autoregressive model (AR) and the moving average model (MA). The AR model describes the relationship between the current value of a time series and its historical values. If it depends on p historical values, the order is denoted as p. The formula defining the p-th order autoregressive process is:

$$y_t = \varphi_1 y_{t-1} + \varphi_2 y_{t-2} + \cdots + \varphi_p y_{t-p} + u_t + \mathrm{c} \tag{4.37}$$

where: y_t represents the current value of the time series. $y_{t\text{-}p}$ represents the historical value at time $t\text{-}p$. φ_i represents the autoregressive coefficients, which are the parameters to be estimated in the model ($i = 1, 2, \ldots, p$). u_t represents the white noise, which is the random fluctuation component in the series. c is the constant term.

The MA model influences the predicted value at the current time point through a linear combination of historical white noise in the time series. This model can effectively eliminate random fluctuations in the predictions. If a segment of the white noise sequence is represented as $\{u_1, u_2, \ldots, u_\mathrm{n}\}$, then the formula defining a q-th order moving average process is:

$$y_t = u_t + \theta_1 u_{t-1} + \theta_2 u_{t-2} + \cdots + \theta_q u_{t-q} + \mathrm{c} \tag{4.38}$$

where: $u_{t\text{-}p}$ represents the white noise sequence at time $t\text{-}q$. θ_i represents the moving average coefficients, which are the parameters to be estimated in the model (for ($i =$ 1, 2, …, q)).

The ARMA model is an integration of the AR model and the MA model, which can simultaneously address the relationship between current data and subsequent data as well as random fluctuations, while maintaining a similar goodness of fit for the data. The defining equation for the ARMA model is shown in Eq. 4.39.

$$y_t = \sum_{i=1}^{p} \varphi_i y_{t-i} + \sum_{i=1}^{q} \theta_i u_{t-i} + u_t + \mathrm{c} \tag{4.39}$$

The ARMA model can handle stationary time series; for non-stationary time series, a differencing term d must be introduced to transform the non-stationary series into a stationary series, resulting in the ARIMA model. Its defining equation is as follows:

$$\left(1 - \sum_{i=1}^{p} \varphi_i L^i\right)(1 - L)^d Y_t = \left(1 + \sum_{i=1}^{q} \theta_i L^i\right) u_t \tag{4.40}$$

where: L is the lag operator. d represents the difference order where ($d \in \mathrm{z}$) and z is a positive integer.

The expression for the differencing operator is:

$$\Delta^d y_t = (1 - L)^d y_t \tag{4.41}$$

Let w_t be equal to:

$$w_t = \Delta^d y_t = (1 - L)^d y_t \tag{4.42}$$

Then the definition of ARIMA model can be transformed into:

$$w_t = \varphi_1 w_{t-i} + \varphi_2 w_{t-2} + \cdots + \varphi_p w_{t-p} + u_t + \theta_1 u_{t-1} + \theta_2 u_{t-2} + \cdots \theta_q u_{t-q} \tag{4.43}$$

4.5.2 Steps for ARIMA Model

The construction of the ARIMA model is based on the ARMA model. The ARIMA model addresses the challenges of handling non-stationary time series by using the differencing method. The specific steps for modeling an ARIMA model are as follows.

Stationarity and non-white noise testing of time series

The purpose of stationarity testing is to determine whether the current time series will continue to develop according to its existing trend in the future. If the current time series exhibits this trend over a future period, the analysis of this series becomes meaningful. The methods for testing stationarity mainly include graphical tests and the Augmented Dickey-Fuller (ADF) test for unit roots. The graphical test involves plotting the time series of observed values to assess whether the series fluctuates randomly around a mean. If the series values exhibit random fluctuations around a mean, it is considered a stationary series. Conversely, if the fluctuations are excessive and there is no fixed trend, the series is classified as non-stationary. The ADF test is a hypothesis test that initially assumes the time series is non-stationary. It then compares the resulting test statistic with critical values corresponding to different significance levels (1, 5, 10%). If the test statistic is less than the critical value corresponding to 1%, the series is determined to be stationary.

Once the non-stationary series has been differenced to obtain a stationary time series, the value of the parameter d can be determined. It is important to note that differencing is a crucial process for achieving stationarity in a time series. However, if the order of differencing is chosen incorrectly or is excessive, it may lead to the loss of valuable information in the time series. After the series is made stationary, a pure randomness test should also be conducted on the observed series to determine whether it is a white noise series. This test assesses whether there is any correlation between the time series and whether historical values influence future values. If the series is a purely random series, this indicates that historical values do not affect the trends of future values, suggesting that there is no significant relationship between the series, and thus the series lacks analytical relevance.

ARIMA model identification

The model construction is based on the differenced stationary time series. The parameter values p and q are initially determined using the autocorrelation coefficients and partial autocorrelation coefficients. The specific criteria for determining the order are shown in Table 4.6.

When selecting the final model, it is necessary to optimize the chosen model by selecting the p and q values that provide the best fit. This evaluation also requires the use of the Akaike Information Criterion (AIC) and the Bayesian Information Criterion (BIC) to determine the final model parameters.

The AIC criterion, known as the Akaike Minimum Information Criterion, serves as an evaluation index for the effectiveness of statistical model fitting. Generally, the definition of the AIC criterion is as follows:

$$AIC = 2k - 2\ln(L) \tag{4.44}$$

where: k represents the number of parameters in the model. L is the likelihood function.

Table 4.6 Order determination of autocorrelation and partial autocorrelation

Applied model	Autocorrelation coefficient (ACF)	Partial autocorrelation coefficient (PACF)
AR(p)	Attenuation tends to zero (geometric or oscillatory)	Post-p-truncation
MA(q)	q-order post-truncation	Attenuation tends to zero (geometric or oscillatory)
ARMA(p, q)	After q order, the attenuation tends to zero (geometric type or oscillating type)	The attenuation tends to zero after p order (geometric or oscillating)

The Bayesian Information Criterion (BIC) measures the quality of a model's fit to the current data from the perspective of model fitting. Its expression is given by:

$$BIC = k\ln(n) - 2\ln(L) \tag{4.45}$$

where: n represents the sample size. k and L have the same meanings as in the AIC equation, where k is the number of parameters in the model and L is the likelihood function.

The BIC criterion is based on the AIC criterion but incorporates the sample size. When the sample size is large, it can allow for greater model complexity. This criterion is more suitable for predicting data series with large sample sizes.

Model significance testing

Significance testing, also known as white noise testing, is generally used to evaluate whether the established model can adequately explain the correlations in the series. It involves determining whether the residual series is a white noise series, which is the primary basis for model significance testing. If the residual series is found to be a white noise series, it indicates that the established model is relatively ideal and suitable for further prediction; conversely, if the residual series is not white noise, the model is regarded as inappropriate, necessitating a re-evaluation of the parameter values and re-modeling.

Model prediction

After establishing the model and conducting the tests mentioned above, it is confirmed that the established model fits well within the confidence interval. This model is considered the optimal model. Under the Anaconda3 runtime environment, the accuracy of the model is validated by assessing the relative error between the experimental values and predicted values, thereby confirming the fitting effect of the established model and evaluating the prediction accuracy of the constructed model.

4.5.3 Prediction of AC Impedance Parameter

The AC impedance parameter values R_s, R_{ct1}, and CPE with different water-to-binder ratios are the subjects of the accuracy validation and prediction of the impedance parameters. The analysis is conducted in the Anaconda3 environment, where all data checks and time series predictions are executed.

Stationarity test of observed value series

The impedance parameter values of cementitious materials with different water-to-binder ratios are treated as the original observation value sequences of the time series. First, these sequences undergo stationarity and non-white noise tests. The ADF unit root test is conducted on the observation value sequences of the impedance parameters for different water-to-binder ratios to check the stationarity of the sequences.

The Ljung-Box test is used to assess the pure randomness of the observation value sequences.

The ADF test results for the observation value sequences of the impedance parameters for different water-to-binder ratios are shown in Table 4.7. All sequences have ADF test statistic T values that exceed the critical values at the 1% significance level, indicating that all sequences are non-stationary and require first-order differencing. The formula for first-order differencing is shown in Eq. 4.46. The ADF unit root test results after first-order differencing are presented in Table 4.8. The test results show that the T statistic values for all observation value sequences are less than the critical values at the 1% significance level, confirming that the observation value sequences have become stationary and therefore can proceed to the next step of testing.

$$\nabla y_t = y_t - y_{t-1} \tag{4.46}$$

where: y_t represents the current value. y_{t-1} represents the historical value at time $t-1$.

Using the Ljung-Box test, the Autocorrelation Function (ACF) is introduced in Anaconda3 to assess the pure randomness of each observation value sequence. When using this function for testing, the calculation program in Anaconda3 will by default return a probability value P. If the P value is greater than 0.05, the sequence is considered to be a white noise series; conversely, it is regarded as a non-white noise series. The test results indicate that the P values for the observation value sequences of the AC impedance parameters with different water-to-binder ratios are well below 0.05, leading to the rejection of the null hypothesis of pure random sequences. This suggests that the first-order differenced sequences are stationary non-white noise series, indicating that the historical information in the observation value sequences has an impact on future values, thus allowing for the next steps in modeling.

Table 4.7 Observation sequence ADF single root test results

Impedance parameter	w/b	T	P	Critical values: (1%)	Critical values: (5%)	Critical values: (10%)
R_s	0.42	−3.331	0.003	−0.3443	−2.867	−2.570
	0.47	−2.376	0.149	−0.3443	−2.867	−2.570
	0.52	−3.273	0.016	−3.443	−2.867	−2.570
R_{ct}	0.42	−2.065	0.024	−3.443	−2.867	−2.570
	0.47	−1.168	0.0118	−3.443	−2.867	−2.570
	0.52	−3.426	0.017	−3.443	−2.867	−2.570
CPE	0.42	−1.464	0.036	−3.455	−2.872	−2.573
	0.47	−1.929	0.319	−3.455	−2.873	−2.573
	0.52	−1.359	0.602	−3.455	−2.873	−2.573

Table 4.8 ADF single root test results of observed value sequence after first order difference

Impedance parameter	w/b	T	P	Critical values: (1%)	Critical values: (5%)	Critical values: (10%)
R_s	0.42	−5.330	4.00E-06	−0.3443	−2.867	−2.570
	0.47	−3.686	4.32E-03	−0.3443	−2.867	−2.570
	0.52	−3.914	1.93E-03	−3.443	−2.867	−2.570
R_{ct}	0.42	−12.065	4.41E-24	−3.443	−2.867	−2.570
	0.47	−5.779	1.18E-21	−3.443	−2.867	−2.570
	0.52	−8.085	1.43E-12	−3.443	−2.867	−2.570
CPE	0.42	−12.009	3.22E-22	−3.455	−2.872	−2.573
	0.47	−3.579	6.17E-03	−3.456	−2.873	−2.573
	0.52	−6.201	5.82E-08	−3.456	−2.873	−2.573

Determination of model parameters

The identification of the model involves the process of establishing the parameters p, d, and q for the ARIMA model. The values of p and q for the observation value sequences of the impedance parameters at different water-to-binder ratios are presented in Table 4.9.

Based on the stationarity test of the observation value sequences, it can be concluded that the first-order differenced sequences of the impedance parameters exhibit stationarity; therefore, the value of d is taken as 1. The other two parameters p and q in the ARIMA model need to be determined. The values of p and q are identified using the autocorrelation function (ACF) and partial autocorrelation function (PACF), which involves a significant degree of subjectivity, and the resulting model is not unique. Therefore, the AIC and BIC criteria can be used as supplementary methods to determine the values of p and q, ensuring that p and q are positive integers. In this study, heatmap analysis is conducted on the observation

Table 4.9 Model selection of impedance parameter observation sequence

Parameter	w/b	p	q	ARIMA(p, d, q)
R_s	0.42	1	1	(1, 1, 2)
	0.47	2	3	(2, 1, 3)
	0.52	1	2	(1, 1, 2)
R_{ct}	0.42	3	4	(3, 1, 4)
	0.47	1	2	(1, 1, 2)
	0.52	1	1	(1, 1, 1)
CPE	0.42	5	5	(5, 1, 5)
	0.47	1	2	(1, 1, 2)
	0.52	4	1	(4, 1, 1)

value sequences of the impedance parameters with different water-to-binder ratios to compare the statistical values under various parameter settings. Generally, the smaller the statistical indicator values, the better the predictive performance of the chosen model. This approach helps to identify the optimal combinations of p and q. The determination of p and q values also indicates the specification of the predictive model for the observation value sequences of the impedance parameters.

Model significance test

The established model undergoes significance testing, which involves performing a white noise test on the residual sequence of the model to ensure its correctness. If the information extracted by the model is sufficiently comprehensive, it indicates that the model is well-fitted. If the fitted residuals contain no relevant information, then the residual sequence is considered to be a white noise series, making such a model significantly effective. The significance testing of the models established for the observation value sequences of the impedance parameters at different water-to-binder ratios shows that the P values are all greater than 0.05, allowing the residual sequences to be regarded as white noise series, thus confirming that the models are significantly valid.

Model validation and prediction

After constructing the model, the observation value sequences of the AC impedance parameters for different water-to-binder ratios are validated for accuracy using the models determined in Table 4.9. The predict function is introduced in the Anaconda3 runtime environment. Tables 4.10, 4.11 and 4.12 list the comparisons between the experimental values and the predicted values of the impedance parameters for different water-to-binder ratios. Since the observation value sequences have been subjected to first-order differencing, the prediction of the AC impedance parameter values starts from the 2-h hydration mark.

Table 4.10 Comparison of experimental and predicted values of AC impedance parameter R_s

w/b	Number (Ω)	Hydration time								
		2 h	4 h	6 h	12 h	1 d	3 d	7 d	14 d	28 d
0.42	Exp. value	26.85	32.42	33.40	35.71	38.02	38.82	39.20	39.56	39.69
	Pred. value	27.93	32.25	33.43	35.78	37.97	38.83	39.41	40.02	40.75
	Rel. error %	4.02	0.52	0.09	0.20	0.13	0.03	0.54	1.16	2.67
0.47	Exp. value	23.22	30.55	30.87	32.12	33.61	33.96	34.44	37.04	37.89
	Pred. value	23.58	29.95	30.88	31.80	32.98	33.67	34.61	37.59	38.44
	Rel. error %	1.55	1.96	0.03	1.00	1.87	0.85	0.49	1.48	1.45
0.52	Exp. value	18.65	23.59	26.92	27.36	30.24	31.23	31.56	33.07	33.39
	Pred. value	19.12	23.12	26.34	27.74	30.47	31.55	32.07	33.92	34.21
	Rel. error %	2.52	1.99	2.15	1.39	0.76	1.02	1.62	2.57	2.46

Table 4.11 Comparison of experimental and predicted values of AC impedance parameter R_{ct1}

w/b	R_{ct1} (Ω)	Hydration time								
		2 h	4 h	6 h	12 h	1 d	3 d	7 d	14 d	28 d
0.42	Exp. value	11,602	18,776	23,881	26,106	31,550	32,184	33,665	34,644	36,097
	Pred. value	11,564	18,023	22,951	25,277	31,437	32,530	34,046	35,060	36,528
	Rel. error %	0.33	4.01	3.89	3.18	0.36	1.07	1.13	1.20	1.19
0.47	Exp. value	9325	12,967	14,135	16,601	19,453	20,902	22,540	23,415	24,269
	Pred. value	9511	12,693	13,705	15,988	19,201	20,854	22,651	23,794	24,959
	Rel. error %	1.99	2.11	3.04	3.69	1.29	0.23	0.49	1.62	2.84
0.52	Exp. value	8168	9126	10,423	11,387	13,966	14,337	15,334	16,143	16,477
	Pred. value	7678	8678	10,111	11,355	14,017	14,507	15,541	16,145	16,906
	Rel. error %	2.33	4.90	2.99	0.28	0.37	1.18	1.34	0.01	2.60

Table 4.12 Comparison of experimental and predicted values of AC impedance parameter *CPE*

w/b	*CPE* (pF)	Hydration time								
		2 h	4 h	6 h	12 h	1 d	3 d	7 d	14 d	28 d
0.42	Exp. value	1397	1288	1288	1269	1208	1200	1188	1182	1188
	Pred. value	1378	1278	1295	1269	1212	1191	1178	1169	1176
	Rel. error %	1.33	0.76	0.55	0.01	0.34	0.75	0.85	1.10	1.02
0.47	Exp. value	1440	1326	1311	1296	1284	1258	1242	1205	1200
	Pred. value	1420	1320	1328	1305	1291	1264	1246	1207	1196
	Rel. error %	1.33	0.49	1.29	0.69	0.53	0.49	0.33	0.25	0.38
0.52	Exp. value	1514	1410	1348	1352	1313	1261	1251	1250	1235
	Pred. value	1525	1424	1357	1344	1327	1271	1255	1246	1219
	Rel. error %	0.74	1.04	0.72	−0.65	1.06	0.74	0.29	0.30	1.33

From the analysis of the data in the tables, it can be observed that for the AC impedance parameter R_s, the average relative errors between the experimental and predicted values for water-to-binder ratios of 0.42, 0.47, and 0.52 are 1.04, 1.19, and 1.83%, respectively. For the AC impedance parameter R_{ct1}, the average relative errors between the experimental and predicted values for the same water-to-binder ratios are 1.82, 1.92, and 1.78%, respectively. For the AC impedance parameter *CPE*,

Table 4.13 Predicted value of AC impedance parameters at 56 days of hydration

w/b	R_s (Ω)	R_{ct1} (Ω)	*CPE* (pF)
0.42	41.38	37,201	1169
0.47	39.32	25,926	1187
0.52	35.07	17,531	1206

the average relative errors for water-to-binder ratios of 0.42, 0.47, and 0.52 are 0.75, 0.64, and 0.76%, respectively.

The average relative errors for all groups of AC impedance parameter values are less than 5%, indicating that the established ARIMA models for the different parameter values are reasonable and effective. Therefore, it is feasible to use these models to predict how the AC impedance parameter values change over time.

Based on the established ARIMA models for different water-to-binder ratios, the impedance parameters for the cementitious materials at a hydration age of 56 days are predicted. The prediction results are shown in Table 4.13, and they align with the actual changes in the parameters. The analysis reveals that for the cementitious materials with water-to-binder ratios of 0.42, 0.47, and 0.52, the R_s values at 56 days of hydration increased by 0.63, 0.88, and 0.86 Ω, respectively, while the R_{ct1} values increased by 673, 967, and 681 Ω, respectively. The CPE values decreased by 7 pF, 9 pF, and 13 pF, respectively, indicating that all impedance parameter values experienced slight changes. This suggests that as hydration time continues to extend, the ionic transfer process within the cementitious material system becomes increasingly difficult, resulting in a continuous increase in impedance values, though the rate of increase becomes progressively smaller. This also indicates that the internal structure of the cementitious material system is becoming more stable.

4.6 Conclusions

(1) Mechanical performance and hydration products of low-carbon cementitious materials prepared using coal gangue, calcium carbide residue, and desulfurized gypsum as raw materials were systematically investigated. Test results revealed that the 3d compressive strength of the cementitious material with a water-to-binder ratio of 0.47 reached 52.52 MPa, accounting for 83.13% of its 28d compressive strength. Mineralogical analysis identified $C_4A_3\hat{S}$, C_2S, and $CaSO_4$ as primary crystalline phases. The hydration of $C_4A_3\hat{S}$ contributed to early strength development, while subsequent hydration of C_2S enhanced later-stage compressive strength.

(2) Alternating current impedance spectroscopy was employed to evaluate cementitious materials with varying water-to-binder ratios at different hydration stages. An equivalent circuit model characterizing hydration degree was established, with systematic analysis of impedance parameter evolution. Combined with hydration heat and chemically bound water measurements, correlations between

impedance parameters and hydration progression were elucidated. Experimental findings demonstrated that the high-frequency semicircular arc in Nyquist plots progressively expanded and shifted rightward with hydration advancement. Resistive parameters R_s and R_{ct1} exhibited monotonic increases, while the constant phase element (*CPE*) progressively decreased. Strong correlations were observed between impedance parameters and key performance indicators including compressive strength, cumulative heat release, and chemically bound water content.

(3) A microstructure-informed equivalent circuit model was developed to investigate impedance parameter variations. Complementary microstructural characterization through morphological analysis and pore structure quantification revealed fundamental relationships between electrical parameters and porosity characteristics. Results indicated inverse trends between R_{CP}/R_{CCP}, C_{DP} parameters and hydration duration, while R_{CP} and R_{CCP} demonstrated contrasting evolutionary patterns. Exponential correlation was established between R_{CCP} and the product of most probable pore diameter/porosity, with linear relationship identified between R_{CP}/R_{CCP} ratio and porosity.

(4) Autoregressive Integrated Moving Average (ARIMA) modeling was implemented for predictive analysis of impedance parameters across different water-to-binder ratios. Model validation showed mean relative errors below 5% between experimental and predicted values. The 56d impedance parameter predictions exhibited satisfactory agreement with actual variation trends, demonstrating the model's effectiveness in characterizing long-term performance evolution of solid waste-based cementitious materials.

References

Chi, L., Li, W., Li, Z., Wang, Z., Lu, S., & Liu, Q. (2022). Investigation of the hydration properties of cement with EDTA by alternative current impedance spectroscopy. *Cement and Concrete Composites, 126*, 104365.

Darquennes, A., Espion, B., & Staquet, S. (2013). How to assess the hydration of slag cement concretes. *Construction and Building Materials, 40*, 1012–1020.

Farzanian, K., Teixeira, K. P., Rocha, I. P., Carneiro, L. D. S., & Ghahremaninezhad, A. (2016). The mechanical strength, degree of hydration, and electrical resistivity of cement pastes modified with superabsorbent polymers. *Construction and Building Materials, 109*, 156–165.

Gu, P., Xie, P., Fu, Y., & Beaudoin, J. J. (1994). AC impedance phenomena in hydrating cement systems: Frequency dispersion angle and pore size distribution. *Cement and Concrete Research, 24*(1), 86–88.

Jain, J., & Neithalath, N. (2011). Electrical impedance analysis based quantification of microstructural changes in concretes due to non-steady state chloride migration. *Materials Chemistry and Physics, 129*(1–2), 569–579.

Liang, N., Miao, Q., Liu, X., & Zhong, Y. (2019). Frost-resistance mechanism of multi-scale PFRC based on NMR. *Magazine of Concrete Research, 71*(13), 710–718.

Li, J. L., Zhou, K. P., Liu, W. J., & Deng, H. W. (2016). NMR research on deterioration characteristics of microscopic structure of sandstones in freeze–thaw cycles. *Transactions of Nonferrous Metals Society of China, 26*(11), 2997–3003.

McCarter, W. J. (1994). A parametric study of the impedance characteristics of cement-aggregate systems during early hydration. *Cement and Concrete Research, 24*(6), 1097–1110.

Mlynárik, V. (2017). Introduction to nuclear magnetic resonance. *Analytical Biochemistry, 529*, 4–9.

Sargam, Y., & Wang, K. (2021). Hydration kinetics and activation energy of cement pastes containing various nanoparticles. *Composites Part b: Engineering, 216*, 108836.

Song, G. (2000). Equivalent circuit model for AC electrochemical impedance spectroscopy of concrete. *Cement and Concrete Research, 30*(11), 1723–1730.

Tambara, L. U. D., Jr., Cheriaf, M., Rocha, J. C., Palomo, A., & Fernández-Jiménez, A. (2020). Effect of alkalis content on calcium sulfoaluminate (CSA) cement hydration. *Cement and Concrete Research, 128*, 105953.

Wu, Y., Duan, P., & Yan, C. (2018). Role of layered double hydroxides in setting, hydration degree, microstructure and compressive strength of cement paste. *Applied Clay Science, 158*, 123–131.

Xie, P., Gu, P., Xu, Z., & Beaudoin, J. J. (1993). A rationalized ac impedence model for microstructural characterization of hydrating cement systems. *Cement and Concrete Research, 23*(2), 359–367.

Zou, Y., Shen, X., Zuo, X., Xue, H., & Li, G. (2022). Experimental study on microstructure evolution of Aeolian sand concrete under the coupling freeze–thaw cycles and carbonation. *European Journal of Environmental and Civil Engineering, 26*(4), 1267–1282.

Chapter 5
Desulfurization Gypsum Regulates Hydration Process of Low-Carbon Cementitious Material

5.1 Introduction

This study developed low-carbon cementitious materials (LCCM) through calcination of industrial solid wastes (coal gangue, calcium carbide slag, and desulfurization gypsum), with particular focus on hydration process regulation via desulfurization gypsum incorporation. A comprehensive investigation was conducted: 1. The effects of desulfurization gypsum on LCCM setting time, strength development, phase evolution of hydration products, chemically bound water content, molecular structure of hydration products, and microstructure of hydration products were studied. 2. The influence of desulfurization gypsum on the porosity and pore size distribution of LCCM was studied. 3. The influences of desulfurization gypsum on Nyquist diagram, conductive path, and impedance parameters of LCCM were studied based on AC impedance method. 4. The relationship between impedance parameters and mechanical properties of LCCM is established by using grey theory. 5. The effect of desulfurization gypsum on the hydration heat release of LCCM was studied, and the hydration kinetics model was established to analyze the hydration reaction degree of LCCM.

5.2 Experimental Overview

5.2.1 Raw Materials

All experimental materials were composed entirely of industrial solid wastes, including coal gangue, carbide slag, and desulfurization gypsum. Coal gangue was sourced from the Jungar City coal mine in Inner Mongolia. It was ground into powder using a ball mill and sieved through a 200-mesh sieve for subsequent use. The desulfurization gypsum and carbide slag were obtained from Gongyi City, Henan Province,

C. Yan et al., *Low-Carbon Cementitious Materials with 100% Solid Wastes*,
https://doi.org/10.1007/978-981-96-8742-8_5

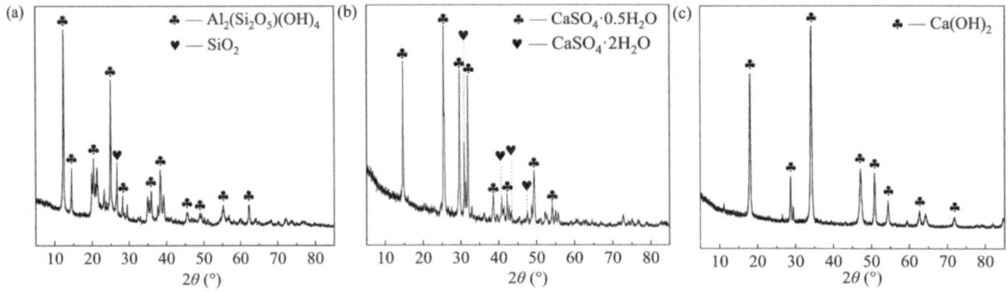

Fig. 5.1 XRD of raw materials. **a** Coal gangue, **b** desulfurization gypsum, **c** carbide slag

Fig. 5.2 Microstructure of raw materials

Fig. 5.3 Particle size distribution of raw materials

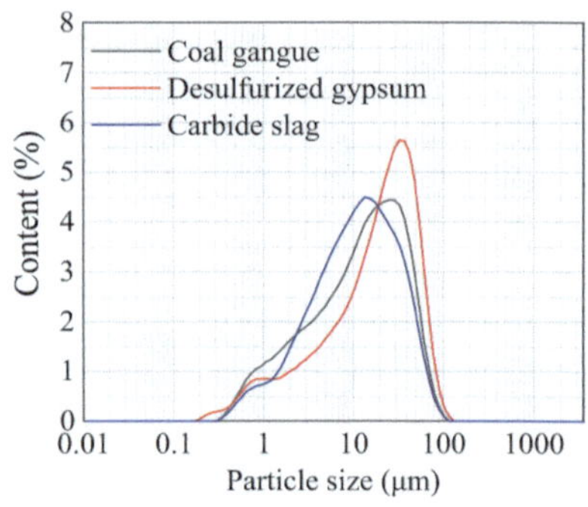

and similarly sieved through a 200-mesh sieve. The phase composition of raw materials, as determined by XRD, is shown in Fig. 5.1. The microstructural morphology of the raw materials is presented in Fig. 5.2, the particle size distribution in Fig. 5.3, and the chemical composition in Table 5.1.

5.2.2 Preparation Process of LCCM

The preparation process of the clinker is illustrated in Fig. 5.4. Coal gangue, desulfurization gypsum, and carbide slag were ground and sieved through a 200-mesh

Table 5.1 Chemical composition of solid waste (wt.%)

Raw material	CaO	SiO_2	SO_3	Al_2O_3	MgO	Fe_2O_3	Na_2O	Ignition loss
Coal gangue	2.23	46.32	1.67	45.34	0.36	0.87	0.02	28.20
Desulfurization gypsum	41.4	1.19	54.9	0.51	0.41	0.30	0.04	12.40
Carbide slag	94.7	3.04	1.67	1.09	0.07	0.28	0.03	22.03

sieve (aperture: 0.075 mm). The materials were weighed according to the calculated proportions, thoroughly mixed using a mixer, and then sprayed with an appropriate amount of water. A portion of the mixture was placed into a specially designed mold and compacted into a round raw material block with a diameter of 6 cm. The specimens were subsequently dried and calcined in a high-temperature furnace under the heating regime shown in Fig. 5.5. The temperature was first raised to 900 °C at a heating rate of 10 °C/min, followed by a further increase to 1280 °C at a rate of 5 °C/min, and held at this temperature for 40 min. After calcination, the specimens were rapidly cooled by removing them from the furnace. Once cooled to room temperature, the specimens were ground and sieved through a 200-mesh sieve to obtain the clinker. The chemical composition and phase composition of the clinker are presented in Table 5.2 and Fig. 5.6, respectively. The primary components were C_2S and $C_4A_3\hat{S}$, along with a small amount of $CaSO_4$.

The clinker was thoroughly mixed with a specified proportion of desulfurization gypsum to prepare LCCM. The samples were labeled as A_0, A_5, A_{10}, A_{15}, and A_{20}, representing desulfurization gypsum mass percentages of 0, 5, 10, 15, and 20% of the clinker, respectively. The phase composition of the prepared LCCM is shown in Fig. 5.7. The LCCM without desulfurization gypsum primarily consisted of $C_4A_3\hat{S}$ and C_2S, along with characteristic peaks of $CaSO_4$, which resulted from the incomplete dehydration of desulfurization gypsum during the calcination of the clinker. As

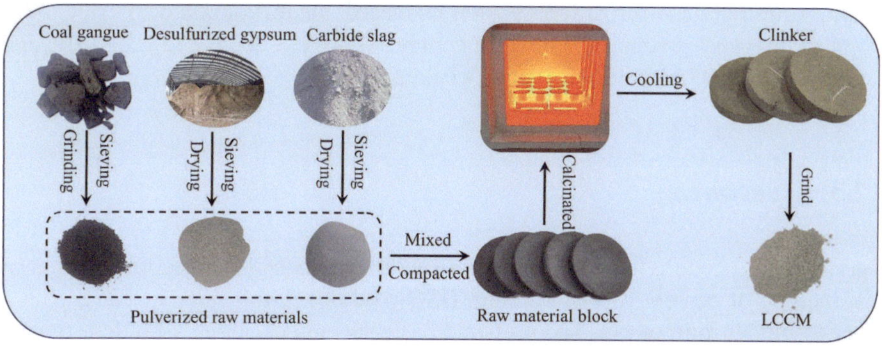

Fig. 5.4 Flow chart of the clinker preparation

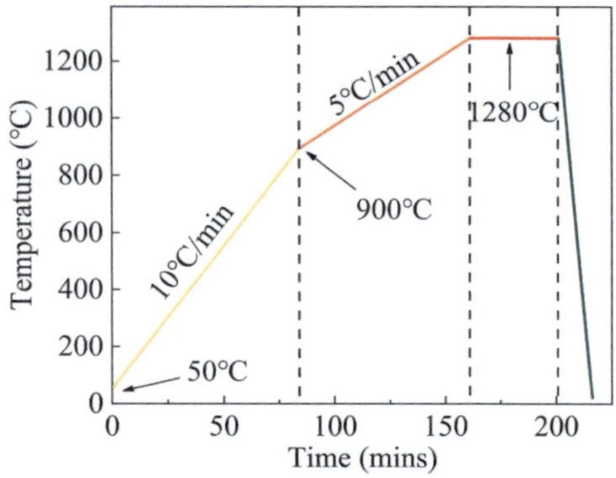

Fig. 5.5 Calcination process of the clinker

Table 5.2 Chemical composition of the clinker (wt.%)

CaO	SiO_2	Al_2O_3	SO_3	Na_2O	MgO	LOSS
44.79	17.68	15.28	17.50	2.25	1.10	1.24

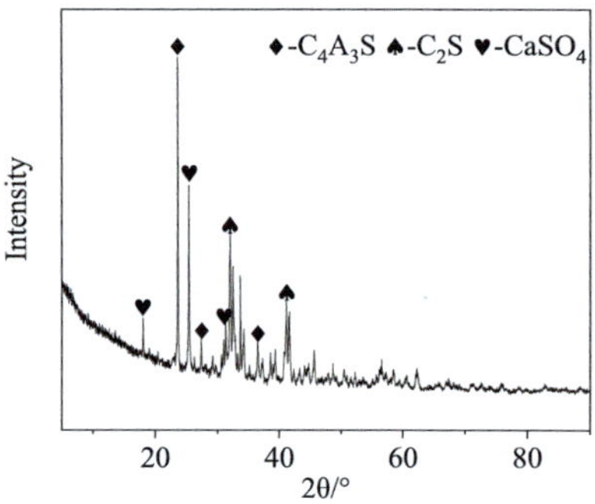

Fig. 5.6 XRD of the clinker

the proportion of desulfurization gypsum increased, the diffraction peaks of $CaSO_4$ began to intensify gradually, while the content of $C_4A_3\hat{S}$ decreased relatively, and the intensity of its characteristic peaks weakened.

5.2.3 *Specimen*

The mixing procedure for preparing the LCCM specimens followed the Method of Test method of cement mortar strength (ISO method) (GB/T 17671-2021), with a water-to-binder ratio of 0.3. The prepared paste specimens, along with their molds, were placed in a curing chamber for standard curing, with the temperature set at 20 °C and the relative humidity at 95%. After 3 h of curing in the standard chamber,

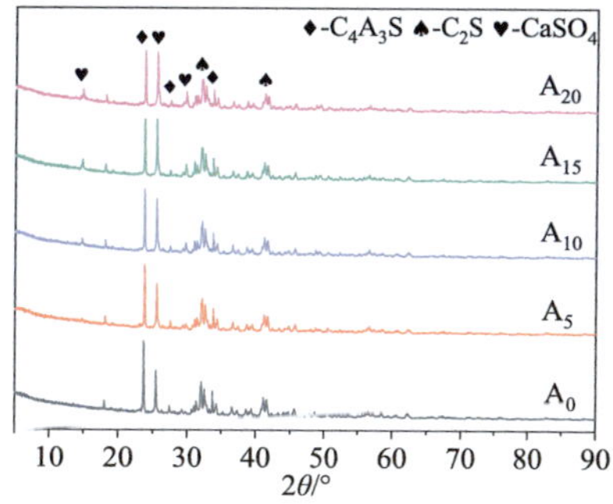

Fig. 5.7 XRD diffraction pattern of LCCM with different desulfurization gypsum content

the specimens were demolded and continued to be cured until the designated testing ages.

For compressive strength testing, the specimens were prepared with dimensions of 20 mm × 20 mm × 20 mm. For nuclear magnetic resonance (NMR) testing, the specimens were prepared with dimensions of 40 mm × 40 mm × 40 mm. For EIS testing, the specimens were prepared with dimensions of 160 mm × 40 mm × 40 mm, and parallel stainless steel mesh sheets were embedded at 20 mm from each end.

Samples for microscopic testing, including XRD, TG-DSC, SEM, and FTIR, were taken from fragments of the specimens after compressive strength testing. Hydration was stopped using anhydrous ethanol. For SEM analysis, the samples were similarly taken from fragments of the compressive strength-tested specimens, with dimensions of less than 5 mm in length and width, and less than 3 mm in thickness, ensuring a flat surface. Hydration was also terminated using anhydrous ethanol.

5.2.4 *Test Methods*

Setting time

The initial and final setting times were tested according to the method specified in Test methods for water requirement of standard consistency, setting time and soundness of the Portland cement (GB/T 1346-2011).

Compressive strength testing

Compressive strength testing was conducted in accordance with the Method of Test method of cement mortar strength (ISO method) (GB/T 17671-2021). The specimens were prepared with a water-to-binder ratio of 0.3 and dimensions of 20 mm × 20 mm × 20 mm. The cement paste specimens were cured in a standard curing chamber until the designated ages, and compressive strength tests were performed using a universal testing machine at a loading rate of 1.25 MPa/s.

Phase analysis

XRD was employed to analyze the phase composition of the LCCM and its hydration products. The samples were prepared with a water-to-binder ratio of 0.3 and cured to the specified ages. After curing, the specimens were crushed, and hydration was stopped using anhydrous ethanol. The samples were then dried at 45 °C and ground to pass through a 200-mesh sieve. XRD analysis was performed using a Rigaku Ultima IV X-ray diffractometer, with a scanning angle range of 5–90° and a scanning speed of 10°/min.

Thermogravimetric analysis

TG-DSC was used to simultaneously measure the mass change and heat flow of LCCM hydration products during heating. The sample preparation method was consistent with that used for XRD testing. Approximately 30 mg of the sample was used for the analysis, which was performed using a NETZSCH STA 449 F5 simultaneous thermal analyzer (Germany). The temperature range was set from 50 to 1000 °C, with a heating rate of 10 °C/min.

Molecular structure analysis

FTIR was conducted using a Shimadzu IR Tracer-100 spectrometer (Japan) to characterize the molecular structure of the LCCM hydration samples that had passed through a 200-mesh sieve. The scanning wavelength range was 400–4000 cm^{-1}, and the number of scans was 45.

Hydration heat analysis

The hydration heat release rate and cumulative heat release of the LCCM over 3 days were measured using a TAM Air eight-channel isothermal calorimeter (TA Instruments, USA). The test was conducted at a constant temperature of 20 $\pm$ 1 °C, with 4 g of cement and 2 g of deionized water. The testing duration was 72 h.

Microstructural morphology analysis

SEM was employed to characterize the microstructural morphology of the LCCM hydration samples. A ZEISS Sigma 500 field-emission scanning electron microscope (Germany) was used for this purpose. The samples, with dimensions of less than 5 mm in length and width and less than 3 mm in thickness, were prepared with a flat surface and coated with gold prior to testing.

Pore structure analysis

NMR technology was used to analyze the pore structure of the specimens. The tests were conducted using a MesoMR23-060H-I NMR imaging analyzer (Suzhou Niuman Analytical Instrument Corporation, China). The pore structure testing process is illustrated in Fig. 5.8. Prior to testing, the specimens were subjected to vacuum saturation with water. To prevent the loss of surface moisture during testing, which could affect the results, the vacuum-saturated specimens were sealed with

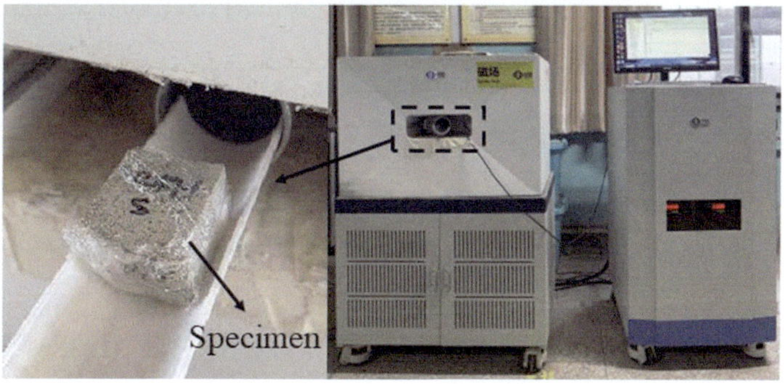

Fig. 5.8 Pore structure test

plastic film and then placed into the testing coil for pore structure analysis using the NMR analyzer.

Electrochemical Impedance Spectroscopy (EIS)

A schematic diagram of the EIS testing process is shown in Fig. 5.9. The EIS test frequency range was set from 0.01 to 1 MHz, with a minimum high-frequency current range of 2 mA and an AC amplitude of 10 mV. The testing mode selected was impedance-frequency scanning.

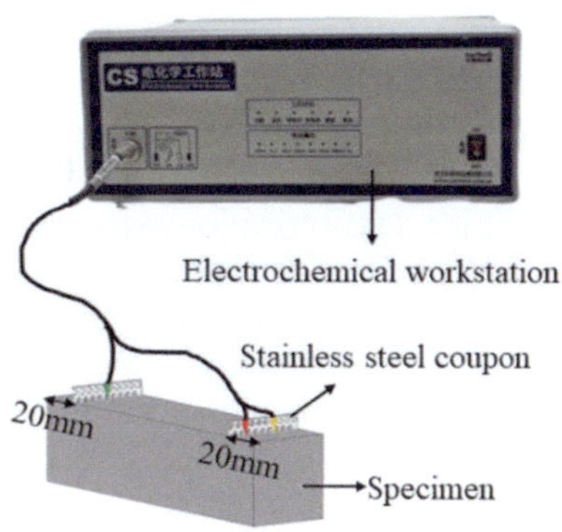

Fig. 5.9 AC impedance test diagram

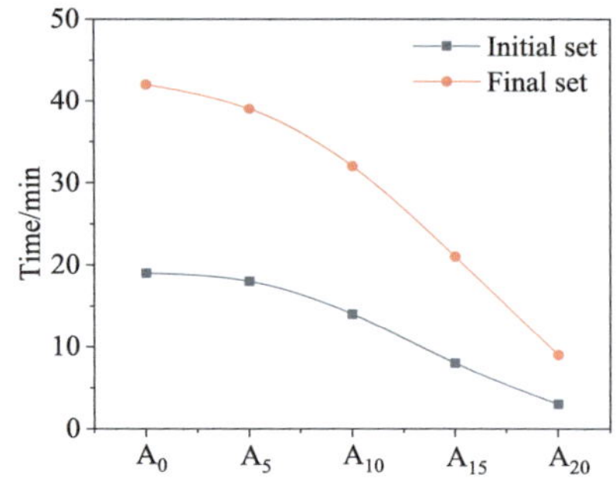

Fig. 5.10 Initial setting time and final setting time of LCCM

5.3 Desulfurization Gypsum Regulates the Strength Development of LCCM

5.3.1 Setting Time

The setting times of LCCM with varying desulfurization gypsum contents are shown in Fig. 5.10. As illustrated, both the initial and final setting times of the samples decreased with increasing desulfurization gypsum content, with the reduction in final setting time being more pronounced than that of the initial setting time. The interval between the initial and final setting times gradually narrowed. This is because the solubility of desulfurization gypsum is higher than that of anhydrite. The incorporation of desulfurization gypsum increased the concentration of calcium ions (Ca^{2+}) and sulfate ions (SO_4^{2-}), accelerating the hydration rate of $C_4A_3\hat{S}$ and the formation of ettringite, thereby shortening the setting time. On the other hand, hemihydrate gypsum in the desulfurization gypsum also reacts with water, consuming moisture. When the desulfurization gypsum content exceeded 10%, the setting time decreased rapidly. The final setting time of A_{20} was only 9 min, representing a 78.6% reduction compared to A_0. This phenomenon, where the mixture loses fluidity and solidifies within minutes after water addition, is referred to as false set. This occurs because, when a large amount of desulfurization gypsum is present in the LCCM, the cement paste contains a low concentration of aluminate ions, which rapidly becomes supersaturated with calcium and sulfate ions, leading to the quick formation of large gypsum crystals. This causes the cement paste to lose its consistency, manifesting as false solidification (Mota et al., 2020).

5.3.2 Strength Development

Figure 5.11 shows the compressive strength of LCCM with varying desulfurization gypsum contents. As illustrated, the compressive strength of all groups at 6 h gradually decreased with increasing desulfurization gypsum content, declining from

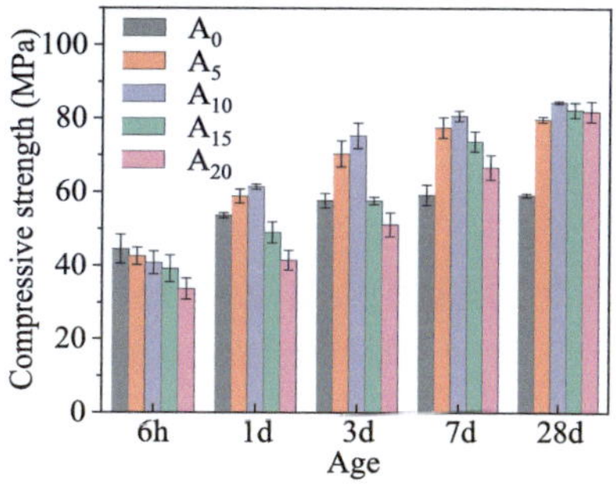

Fig. 5.11 Compressive strength of LCCM under different hydration ages

44.48 MPa for A_0 to 33.66 MPa for A_{20}. Based on the previous analysis, this is attributed to the fact that the addition of desulfurization gypsum slowed the early hydration rate of the LCCM and reduced the formation of ettringite (AFt), leading to a gradual decrease in the strength of the hardened paste. From 6 h to 3 days, A_5 and A_{10} exhibited rapid compressive strength growth, increasing by 65.35 and 84.47%, respectively. During this stage, the sufficient $CaSO_4$ content facilitated the hydration of $C_4A_3\hat{S}$, generating a large amount of AFt and aluminum hydroxide gel (AH_3), which formed a dense skeletal structure and became the primary reason for the strength increase. For A_{15} and A_{20}, compressive strength began to increase rapidly after 3 days, with the entire strength development cycle being prolonged. Notably, A_{20} showed a 22.8% increase in compressive strength from 7 to 28 days, indicating that desulfurization gypsum significantly extended the hydration cycle of the LCCM. At 28 days, A10 achieved the highest compressive strength of 84.36 MPa, and the compressive strength ranking of the five groups was $A_{10} > A_{15} > A_{20} > A_5 > A_0$.

It is generally accepted that $C_4A_3\hat{S}$ exhibits rapid early hydration, and its strength develops quickly as hydration progresses. Observing the compressive strength development of LCCM reveals that, in the absence of desulfurization gypsum, the strength development aligns with this pattern. However, after the addition of desulfurization gypsum, the early hydration rate is significantly slowed, and the hydration cycle is prolonged with increasing gypsum content. Even after 7 days of hydration, A_{15} and A_{20} samples still show substantial strength growth. As discussed earlier, this is attributed to two factors: (1) the addition of desulfurization gypsum delays early strength development, allowing strength potential to be released in the later hydration stages, and (2) as shown in Eqs. 5.1–5.2 (Hargis et al., 2014; Telesca et al., 2014), the continuous reaction between desulfurization gypsum and AH_3 generates AFt, which supplements strength development. Additionally, this reaction consumes $Ca(OH)_2$, a hydration product of C_2S, promoting further hydration of C_2S to produce more C-S-H gel, thereby supporting significant strength improvement in the later stages. Therefore, although high desulfurization gypsum content negatively impacts the early hydration and hardening of LCCM, the abundant $CaSO_4$ ensures sustained formation of hydration products, ultimately benefiting long-term strength development.

$$3CH + 3C\overline{S} + AH_3 + 26H \rightarrow C_6A\overline{S}_3H_{32} \tag{5.1}$$

$$C_2S + H \rightarrow CH + C\text{-}S\text{-}H(gel) \quad (5.2)$$

5.4 Desulphurization Gypsum Regulates the Microstructure of LCCM

5.4.1 Phase Transition

Figure 5.12 shows the XRD patterns of hydration products for LCCM with varying desulfurization gypsum contents at different ages. As shown in Fig. 5.12a, compared to unhydrated sample, the characteristic peaks of $C_4A_3\hat{S}$ and $CaSO_4$ decreased rapidly at 6 h, with the $CaSO_4$ peak almost completely disappearing, while the AFt peak intensity increased significantly. Subsequently, the AFt peak intensity gradually increased, but remained relatively unchanged from 3 to 28 days. This indicates that, in the absence of desulfurization gypsum, $C_4A_3\hat{S}$ and $CaSO_4$ in the LCCM reacted rapidly during the early hydration stage, generating a large amount of AFt, as described by Eq. 5.3. However, after 3 days, due to the limited $CaSO_4$ content, the remaining $C_4A_3\hat{S}$ hydrated to form AFm, resulting in negligible growth in the AFt peak intensity. The hydration process during this stage is represented by Eq. 5.4.

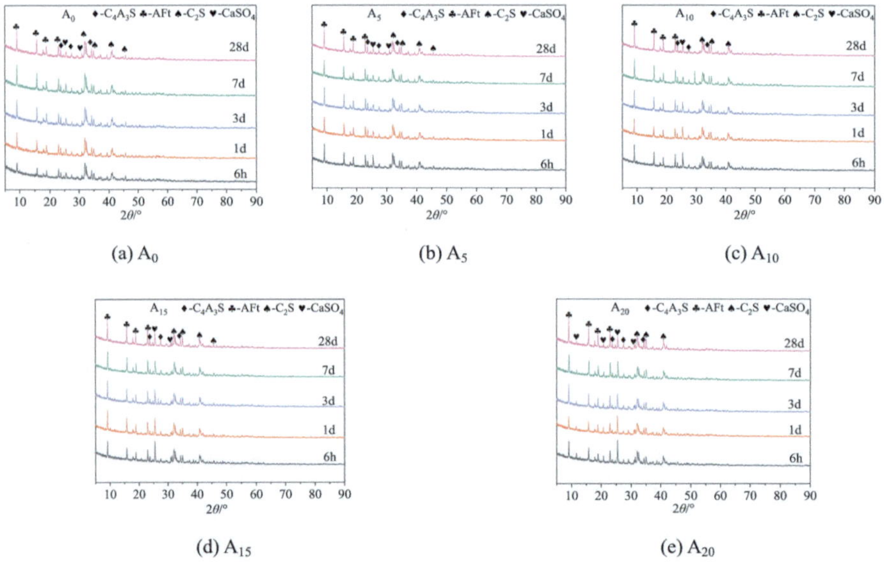

Fig. 5.12 XRD diffraction pattern of LCCM

Observing the phase changes of LCCM at different hydration ages, the rapid decrease in $CaSO_4$ peak intensity is progressively delayed with increasing desulfurization gypsum content. When the desulfurization gypsum content is 0%, $CaSO_4$ is rapidly consumed within 0–6 h, and its content stabilizes after 1 day. However, when the desulfurization gypsum content increases to 20%, the $CaSO_4$ content decreases rapidly only during 3–7 days and continues to decline until 28 days. Correspondingly, at low desulfurization gypsum content, AFt forms in large quantities during the early stage, with minimal growth in the later stage. As the desulfurization gypsum content increases, the early-stage formation of AFt slows, while its growth continues in the later stages. The higher the desulfurization gypsum content, the more pronounced the delayed formation of AFt. This indicates that desulfurization gypsum affects the hydration of LCCM in two ways: (1) it slows the early hydration rate, and (2) it extends the hydration cycle, preventing excessive concentration of hydration progress. This is because the addition of desulfurization gypsum ensures sufficient $CaSO_4$ content, allowing $C_4A_3\hat{S}$ to hydrate rapidly, forming AFt and AH_3 that coat the LCCM particles. This creates a relatively dense layer that slows water penetration and further hydration of $C_4A_3\hat{S}$, resulting in slower early hydration and sustained hydration in the later stages.

$$C_4A_3\overline{S} + 2C\overline{S} + 38H \rightarrow C_6A\overline{S}_3H_{32} + 2AH_3 \tag{5.3}$$

$$C_4A_3\overline{S} + 18H \rightarrow C_4A\overline{S}H_{12} + 2AH_3 \tag{5.4}$$

5.4.2 Comprehensive Thermal Analysis

At high temperatures, hydration products of LCCM undergo endothermic or exothermic reactions. Different hydration products experience endothermic dehydration or decomposition within specific temperature ranges. The types and quantities of hydration products can be determined based on the temperature ranges and peak sizes of the endothermic peaks. The TG-DSC curves of LCCM at different hydration ages and the mass loss rates of their major hydration products are shown in Fig. 5.13 and Table 5.3, respectively. By analyzing Fig. 5.13 and Table 5.3, it is evident that the primary hydration products of LCCM are AFt and AH_3, with their endothermic dehydration peaks occurring between 100–160 °C and 220–280 °C (Wang et al., 2017), respectively. For A_0, the endothermic peaks of AFt and AH_3 increased slowly from 6 h to 28 days. At 6 h, the mass loss contribution of AFt was 12.39%, reaching 90.5% of the 28-day value, indicating that most of the $C_4A_3\hat{S}$ in A_0 had already hydrated within the first 6 h, generating a large amount of AFt and AH_3, with rapid and concentrated hydration progress.

As the desulfurization gypsum content increased, the endothermic peak of AFt no longer concentrated in the early stage. For A_{10}, AFt began to form in large quantities

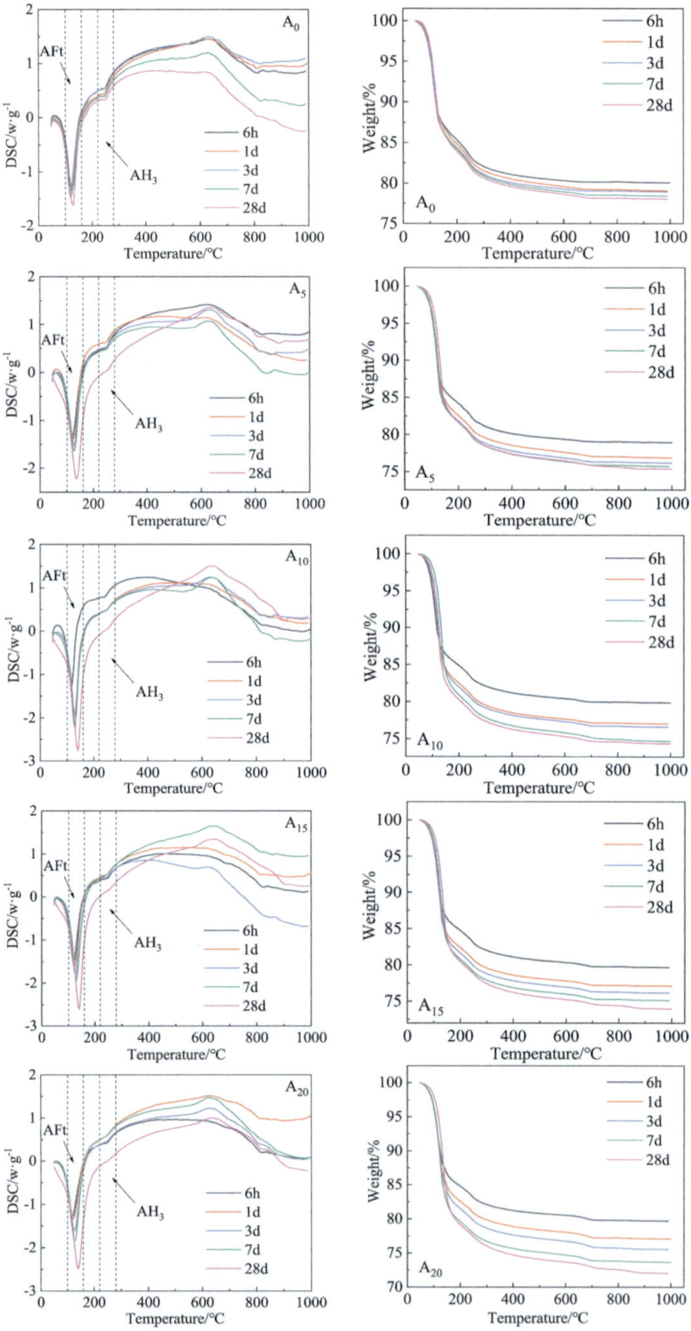

Fig. 5.13 TG-DSC spectra of LCCM at different hydration ages

Table 5.3 Mass loss rate of main hydration products of LCCM at different ages (%)

Specimen	Phase	6 h	1 d	3 d	7 d	28 d
A_0	AFt	12.391	13.005	13.252	13.388	13.687
	AH_3	2.189	2.201	2.203	2.206	2.212
A_5	AFt	13.746	15.283	15.903	15.957	15.932
	AH_3	1.958	1.983	1.995	2.156	2.064
A_{10}	AFt	13.329	15.561	16.154	16.987	17.661
	AH_3	1.839	1.894	1.914	1.953	1.857
A_{15}	AFt	13.284	15.322	16.123	16.835	17.311
	AH_3	1.822	1.824	1.827	1.879	1.831
A_{20}	AFt	13.118	15.012	15.851	16.517	17.215
	AH_3	1.786	1.770	1.836	1.849	1.821

after 1 day, while for A_{20}, the rapid increase in AFt content occurred only after 3 days, demonstrating the increasingly significant role of desulfurization gypsum in delaying the hydration of $C_4A_3\hat{S}$. At 28 days, the endothermic peak values of AFt for the five groups showed an initial increase followed by a decrease, with mass loss rates of 13.687, 15.932, 17.661, 17.311, and 17.215% for A_0–A_{20}, respectively. This is primarily because the increase in desulfurization gypsum content reduced the $C_4A_3\hat{S}$ content, leading to a decrease in the total AFt formation. The 28-day AFt content was highest in the LCCM with 10% desulfurization gypsum. Further increasing the desulfurization gypsum content reduced AFt formation, but it remained higher than that of A_0 and A_5.

Additionally, the mass loss rate of AH_3 at the same hydration age shows a decreasing trend with increasing desulfurization gypsum content, declining from 2.218% for A_0 to 1.821% for A_{20} at 28 days. Furthermore, comparing the mass loss rates of AH_3 at different ages for each group, the AH_3 content gradually increased within the first 7 days after the addition of desulfurization gypsum but decreased at 28 days, with more pronounced changes observed at higher desulfurization gypsum contents. This is because AH_3 reacts with $CaSO_4$ to form AFt. The continuous increase in AH_3 content within the first 7 days indicates ongoing hydration of $C_4A_3\hat{S}$ during this period. After 7 days, $C_4A_3\hat{S}$ is almost completely hydrated, and the consumption of AH_3 exceeds its generation, leading to a decrease in AH_3 content at 28 days. This also explains why LCCM with high desulfurization gypsum content, despite having lower $C_4A_3\hat{S}$ content and faster hydration, can still generate a significant amount of AFt in the later hydration stages.

If the total mass loss rate of LCCM at different hydration ages is used to characterize the degree of hydration, with the 28-day hydration degree defined as 1, the development of hydration degree over time is shown in Fig. 5.14. It can be observed that the hydration degree at 6 h decreases with increasing desulfurization gypsum content, indicating that desulfurization gypsum slows the early hydration of the cement. As the hydration age increases, the hydration degree of A_5 exceeds that

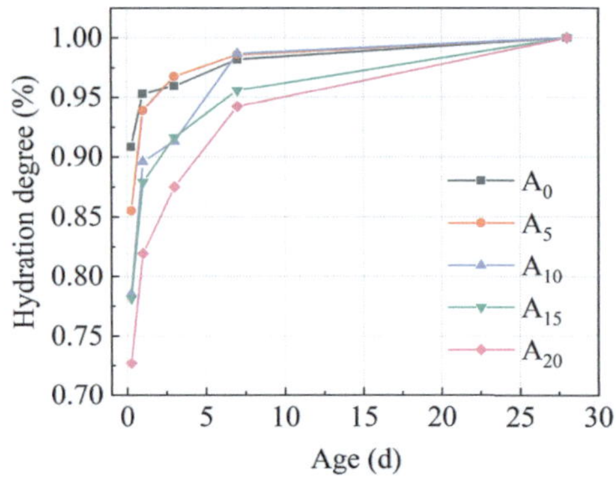

Fig. 5.14 Hydration degree of LCCM at different ages

of A_0 at 3 days, followed by A_{10} surpassing A_0 at 7 days. However, the hydration degrees of A_{15} and A_{20} remain lower than those of the first three groups throughout the process. This suggests that an appropriate amount of desulfurization gypsum can slow the early hydration rate of the cementitious material, but the desulfurization gypsum content should not exceed 10%, as higher contents significantly increase the inhibitory effect on hydration.

5.4.3 Molecular Characteristics Analysis

Figure 5.15 shows the FTIR spectra of hydration products of LCCM from 6 h to 28 days. Table 5.4 lists the characteristic infrared absorption bands of minerals and hydration products in LCCM (Sun et al., 2022; Yin et al., 2018; Trezza & Lavat, 2001). As shown in Fig. 5.15a, A0 exhibits three distinct AFt absorption peaks at 1110, 3422, and 3635 cm^{-1}, primarily caused by the asymmetric stretching of $[SO_4^{2-}]$, H_2O, and $[OH^-]$. The increase in AFt absorption peaks from 6 h to 28 days is minimal. Additionally, the absorption peaks of $C_4A_3\hat{S}$ at 616 and 882 cm^{-1} and $CaSO_4$ at 667 cm^{-1} are less pronounced, indicating rapid hydration of the LCCM in the absence of desulfurization gypsum, with hydration largely completed within the first 6 h. The C_2S absorption peak at 28 days shows little change compared to 6 h, suggesting minimal hydration of C_2S during the first 28 days.

Comparing Fig. 5.15a–e, the AFt absorption peaks at 6 h gradually decrease with increasing desulfurization gypsum content, particularly the peak at 3635 cm^{-1}, which becomes smoother. This indicates that desulfurization gypsum slows the early hydration of LCCM, reducing AFt formation. After 1 day, the AFt absorption peaks initially increase and then decrease with increasing desulfurization gypsum content. At 1 day, A_5 exhibits a larger AFt absorption peak than the other groups, indicating rapid hydration and significant AFt formation between 6 h and 1 day. In contrast, LCCM with more than 5% desulfurization gypsum content still hydrate slowly. From 3 to 28 days, A_{10} shows the largest AFt absorption peak, and at 28 days, the order

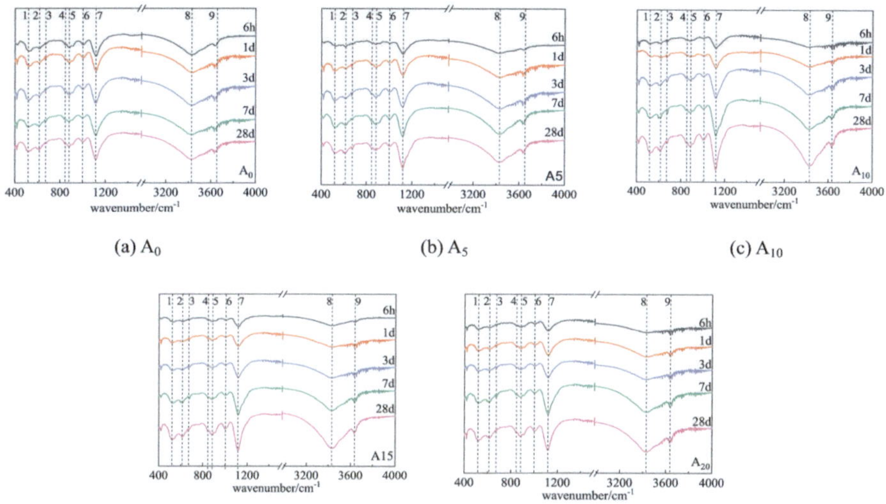

(a) A_0 (b) A_5 (c) A_{10}

(d) A_{15} (e) A_{20}

Fig. 5.15 FTIR spectra of LCCM at different hydration ages

Table 5.4 Infrared absorption characteristics of LCCM minerals and their hydration products

Phase	Functional group	Wave number (cm^{-1})	No
AFt	$[SO_4^{2-}]$	1110	7
	H_2O	3422	8
	$[OH^-]$	3635	9
AH_3	Al-O	520	1
$C_4A_3\overline{S}$	$[AlO_4]$	616, 882	2, 5
C_2S	$[SiO_4^{4-}]$	847, 1000	4, 6
$CaSO_4$	$[SO_4^{2-}]$	667	3

of AFt peak intensity is $A_{10} > A_{15} > A_{20} > A_5 > A_0$. This suggests that LCCM with 10% desulfurization gypsum content begin rapid hydration and generate substantial AFt between 1 and 3 days, with the hydration process continuing and AFt formation persisting. When the desulfurization gypsum content exceeds 10%, the hydration rate of LCCM is extremely slow in the initial 3 days but accelerates afterward, generating a large amount of AFt, ultimately exceeding the AFt formation in materials with lower or no desulfurization gypsum content.

In summary, the FTIR spectra of LCCM at different ages exhibit the same trends as the XRD and TG-DSC results. Together, these methods confirm that desulfurization gypsum slows the early hydration rate, extends the hydration cycle, and increases AFt formation in LCCM.

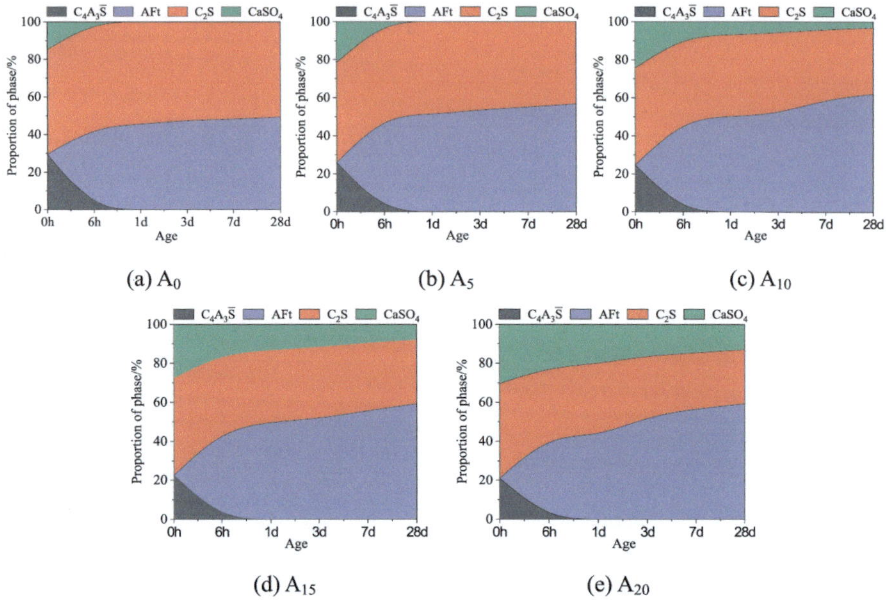

Fig. 5.16 Main phase content change diagram of LCCM cement paste

5.4.4 *Quantitative Analysis of Phase Change*

Based on the XRD patterns, the Rietveld method was used to quantitatively analyze the main phases of LCCM samples at different ages, resulting in the phase content variation graph shown in Fig. 5.16. It can be observed that, due to the absence of desulfurization gypsum, the calcium sulfate and $C_4A_3\hat{S}$ in A_0 and A_5 samples were almost completely consumed within the first day. Combined with the XRD patterns, a small amount of residual calcium sulfate remained, but its concentration was insufficient to sustain the reaction process described by Eq. 5.3, resulting in negligible growth in AFt content after 1 day. When the desulfurization gypsum content reached 10% or higher, the AFt content in the samples continued to increase significantly after 3 days, accompanied by a decrease in calcium sulfate. Since anhydrous calcium sulfate was almost entirely consumed by this stage, it is speculated that the increase in AFt content during this period was primarily due to the reaction between calcium sulfate and aluminum AH_3. On the other hand, residual calcium sulfate began to appear after 28 days of hydration, with A_{10} containing 3.2% and A_{20} containing 12.9% calcium sulfate at 28 days. At 28 days, the LCCM with 10% desulfurization gypsum content exhibited the highest AFt content at 61.9%, which decreased with further increases in desulfurization gypsum content. This indicates that the 10% desulfurization gypsum sample achieved the most complete hydration, while excessive desulfurization gypsum reduced the proportion of C_2S, adversely affecting long-term strength development.

5.4.5 *Microscopic Morphology Development Process*

Figure 5.17 shows the SEM images of A_0 hydration samples at 6 h, 3 days, and 28 days. At 6 h, the A_0 hydration sample contains a small amount of incompletely hydrated LCCM particles, with some needle-like AFt and flocculent AH_3 distributed on the particles. Additionally, a large number of rod-like AFt are observed. These AFt structures are robust and well-developed, interconnecting to form a skeletal structure and nesting with AH_3 to create a dense network. By 3 days, the cement undergoes further hydration, with a slight increase in the size and quantity of AFt, resulting in a denser structure. A small amount of layered AFm also appears. At 28 days, due to insufficient $CaSO_4$ content, $C_4A_3\hat{S}$ continues to hydrate, forming AFm. Additionally, a small amount of grass-like and fibrous C-S-H is observed, indicating the onset of C_2S hydration. The quantity and size of AFt show little change compared to 6 h, suggesting that A_0 had already completed most of its hydration process within the initial 6 h.

Figure 5.18 shows the SEM images of A_{10} hydration samples at 6 h, 3 days, and 28 days. At 6 h, the A_{10} contains more unhydrated LCCM particles and desulfurization gypsum particles compared to A_0. The surfaces of the LCCM particles are irregular, mostly coated with hydration products, and a small portion has hydrated to form short rod-like AFt, which fails to form a complete skeletal structure with AH_3. By 3 days, the LCCM particles decrease rapidly, generating a large amount of AFt and flaky AH_3. The AFt crystals further grow in size and aggregate to form columnar structures, while AH_3 links and fills the gaps between AFt columns, resulting in a denser structure. At 28 days, clusters of AFt form within microcracks and pores, distributed in a disordered manner. The interconnected spatial structure becomes more compact, effectively reducing crack width and length. Combined with a small amount of C-S-H generated from C_2S hydration, the overall matrix density is further enhanced.

Figure 5.19 shows the SEM images of A_{20} hydration samples. At 6 h, the A_{20} sample contains a large number of unhydrated LCCM particles, along with a small amount of needle-like AFt and flocculent AH_3, without forming a dense skeletal structure. The hydration degree of the sample is low, indicating a significant inhibitory effect of desulfurization gypsum on the early hydration rate of LCCM. By 3 days, the number and size of unhydrated LCCM particles decrease, but compared to A_{10} samples of the same age, the quantity remains higher. Meanwhile, AFt crystals gradually grow longer and thicker, but their quantity is insufficient to form a dense structure. This demonstrates that increasing the desulfurization gypsum content extends the hydration cycle, with a more pronounced delay in hydration as the desulfurization gypsum content increases. At 28 days, the LCCM particles in the sample significantly decrease, and a large number of columnar and prismatic AFt begin to appear, with both quantity and size surpassing those of the A_0 sample without desulfurization gypsum. Some desulfurization gypsum particles attached to AFt react with AH_3, causing the AFt crystals to continuously grow and aggregate, interweaving with AH_3. Eventually, the two interconnect, resulting in a denser structure. Additionally,

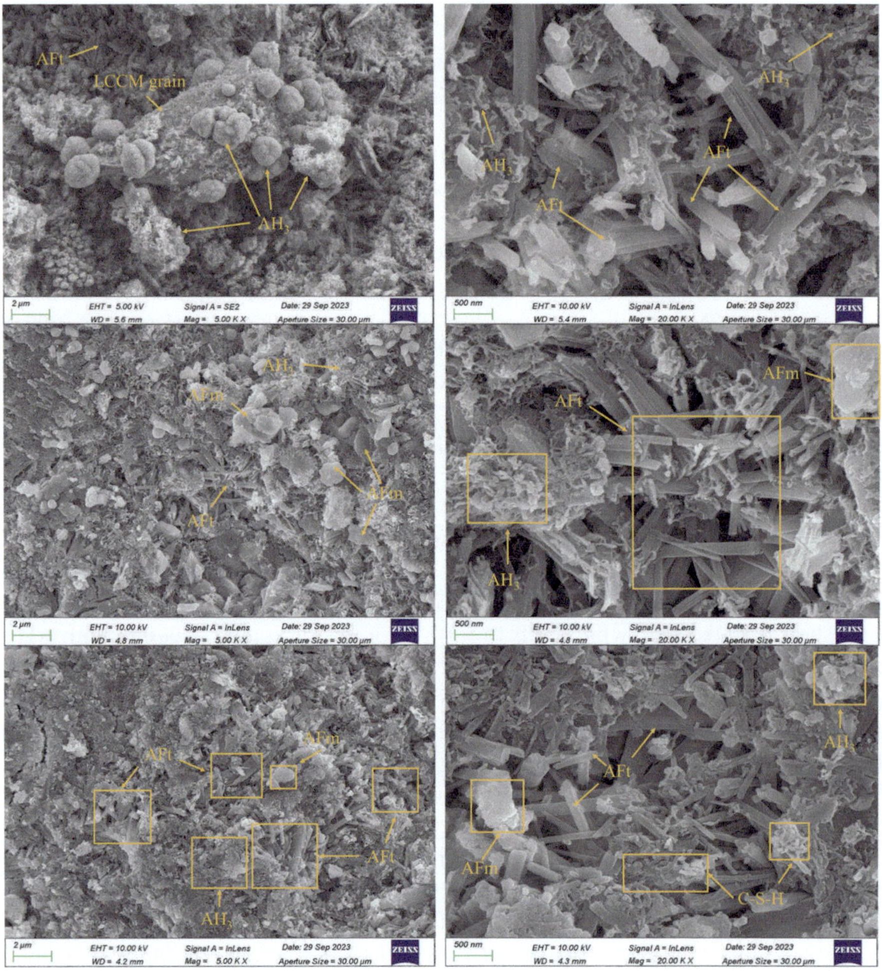

Fig. 5.17 SEM diagram of A_0 hydration sample

a significant amount of grass-like C-S-H gel is observed in the 28-day hydration sample. As described by Eqs. 5.3 and 5.4, the reaction between sufficient calcium sulfate and aluminum AH_3 to form AFt consumes $Ca(OH)_2$, reducing its concentration in the hardened paste and promoting the hydration of C_2S. This leads to the formation of a large amount of C-S-H gel in the hardened paste by 28 days.

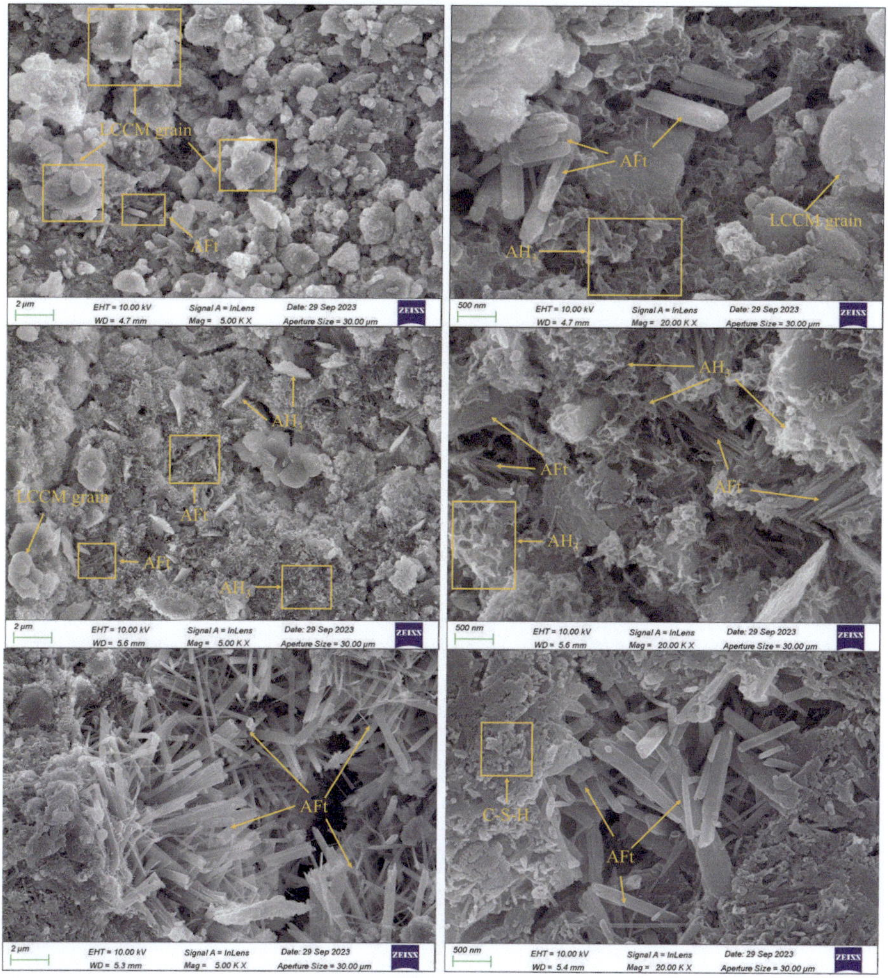

Fig. 5.18 SEM diagram of A_{10} hydration sample

5.5 Desulfurization Gypsum Regulates the Pore Structure of LCCM

5.5.1 Porosity

Figure 5.20 illustrates the variation in porosity of hardened pastes of LCCM. It can be observed that, during the early hydration stage (within 6 h), the porosity of the LCCM samples increases with higher desulfurization gypsum content. Based on previous analysis, this is attributed to the fact that high desulfurization gypsum content slows the early hydration rate, resulting in fewer hydration products and thus

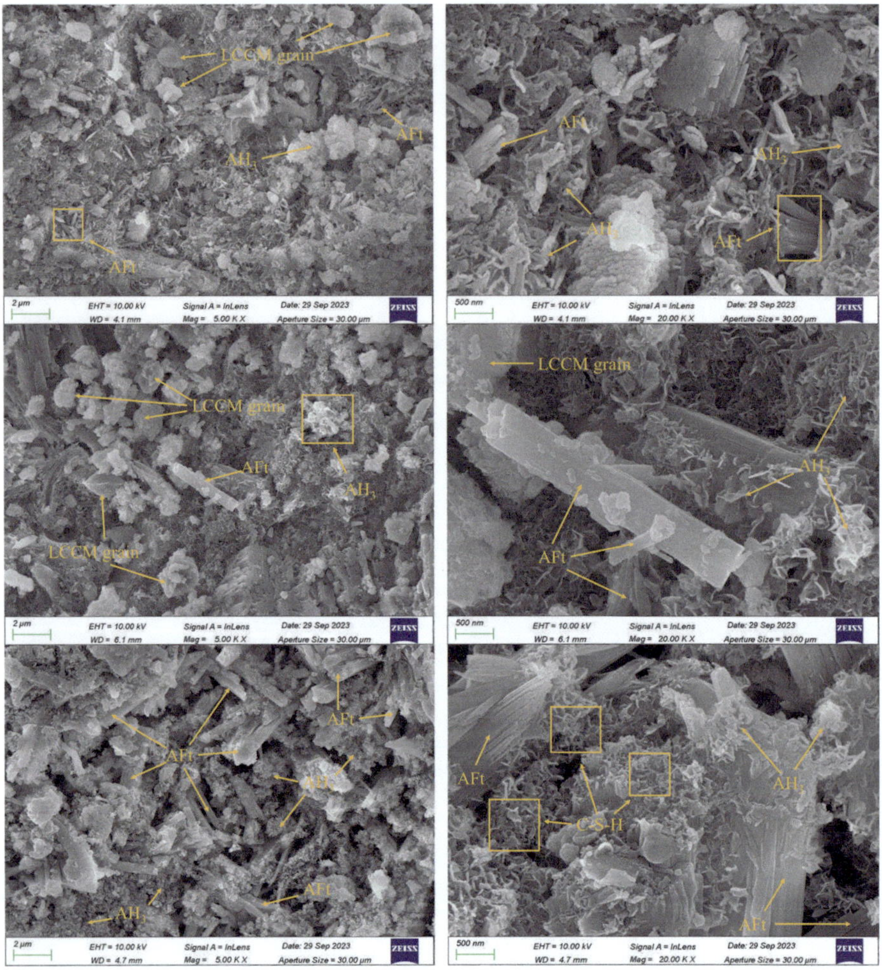

Fig. 5.19 SEM diagram of A_{20} hydration sample

higher porosity. From 3 to 28 days, the porosity of A_0 and A_{10} samples changes minimally, while the porosity of A_{15} and A_{20} samples decreases rapidly as hydration progresses. This indicates that LCCM with little or no desulfurization gypsum complete most of their hydration within the initial 3 days, whereas high desulfurization gypsum content delays the early hydration rate and extends the hydration cycle. At 28 days, the porosity of the samples is inversely proportional to the desulfurization gypsum content, with the porosities of the five groups being 3.09, 2.66, 2.43, 1.61, and 1.52%, respectively. The higher the desulfurization gypsum content, the more pronounced the decrease in porosity. This is because, under sufficient calcium sulfate conditions, the size and quantity of AFt in the hydration products increase, leading to a denser structure and reduced porosity.

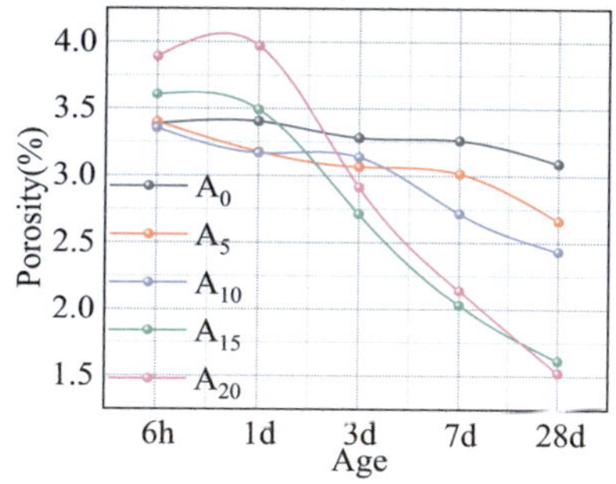

Fig. 5.20 Porosity of hardened paste of LCCM

5.5.2 *Pore Size Distribution*

Figure 5.21 shows the pore size distribution of the samples at different ages. As observed in Fig. 5.21a, the pore size distribution of the hardened paste of LCCM is primarily concentrated in the range of 0.01–0.1 μm. In the absence of desulfurization gypsum, the pore size distribution curves of the A_0 sample at various hydration ages are relatively concentrated. As the desulfurization gypsum content increases, the peak of the pore size distribution curve shows a more pronounced downward trend with increasing age. This is consistent with the porosity variation pattern, indicating that porosity is mainly influenced by changes in pores within this range. A small number of pores exist in the range of 1–10 μm, and the pore size distribution curves in this range exhibit a shift toward smaller pore sizes with increasing desulfurization gypsum content. This suggests that desulfurization gypsum can reduce the size of larger pores and improve the pore size distribution. This phenomenon may be attributed to the fact that the addition of desulfurization gypsum increases the quantity and size of AFt, promotes the hydration of C_2S to form C-S-H gel in the later stages, and the remaining unreacted desulfurization gypsum also contributes to pore filling, resulting in a reduction in the size of these pores.

Based on pore size, the pores in the hardened paste of LCCM are classified into unharmful pores ($r < 20$ nm), less harmful pores (20 nm $\leq r < 50$ nm), harmful pores (50 nm $\leq r < 200$ nm), and more harmful pores ($r \geq 200$ nm). The proportions of these different types of pores are shown in Fig. 5.21b. For A_0, A_5, and A_{10}, the proportions of different pore types show little change with age. Overall, the proportions of highly harmful and harmful pores decrease with increasing desulfurization gypsum content, indicating that desulfurization gypsum promotes the development of smaller pores. For A_{15} and A_{20}, the proportions of harmful and highly harmful pores decrease significantly with age. At 28 days, A_{15} has the lowest proportion of harmful and highly harmful pores among the five groups, accounting for only 8%. However, when the desulfurization gypsum content is further increased, the proportions of harmful and highly harmful pores in A_{15} and A_{20} rise. This may be due to the excessive AFt content caused by high desulfurization gypsum content, which, due to its high expansiveness, leads to expansion cracks in the paste, thereby increasing the proportion of highly harmful pores.

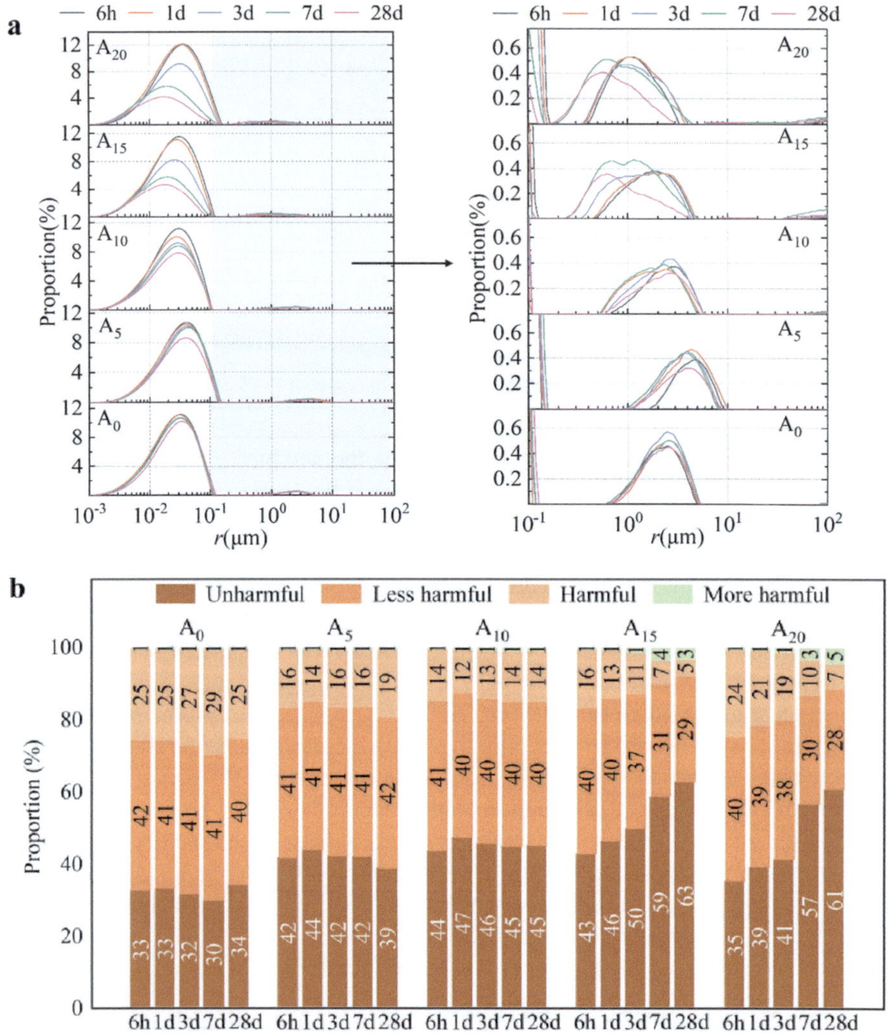

Fig. 5.21 Pore size distribution of hardened paste of LCCM

5.6 AC Impedance Characteristics of LCCM Regulated by Desulfurization Gypsum

5.6.1 Nyquist Curves

For the analysis of electrochemical impedance spectroscopy (EIS) results, the most common approach is to establish an equivalent circuit model by combining different circuit elements to simulate the hydration reaction process within the cementitious material (Gu et al., 1994; McCarter, 1994). Figure 5.22 illustrates the conventional equivalent circuit model for the hardened paste of LCCM. In this model, R_0 represents the pore solution resistance of the hardened paste, R_1 represents the resistance generated by the charge transfer process within the cementitious material, and R_2 represents the resistance arising from the charge transfer process at the interface between the hardened paste and the electrode. CPE_1 representing the electric double layer properties at the solid–liquid interface within the hardened paste, while CPE_2 representing the electric double layer properties at the interface between the hardened paste and the electrode.

The fitting results of the Nyquist curves for the hydration process of LCCM hardened paste, based on the conventional equivalent circuit model, are shown in Fig. 5.23. The Nyquist curve consists of two parts: a semicircular curve in the high-frequency region and a straight line in the low-frequency region (Macphee et al., 2000). In the Nyquist plot, the vertical axis represents the imaginary part of the impedance, while the horizontal axis represents the real part of the impedance. The frequency decreases as the horizontal coordinate value increases. At the intersection of the high-frequency semicircle and the low-frequency straight line, the horizontal coordinate corresponds to the bulk resistance (R_1) of the hardened paste of LCCM.

It can be observed that the diameter of the Nyquist curves for all samples gradually increases with age. At 6 h of hydration, the bulk resistance of the hardened paste decreases with increasing desulfurization gypsum content, with values of 11,400 Ohms, 11,081 Ohms, 10,129 Ohms, 8,961 Ohms, and 7,891 Ohms for the five groups

Fig. 5.22 Conventional equivalent circuit model

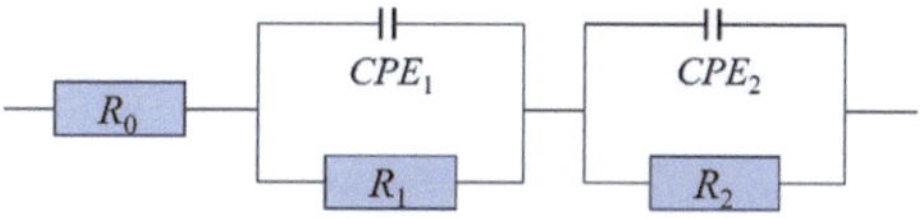

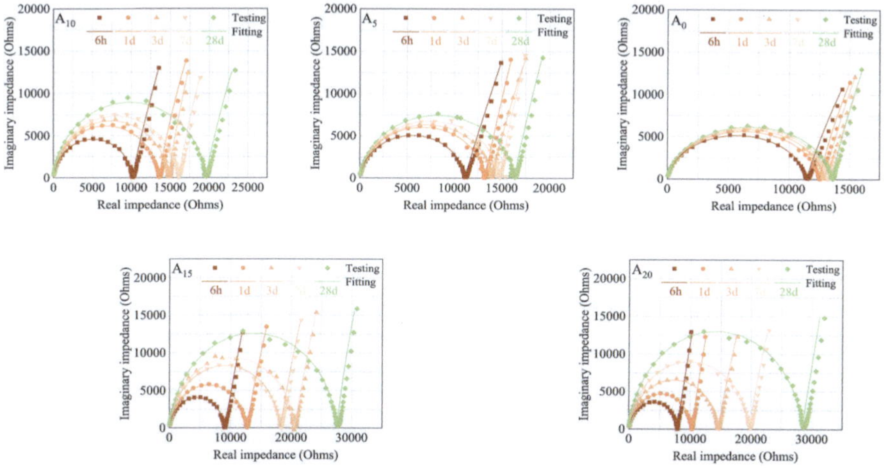

Fig. 5.23 Nyquist curve of LCCM hardened paste

of samples, respectively. As hydration progresses, the higher the desulfurization gypsum content, the greater the change in bulk resistance over the hydration cycle. At 28 days of hydration, the bulk resistance of the A_0 sample (without desulfurization gypsum) is 13,482 Ohms, representing only an 18.3% increase compared to 6 h. In contrast, the A_{20} sample reaches 28,317 Ohms, which is 3.6 times its value at 6 h.

The incorporation of desulfurization gypsum reduces the bulk resistance of the paste in the early stages and significantly increases it in the later hydration stages. This is because, after adding desulfurization gypsum, the hardened paste of LCCM releases more Ca^{2+} and SO_4^{2-} ions within the initial 6 h of hydration, enhancing the paste's conductivity. Additionally, as previously discussed, the addition of desulfurization gypsum slows the early hydration rate of LCCM, resulting in more pores and a less dense structure, which also facilitates current flow. In the later hydration stages, due to the sufficient desulfurization gypsum content, the hydration of the cementitious material continues, generating hydration products continuously. This reduces the content of free ions in the pores of the cementitious material, leading to a decrease in ion concentration. As the paste structure becomes denser, the conductivity of the internal system of the cementitious material deteriorates, making the transport of free ions increasingly difficult.

5.6.2 An Equivalent Circuit Model Based on Microstructure Change

Conductive path

The hardened paste of LCCM is a solid–liquid-gas three-phase porous system composed of a solid matrix (including hydration products and unhydrated cementitious material particles), water, and gas present in the pores. Its electrical conductivity is closely related to changes in its microstructure. While the conventional equivalent circuit model effectively reflects the overall resistance changes in the hardened paste of LCCM, it cannot accurately express the microstructural changes during the hardening process. Therefore, this chapter introduces a microstructure-based equivalent circuit model to analyze the electrochemical impedance changes in the hardened paste of LCCM.

According to the microstructural model proposed by Song (2000), the conductive paths in the hardened paste of LCCM can be divided into three types: (1) continuous conductive paths (CCP), formed by continuously connected conductive pores; (2) discontinuous conductive paths (DCP), composed of discontinuous points (DP) formed by the solid phase of the hardened paste alternating with conductive pores; and (3) insulating paths (ICP), which include all paths other than DCP and CCP. The three conductive paths in the hardened paste of LCCM are illustrated in Fig. 5.24.

The conductive paths in the hardened paste of LCCM can be simplified into the model shown in Fig. 5.25. Figure 5.25a illustrates the simplified conductive paths at the initial stage of hydration. At this stage, due to the limited hydration progress, the

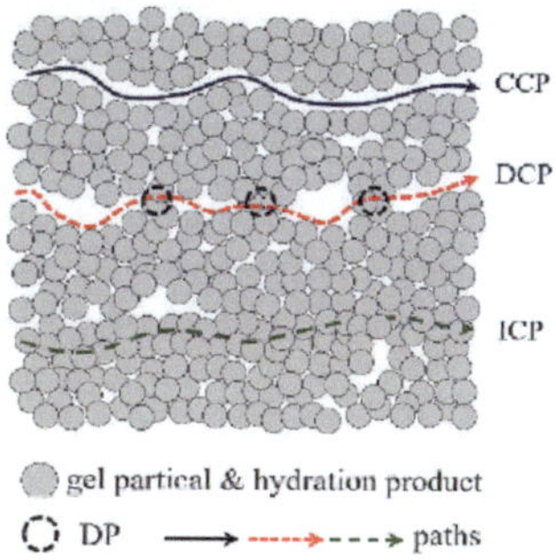

Fig. 5.24 Conductive path of hardened paste of LCCM

system contains fewer hydration products and more water, allowing ions to move freely. The conductive paths are predominantly CCP. As hydration progresses, the conductive paths evolve as shown in Fig. 5.25b. The LCCM gradually hydrates, generating a large amount of hydration products. The CCP paths begin to exhibit discontinuities, transitioning into DCP paths, while some pores in the DCP paths are filled with hydration products, forming ICP paths. Figure 5.25c depicts the simplified conductive paths in the later stages of hydration. At this stage, due to the extensive filling by hydration products, the structure becomes denser, and the conductive paths in the hardened paste system are primarily ICP, with a small number of DCP paths remaining.

Equivalent circuit model

The electrical conductivity of the hardened paste of LCCM depends on the impedance of the three conductive paths. Accordingly, the total impedance Z of the hardened paste is given by (Bai et al., 2023; Hu et al., 2019):

$$\frac{1}{Z} = \frac{1}{Z_{CCP}} + \frac{1}{Z_{DCP}} + \frac{1}{Z_{ICP}} \tag{5.5}$$

where: Z_{CCP} is the impedance of CCP. Z_{DCP} is the impedance of DCP. Z_{ICP} is the impedance of ICP.

Since the current conduction through CCP is caused by the migration of ions in the pore solution, the impedance of CCP can be expressed as:

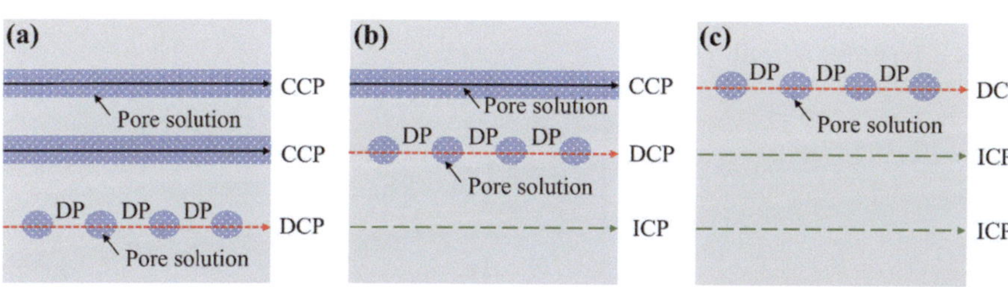

Fig. 5.25 Simplified model of conductive path

$$Z_{CCP} = R_{CCP} \tag{5.6}$$

where: R_{CCP} is the resistance in CCP.

The impedance of the DCP path consists of two parts: the continuous part impedance (Z_{CP}) and the discontinuous point impedance (Z_{DP}) of the DCP. The impedance expression of the DCP path is shown in Eq. 5.7:

$$Z_{DCP} = Z_{CP} + Z_{DP} \tag{5.7}$$

The continuous partial impedance (Z_{CP}) and discontinuous point impedance (Z_{DP}) of the DCP can be expressed as follows:

$$Z_{DP} = \frac{1}{j\omega C_{DP}} \tag{5.8}$$

$$Z_{CP} = R_{CP} \tag{5.9}$$

where: j is an imaginary unit, $j^2 = -1$. ω is the alternating current frequency. C_{DP} is the double parallel plate capacitance of the discontinuous point of DCP. R_{CP} is the resistance of the continuous part of DCP.

The impedance of Z_{ICP} can be expressed by the Eq. 5.10:

$$Z_{ICP} = \frac{1}{j\omega C_{mat}} \tag{5.10}$$

where: C_{mat} represents the double parallel plate capacitance of ICP.

The microstructure-based equivalent circuit model can be represented as Fig. 5.26a. Since ICP can be regarded as an insulator, C_{mat} approaches 0 compared with CDP. In this case, the circuit can be further simplified as shown in Fig. 5.26b. Therefore, the total impedance Z of hardened paste of LCCM can be expressed as:

$$Z = \frac{1}{\left(\frac{1}{R_{CCP}} + \frac{1}{R_{CP} + \frac{1}{j\omega C_{DP}}}\right)} \tag{5.11}$$

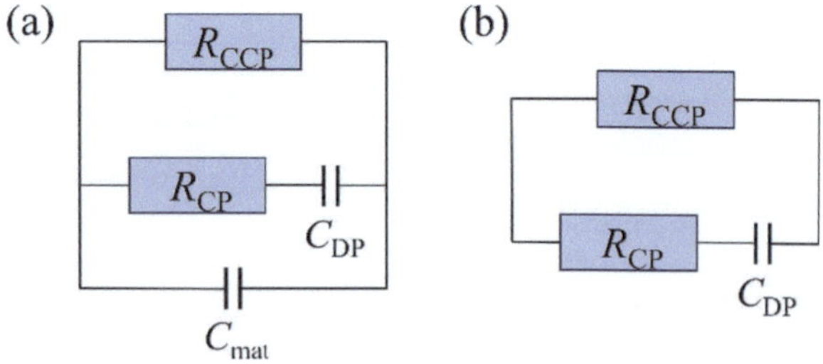

Fig. 5.26 Equivalent circuit model based on microstructure

By comparing the conventional circuit model with the microstructure-based equivalent circuit model, the relationship between the parameters of different equivalent circuit elements can be established:

$$R_{CCP} = R_0 + R_1 \tag{5.12}$$

$$R_{CP} = \frac{(R_0 + R_1)R_0}{R_1} \tag{5.13}$$

$$C_{DP} = CPE_1\left(\frac{R_1}{R_0 + R_1}\right)^2 \tag{5.14}$$

CPE_1 can be calculated by the Eq. 5.15:

$$CPE_1 = \frac{1}{\omega^* R_1} \tag{5.15}$$

where: ω^* is the frequency value corresponding to the highest point of the high-frequency semi-circular arc in the Nyquist diagram.

Impedance parameter

The impedance parameters R_{CCP}, R_{CP}, and C_{DP} at different hydration ages can be calculated using Eqs. 5.12–5.14, as shown in Fig. 5.27a–c, respectively. From the perspective of hydration age, the R_{CCP} values of all five groups of samples exhibit an increasing trend as hydration progresses. At 6 h of hydration, the R_{CCP} values decrease with increasing desulfurization gypsum content, with values of 11,246 Ω, 11,108 Ω, 10,157 Ω, 8,987 Ω, and 7,917 Ω for the five groups, respectively. When the samples continue to hydrate up to 28 days, the R_{CCP} values increase with higher desulfurization gypsum content, reaching 13,514 Ω, 16,290 Ω, 19,523 Ω, 27,505 Ω, and 28,355 Ω for the five groups, respectively. From the perspective of desulfurization gypsum content, the R_{CCP} values of A_0, A_5, and A_{10} increase primarily within the initial 3 days of hydration, while A_{15} and A_{20} show significant growth after 3 days, with their R_{CCP} values increasing by 51.3 and 95.3%, respectively, from 3 to 28 days. The addition of desulfurization gypsum facilitates the dissolution of more Ca^{2+} and SO_4^{2-} ions into the pore solution, enhancing conductivity. Additionally, as previously discussed, desulfurization gypsum slows the formation of hydration products in the early stages, promoting the development of wider connected pore paths, thereby reducing the R_{CCP} values in the early hydration stages. In the later hydration stages, the sufficient desulfurization gypsum content allows the hydration reaction to continue, consuming ions in the pore solution to form hydration products. This narrows the connected pore paths, further reducing the ability of current to pass through, leading to a continuous increase in R_{CCP} values.

The R_{CP} value reflects the conductivity of disconnected pores in the DCP, while the C_{DP} value represents the equivalent cumulative thickness of all DP aligned parallel to

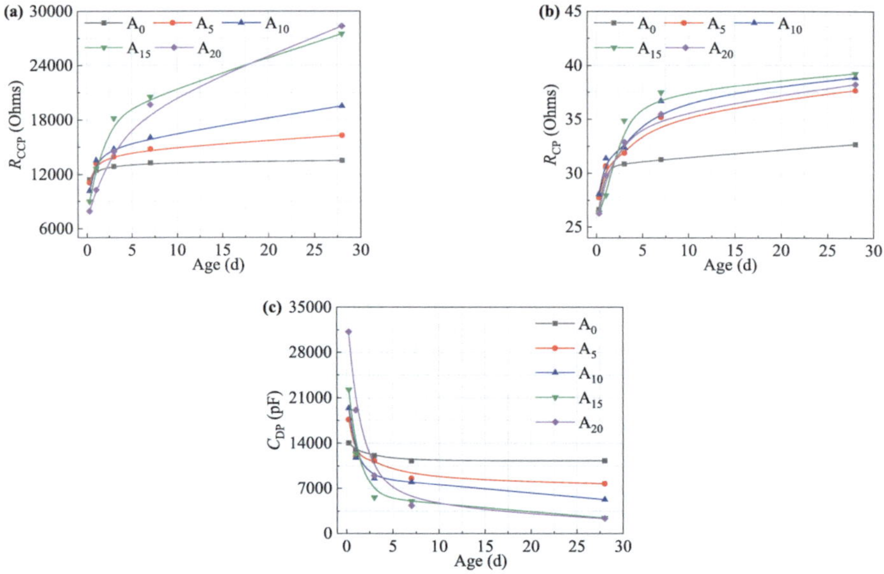

Fig. 5.27 Calculated values of impedance parameters. **a** R_{CCP}, **b** R_{CP}, **c** C_{DP}

the electric field in the DCP. The relationship between the two is that the increase of one side leads to the decrease of the other. The increase in R_{CP} and the decrease in C_{DP} are primarily concentrated within the first 3 days of hydration. This is mainly due to the rapid progress of the hydration reaction, which gradually reduces the pores in the DCP and continuously increases the cumulative thickness of DP, thereby weakening the ability of current to pass through. Compared to the changes in R_{CCP}, the influence of hydration age and desulfurization gypsum content on R_{CP} and C_{DP} values is relatively small. This is because the formation of hydration products not only narrows the disconnected pore paths but also increases the number of discontinuous points, causing more connected pores in the CCP to transform into disconnected pores. This, to some extent, offsets the increase in R_{CP} and the decrease in C_{DP}.

5.6.3 Relationship Between Impedance Parameters and Compressive Strength

Grey entropy correlation grade

The grey relational entropy analysis method is a technique that quantifies the influence of various factors within a grey system to determine their significance on the system (Zhao et al., 2020). It effectively distinguishes the impact of primary and secondary factors on the overall system. By selecting the bulk resistance (R_1) and equivalent circuit model parameters (R_{CCP}, R_{CP}, and C_{DP}) as influencing factors, the grey entropy method is used to analyze the impact of changes in different electrochemical impedance parameters on the compressive strength of hardened paste of LCCM.

Taking the compressive strength of hardened paste of LCCM at different hydration ages as the reference sequence, it can be defined as:

$$Y = \left[y(1), y(2), \cdots, y(k)\right] \quad (5.16)$$

where: Y is the compressive strength of hardened pastc of LCCM, $y(\mathrm{k})$ is the compressive strength of hardened paste of LCCM at the kth hydration age.

The volume resistance $R_1(X_1)$, $R_{CCP}(X_2)$, $R_{CP}(X_3)$, and $C_{DP}(X_4)$ of hardened paste of LCCM are used as comparison sequences, which can be set as:

$$X_i = [x_i(1),\ x_i(2),\ \cdots,\ x_i(k)]\, i = 1,\ 2,\ 3,\ 4 \quad (5.17)$$

where: X_i is the different influencing factors. $x_i\ (k)$ is the parameter value of the kth hydration age.

The original sequence composed of reference sequence and comparison sequence is calculated according to the following steps, and the grey entropy correlation degree between different impedance parameters and compressive strength of hardened paste of LCCM can be obtained. The results are shown in Fig. 5.28.

(1) Meaning processing:

$$\begin{cases} x_i' = \dfrac{X_i}{x_i(a)} = (x_i'(1),\ x_i'(2), \ldots,\ x_i'(k))\ \ i = 1,\ 2, \ldots,\ n \\ y_j' = \dfrac{Y_j}{y_j(a)} = (y_j'(1),\ y_j'(2), \ldots,\ y_j'(k))\ \ j = 1,\ 2, \ldots,\ m \end{cases} \quad (5.18)$$

where: $x_i'(a)$ is the mean value of $x_i(1), x_i(2), \cdots, x_i(k)$. $y_i'(a)$ is the mean value of $y_i(1), y_i(2), \cdots, y_i(k)$.

(2) Difference between reference sequence and comparison sequence:

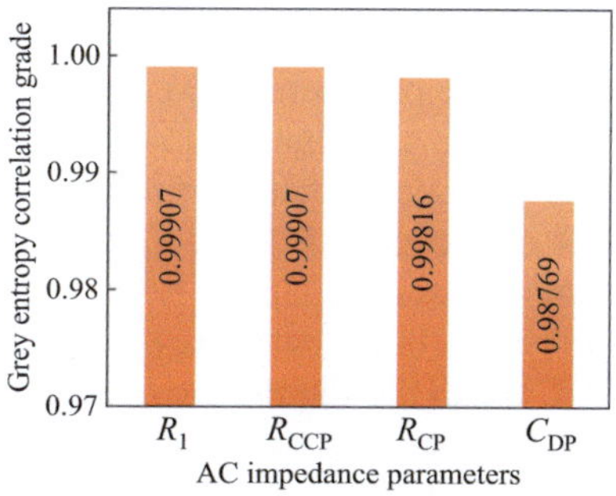

Fig. 5.28 Grey entropy correlation degree between different AC impedance parameters and compressive strength

$$\begin{cases} \Delta_{ij}(k) = |y_j'(k) - x_i'(k)| \\ \Delta_{ij} = (\Delta_{ij}(1), \Delta_{ij}(2), \ldots, \Delta_{ij}(k) \\ (i = 1, 2, \ldots, n; j = 1, 2, \ldots, m; k = 1, 2, \ldots, p) \end{cases} \tag{5.19}$$

(3) Find the maximum difference and minimum difference of the sequence:

$$\begin{cases} M = \max_i \max_k \Delta_i(k) \\ m = \min_i \min_k \Delta_i(k) \end{cases} \tag{5.20}$$

(4) Correlation coefficient:

$$\gamma_{ij} = \frac{m + \xi M}{\Delta_i(k) + \xi M} \tag{5.21}$$

where: ξ is the resolution coefficient, generally 0.5.

(5) Grey entropy correlation density:

$$P_h = \frac{\gamma[y_i(h), x_i(h)]}{\sum_{k=1}^{n} \gamma[y_i(h), x_i(h)]} \quad (h = 1, 2, \ldots, n) \tag{5.22}$$

(6) Grey correlation entropy and grey entropy correlation degree:

Grey correlation entropy:

$$H = -\sum_{k=1}^{n} P_n \ln P_n \tag{5.23}$$

Grey entropy correlation degree:

$$E = \frac{1}{n} H_{\max} \tag{5.24}$$

Figure 5.28 shows the grey entropy correlation between different electrochemical impedance parameters and compressive strength. A higher grey entropy correlation between an impedance parameter and compressive strength indicates a greater influence of that parameter on compressive strength. It can be observed that among the four impedance parameters, R_1 and R_{CCP} have the highest grey entropy correlation, both at 0.99907. This is because R_{CCP} is the sum of the bulk resistance (R_1) and the pore solution resistance (R_0). Due to the low porosity of the hardened paste of LCCM and the significantly higher ionic conductivity of the pore solution compared to the hardened paste matrix, the value of R_1 is much larger than that of R_0, resulting in identical grey entropy correlations for R_1 and R_{CCP}. The grey entropy correlations of R_{CP} and C_{DP} with compressive strength are relatively small, at 0.99816 and 0.98769, respectively. These parameters have a lesser influence on compressive strength, which is consistent with the conclusions drawn from the equivalent circuit model.

Compressive strength prediction formula

Since the bulk resistance R_1 has a high correlation with the compressive strength of the hardened paste of LCCM and is relatively easy to obtain experimentally, R_1 is selected as the independent variable to characterize compressive strength.

Figure 5.29 illustrates the relationship between compressive strength and bulk resistance for hardened paste of LCCM with different desulfurization gypsum contents at various hydration ages. It can be observed that the relationship between compressive strength and bulk resistance varies for samples with different desulfurization gypsum contents. The data in the figure can be divided into two regions, corresponding to samples A_0, A_5, and A_{10}, and samples A_{15} and A_{20}, respectively. Within both regions, compressive strength and bulk resistance exhibit a strong linear correlation. By fitting the data, the relationship between compressive strength $f_c(t)$ and bulk resistance R_1 (t) can be expressed as shown in Eq. 5.25.

$$\begin{cases} f_c(t) = 0.00652R_1(t) - 40.384 \ \ w_{DG} \leq 10\% \\ f_c(t) = 0.00286R_1(t) + 0.902 \ \ w_{DG} > 10\% \end{cases} \tag{5.25}$$

As can be seen from the above formula, the desulfurization gypsum content has a significant impact on the electrochemical impedance characteristics and compressive strength of the hardened paste of LCCM. Excessive desulfurization gypsum content ($w_{DG} > 10\%$) reduces the slope of the compressive strength expression, indicating that, at the same bulk resistance, the compressive strength of the hardened paste with high desulfurization gypsum content decreases. This is because excessive desulfurization gypsum content leads to unreacted desulfurization gypsum particles filling the pores of the paste, reducing porosity and thereby increasing bulk resistance. However, too much desulfurization gypsum decreases the proportion of reactive LCCM, resulting in fewer hydration products and hindering the growth of compressive strength.

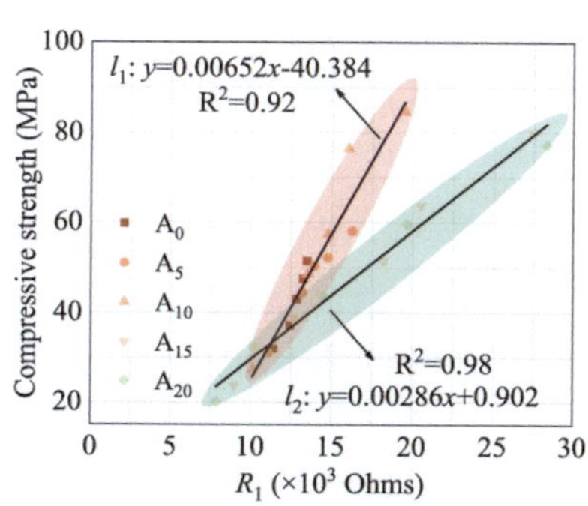

Fig. 5.29 The relationship between R_1 and the compressive strength of LCCM hardened paste compressive strength

The formula indicates that the bulk resistance R_1 can be effectively used to predict changes in compressive strength, providing a feasible method for characterizing compressive strength. Using this method to predict compressive strength allows for real-time monitoring without damaging the specimens. Additionally, due to the sensitivity of electrochemical testing, it can reflect subtle changes in the microstructure of the paste, making this prediction method both precise and capable of continuous monitoring.

5.7 Hydration Kinetics of LCCM Regulated by Desulfurization Gypsum

5.7.1 Hydration Heat

The hydration heat release profile provides a direct reflection of the hydration rate and reaction extent of cement minerals during hydration. Figure 5.30 shows the hydration heat release rate of LCCM over 72 h.

As illustrated, the heat release of LCCM is primarily concentrated within the initial 6 h, exhibiting two exothermic peaks. The first exothermic peak occurs between 0 and 1 h and is mainly attributed to the heat release from the dissolution of various minerals in the LCCM and the partial hydration of $C_4A_3\hat{S}$. Upon contact with water, these LCCM particles react rapidly, releasing a significant amount of heat, which is reflected in the figure as a linear increase in the heat release rate curve during the initial hydration stage. At this stage, A_0 exhibits the highest heat release rate, reaching 57 mW·g^{-1}, due to its highest $C_4A_3\hat{S}$ content and fastest reaction rate. After the addition of gypsum, the hydration heat release rate gradually slows. However, when the desulfurization gypsum content reaches 20%, the presence of a large amount of hemihydrate gypsum leads to the release of substantial heat during its absorption of water to form $CaSO_4 \cdot 2H_2O$, resulting in a higher heat release rate for A_{20} compared to LCCM with lower desulfurization gypsum content.

As hydration progresses, the second exothermic peak appears between 1 and 3 h. This is primarily due to the reaction of $C_4A_3\hat{S}$ with desulfurization gypsum and

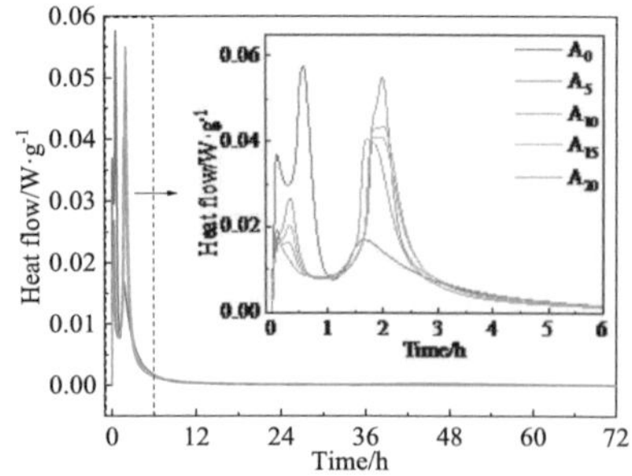

Fig. 5.30 Hydration heat release rate curve of LCCM

water to form AFt and AH_3, releasing a significant amount of heat. At this stage, A_0 exhibits the lowest heat release rate, indicating that most of its $C_4A_3\hat{S}$ has already been consumed during the rapid hydration in the first stage. As the desulfurization gypsum content increases, the heat release rate of LCCM gradually decreases. However, the heat release rate of A_{20} remains at 0.041 $W{\cdot}g^{-1}$, significantly higher than that of A_0 at 0.015 $W{\cdot}g^{-1}$. The heat release rate pattern demonstrates that the addition of desulfurization gypsum slows the early hydration rate and delays the hydration process.

As shown in Fig. 5.31, the cumulative hydration heat release curves of LCCM with different desulfurization gypsum contents indicate that the hydration heat is primarily concentrated within the first 6 h. The addition of desulfurization gypsum significantly increases the total hydration heat, which initially rises and then decreases with increasing desulfurization gypsum content. This suggests that increasing the desulfurization gypsum content effectively promotes the hydration degree of $C_4A_3\hat{S}$ and the formation of AFt, releasing a substantial amount of heat. However, excessive desulfurization gypsum content inhibits the overall hydration degree of LCCM. The total hydration heat of A_0 and A_5 shows little increase after 12 h, while that of A_{10}, A_{15}, and A_{20} continues to grow, significantly exceeding A_0 at 72 h. This is because, when $CaSO_4$ content is sufficient, AFm reacts with $CaSO_4$ and water to transform into AFt, as shown in Eq. 5.26, a process that also releases heat. This further indicates that adding more than 10% desulfurization gypsum, while reducing the early hydration rate of LCCM, extends the later hydration period, allowing the continuous transformation of AFm into AFt, thereby promoting the development of later hydration.

$$C_4A_3\overline{S}H_{12} + 2C\overline{S} + 20H \rightarrow C_6A\overline{S}_3H_{32} \tag{5.26}$$

Fig. 5.31 Accumulative hydration exothermic curve

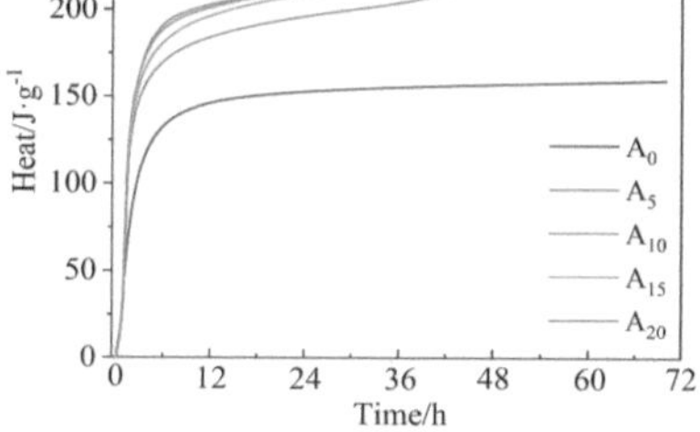

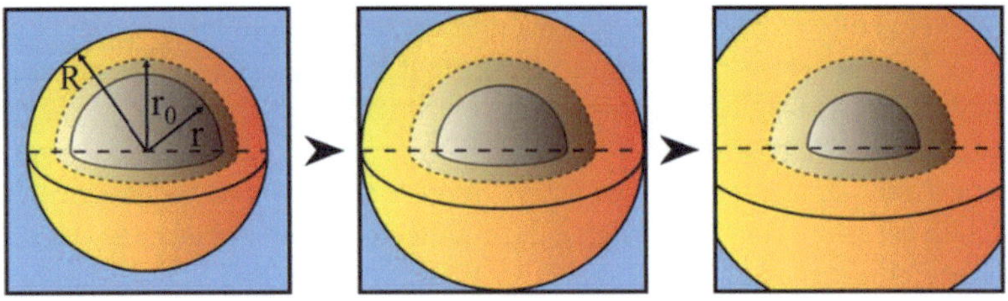

Fig. 5.32 Hydration model diagram

5.7.2 Hydration Kinetic Model of LCCM

It is assumed that the cement paste is composed of countless identical basic cubic hydration units, each with a side length of *L*. Each hydration unit includes an initial cement particle with a radius of r_0 and surrounding water molecules. When the initial cement particle dissolves in water, a thin film of hydration products first forms on the surface of the cement particle. Water molecules penetrate this film to react with the unhydrated cement inside, generating new hydration products that continuously diffuse outward and uniformly cover the surface of the unhydrated cement particle, maintaining the overall spherical shape of the particle. The radius of the cement particle, including the hydration products, is denoted as *R*, while the radius of the unhydrated cement particle is denoted as *r*. As hydration progresses, *r* continuously decreases, while *R* continuously increases. A schematic diagram of the cement hydration model is shown in Fig. 5.32.

According to the hydration model, the volume of the initial cement particles is:

$$V_c = \frac{1}{\frac{w}{c}\rho_c + 1} \tag{5.27}$$

where: *w* is the mass of water. *c* is the mass of cement. ρ_c is the density of cement.

The initial cement particle radius r_0 is:

$$r_0 = \sqrt[3]{\frac{3V_c}{4\pi}} \tag{5.28}$$

The hydration degree α of cement particles is:

$$\alpha = 1 - \left(\frac{r}{r_0}\right)^3 \tag{5.29}$$

According to the minimum theoretical water-cement ratio, the relationship between the degree of hydration α and the radius of cement particles after hydration *R* can be deduced:

$$\alpha(R) = \left(\frac{V_R\left(1 + \frac{w}{c}\rho_c\right)}{L^3} - 1\right) \cdot \frac{1}{\left(\frac{w}{c}\right)_{\min}\rho_c} \tag{5.30}$$

where: V_R is the total volume of cement particles (including hydration products and unhydrated particles) after hydration. $(\frac{w}{c})_{min}$ is the theoretical minimum water-binder ratio.

$$V_R = \frac{4}{3}\pi R^3, R < \frac{L}{2} \tag{5.31}$$

$$V_R = \frac{4}{3}\pi R^3 - 6\pi\left(\frac{2R^3}{3} + \frac{L^3}{24} - \frac{LR^2}{2}\right), \frac{L}{2} < R < \frac{\sqrt{2}}{2}L \tag{5.32}$$

In the process of hydration, without considering the evaporation of water, each kind of cement has a minimum theoretical water cement ratio, which varies according to the different composition of cement, which can be expressed as:

$$\left(\frac{w}{c}\right)_{\min} = \sum\left(\frac{w}{c}\right)_i \varphi_i \tag{5.33}$$

where: $(w/c)_i$ is the minimum theoretical water-binder ratio of a mineral in cement. φ_i is the proportion of the mineral in cement. The minimum theoretical water-cement ratio of C_2S is 0.178. The minimum theoretical water-cement ratio of $C_4A_3\hat{S}$ in the presence of $CaSO_4$ is 0.718.

Cement hydration primarily consists of three fundamental processes: nucleation and crystal growth (NG), phase boundary reaction, and diffusion reaction. These processes can occur simultaneously, but the progress of cement hydration depends on the slowest one. Since the NG process mainly occurs in the early stages of hydration and has a relatively short duration, only the phase boundary reaction and diffusion reaction processes are discussed here.

In the phase boundary reaction process, dissolved cement particles are in direct contact with water. At this stage, the hydration rate is determined by the chemical reaction rate, and the relationship between the thickness of the hydration products and time is given by:

$$\begin{cases} \frac{dh}{dt} = k_1 \\ h = k_1 t \end{cases} \tag{5.34}$$

where: h is the thickness of hydration products, $h = R\text{-}r$. k_1 is the chemical reaction rate, which can be approximately considered as a constant.

From Eq. 5.29, we can get $r = r_0\sqrt[3]{1-\alpha}$. The relationship between hydration degree α and hydration time in phase boundary reaction process can be obtained by introducing Eq. 5.35:

$$R - r_0\sqrt[3]{1-\alpha} = k_1 t \tag{5.35}$$

In the diffusion reaction process, water molecules and ions in the solution penetrate through the hydration product layer to react with the unhydrated cement. As hydration time increases, the thickness of the hydration product layer gradually increases, leading to greater diffusion resistance and a slower diffusion rate. Consequently, the reaction rate decreases as the thickness of the hydration product layer increases. Assuming the thickness of the hydration product layer at this stage is h and the diffusion rate is k_2, the relationship between the thickness of the hydration products and time is given by:

$$\begin{cases} \dfrac{dh}{dt} = \dfrac{k_2}{h} \\ h^2 = 2k_2 t \end{cases} \tag{5.36}$$

Similarly, the relationship between hydration degree α and hydration time in the diffusion reaction process can be obtained by combining Eq. 5.37.

$$\left(R - r_0\sqrt[3]{1-\alpha}\right)^2 = k_2 t \tag{5.37}$$

Suppose that the hydration unit side length L is 10 μm, ρ_c is 3.1 g/cm^3, and the water-binder ratio is 0.5, the hydrate phase content and the minimum water-binder of each group of LCCM samples are as shown in the Table 5.5.

The LCCM sample has completed most of the hydration heat release at 12 h, and the hydration rate is basically unchanged. The hydration degree of low carbon cementitious material at 12 h is taken to calculate the chemical reaction rate k_1 and diffusion rate k_2, and the results are shown in Table 5.6.

Based on the Eqs. 5.34–5.36, the change of hydration degree of low carbon cementitious material samples with time in different reaction processes can be obtained.

Table 5.5 Hydrated phase content and minimum water-binder ratio of LCCM samples

		A_0	A_5	A_{10}	A_{15}	A_{20}
Hydrate phase	$C_4A_3\overline{S}$&$CaSO_4$	0.427	0.376	0.359	0.325	0.305
	C_2S	0.558	0.526	0.511	0.498	0.486
Minimum water-binder ratio		0.43	0.39	0.37	0.34	0.32

Table 5.6 Parameters of different reaction processes

Sample	A_0	A_5	A_{10}	A_{15}	A_{20}
α	0.8963	0.8734	0.8104	0.7578	0.7417
k_1(μm/h)	0.2071	0.1950	0.1678	0.1495	0.1461
k_2(μm/h)	0.2577	0.2282	0.1691	0.1342	0.1282

At the same time, based on the cumulative hydration heat release, according to the Eqs. 5.37–5.38, the hydration degree test value α_T at different time can be obtained.

$$\frac{1}{Q} = \frac{1}{Q_{max}} + \frac{t_{50}}{t \cdot Q_{max}} \tag{5.38}$$

$$\alpha_T = \frac{Q}{Q_{max}} \tag{5.39}$$

where: Q is the cumulative hydration heat. Q_{max} is the maximum cumulative hydration heat when the hydration time is infinitely long. t_{50} is the hydration time when the cumulative hydration heat reaches 50% of the maximum cumulative heat release.

Q_{max} and t_{50} can be obtained by fitting the heat release Q and hydration time t, and then the experimental value of hydration degree can be obtained. The calculated value of hydration degree is compared with the experimental value, as shown in the following figure.

As shown in the Fig. 5.33, in the absence of desulfurization gypsum, the hydration degree curve under the phase boundary reaction process aligns more closely with the experimental values, indicating that the hydration process of LCCM is primarily governed by the chemical reaction rate at this stage. With the addition of desulfurization gypsum, the hydration process of LCCM samples transitions from being dominated by the phase boundary reaction process to the diffusion reaction process. This is because, on one hand, desulfurization gypsum increases the concentration of calcium ions and sulfate ions in the solution, leading to the rapid formation of a hydration layer covering the unhydrated cement particles in the early stages. On the other hand, the addition of desulfurization gypsum results in unhydrated gypsum crystals surrounding the cement particles, reducing the direct contact area between the liquid phase and the cement. Together, these effects cause the liquid phase to penetrate and diffuse into the interior to react with the cement. In the early hydration stages, the hydration degree of LCCM samples is slightly higher than the calculated values. This is because, for simplicity, the model assumes calcium sulfate ($CaSO_4$) as the mineral participating in hydration, whereas desulfurization gypsum primarily consists of $CaSO_4 \cdot 0.5H_2O$, which has better solubility and reacts faster than $CaSO_4$. When the desulfurization gypsum content is high, the hydration degree of LCCM samples in the later stages (hydration degree > 0.8) is slightly lower than the calculated values. This is due to the excessive desulfurization gypsum content, where some desulfurization gypsum remains unreacted after the complete consumption of anhydrous calcium sulphoaluminate. Overall, the model shows a high degree of fit with the experimental values and effectively reflects the regulatory effect of desulfurization gypsum on the hydration process of LCCM.

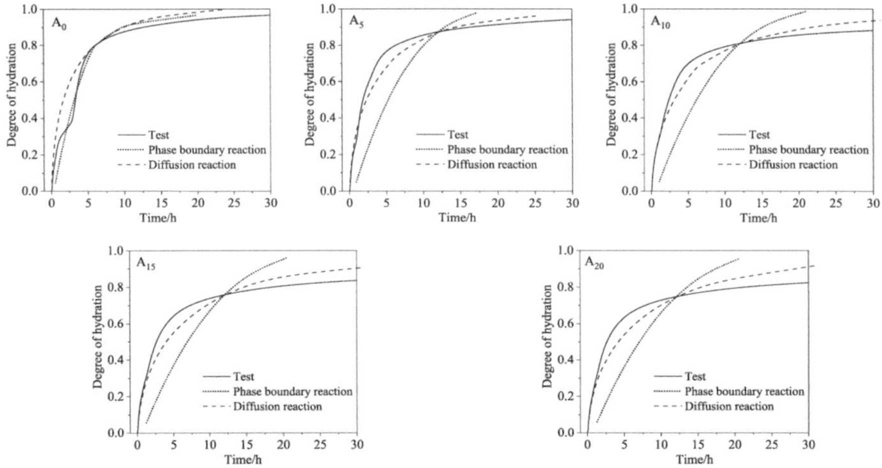

Fig. 5.33 Hydration degree change with time

5.8 Hydration Mechanism of LCCM Regulated by Desulfurization Gypsum

Analysis of the hydration process of LCCM reveals that desulfurization gypsum delays the early hydration and extends the hydration cycle, which is directly related to the distribution characteristics of the hydration reactant $CaSO_4$ in LCCM.

From the perspective of the particle scale of LCCM, the hydration reaction process without desulfurization gypsum is illustrated in Fig. 5.34. After the solid waste raw materials are calcined at high temperatures to form minerals such as $C_4A_3\hat{S}$ and C_2S, a small amount of residual $CaSO_4$ remains uniformly dispersed on the surface and inside of the clinker. At the onset of hydration, the surface layer of the clinker hydrates first. However, due to the relatively low $CaSO_4$ content, the amount of hydration products formed is also limited, and the hydration product layer covering the surface of the LCCM particles is not dense. This provides favorable conditions for water to penetrate into the interior of the clinker. Subsequently, water enters the interior of the LCCM particles, reacting with $C_4A_3\hat{S}$ and $CaSO_4$ to form hydration products. The volume expansion of AFt generated inside causes the outer layer to peel off, further facilitating water penetration into the clinker. This repeated generation and peeling of the hydration layer creates a positive feedback loop, accelerating the hydration process. Overall, $CaSO_4$ can act both on the surface and inside the clinker. The hydration reaction in the core of the clinker is controlled by the chemical reaction rate, resulting in a faster hydration speed, which belongs to the phase boundary reaction process.

After the addition of desulfurization gypsum, the hydration reaction process of LCCM is illustrated in Fig. 5.35. Desulfurization gypsum is dispersed around the clinker. Once water comes into contact with the clinker and desulfurization gypsum,

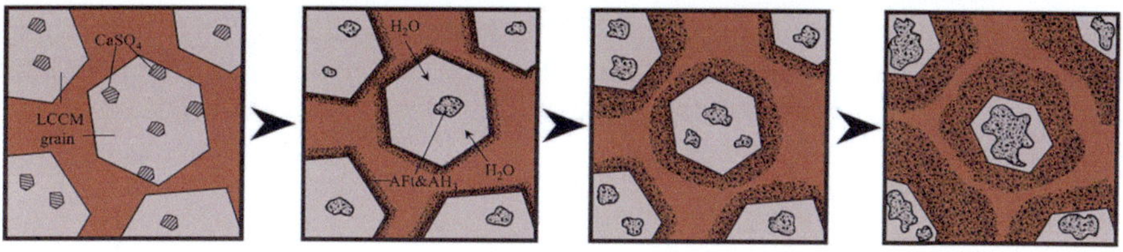

Fig. 5.34 Hydration reaction model of LCCM without desulfurization gypsum

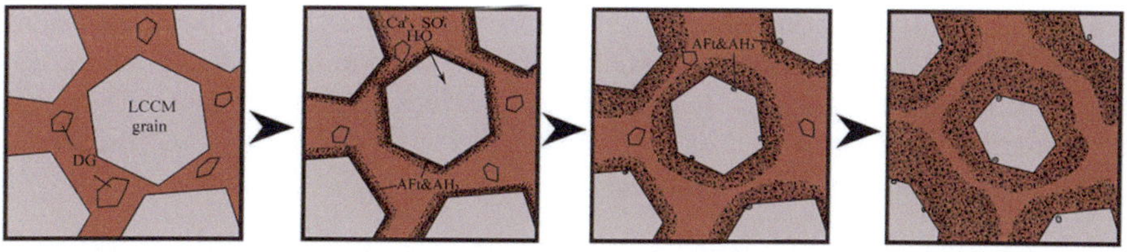

Fig. 5.35 Hydration reaction model of LCCM mixed with desulfurization gypsum

rapid hydration occurs on the surface of the clinker, forming hydration products such as AFt and AH_3. These products accumulate and form a dense layer on the surface of the clinker, hindering further contact between water, clinker, and gypsum, thereby slowing down the hydration reaction. As hydration progresses, the paste structure becomes denser, and only a small amount of water carrying Ca^{2+} and SO_4^{2-} ions penetrates through the pores into the interior of the clinker to react with the clinker minerals. The amount of AFt generated inside the clinker is limited, insufficient to cause the surface layer of the clinker particles to crack or peel off. This delays the early hydration rate of LCCM, extends the hydration cycle, and results in a slow and prolonged hydration process. Overall, desulfurization gypsum primarily participates in hydration starting from the surface of the clinker and gradually progresses inward. The hydration reaction of cement is mainly determined by the ion diffusion rate, leading to a slower early hydration rate and a longer hydration cycle, which belongs to the diffusion reaction process.

Through the analysis of the regulatory mechanism, the influence of different desulfurization gypsum contents on the hydration and hardening process of LCCM can be explained. Simply put, the content of desulfurization gypsum affects the state of the hydration layer on the surface of cement particles, alters the transport mechanism of water into the interior of the cement particles, and thereby influences the hydration reaction rate. The higher the desulfurization gypsum content, the denser the hydration layer on the surface of the cement particles in the early stages, which hinders the progress of hydration. This results in two effects: on one hand, high desulfurization gypsum content slows the early hydration rate of LCCM. On the other hand, it provides a foundation for the continuous and slow hydration of cement particles.

Additionally, sufficient calcium sulfate not only reacts with $C_4A_3\hat{S}$ but also continuously generates AFt through reactions with AH_3, further enhancing the hydration degree of LCCM and promoting the development of their compressive strength.

5.9 Conclusions

In this chapter, the setting time, phase change, microstructure and compressive strength of desulfurization gypsum in the process of hydration and hardening of all solid waste LCCM hardened paste were studied. The hydration kinetic model was established, and the regulation mechanism of desulfurization gypsum was discussed. The hardening process of LCCM mortar was studied by conducting AC impedance spectroscopy measurements. The specific conclusions are as follows:

(1) The setting time of LCCM decreases with the increase of desulfurization gypsum content. The final setting time of A20 is 9 min, which is 78.6% shorter than that of A0, showing a false setting phenomenon. desulfurization gypsum is not conducive to the development of compressive strength in the early stage of hydration, but it can promote the improvement of compressive strength in the later stage. The compressive strength of LCCM decreased from 44.48 MPa of A0 to 33.66 MPa of A20 after hydration for 6 h, which decreased by 23.2%. The compressive strength of A10 was the highest at 28 d, reaching 84.36 MPa, which was 42.7% higher than that of A0.
(2) The incorporation of desulfurization gypsum delays the early hydration rate of LCCM and prolongs the hydration cycle. Quantitative analysis of hydration products showed that there was a positive linear relationship between AFt content and compressive strength. The AFt content of A10 was the highest at 28 d, which was 61.9%, indicating that 10% desulfurization gypsum had the most obvious effect on the hydration of LCCM.
(3) Desulfurization gypsum is beneficial to promote the development of AFt. With the increase of desulfurization gypsum content and the growth of hydration age, the size of AFt gradually increases, showing a trend of needle-rod-column.
(4) The cubic hydration kinetics model shows that the LCCM hydration reaction without desulfurization gypsum is controlled by the chemical reaction rate and the maximum reaction rate was 0.057 $W \cdot g^{-1}$, belongs to the phase boundary reaction process. After adding desulfurization gypsum, the hydration reaction is mainly determined by the diffusion rate of ions through the hydration layer, which slows down the reaction rate and belongs to the diffusion reaction process.
(5) Desulfurization gypsum improves the pore structure of the LCCM mortar. After 28 days of hydration, the porosity of LCCM mortar decreased from 3.09% in A0 to 1.52% in A20; additionally, desulfurization gypsum assists in reducing the pore diameter, significantly lowering the proportion of mesopores (50 nm $\leq r < 200$ nm) and air pores ($r \geq 200$ nm).

(6) The volume resistivity of LCCM mortar gradually increased with the hydration age; a higher amount of desulfurization gypsum decreased the volume resistivity in the early stages of hydration but contributed to its increase in the later stages. An equivalent circuit model based on the AC impedance results of LCCM mortar was established, where an increase in the hydration age and desulfurization gypsum is conducive to the transition from CCP to ICP.
(7) Grey entropy analysis was conducted on the equivalent circuit impedance parameters and compressive strength. R_1 and R_{CCP} exhibited the highest correlation with the compressive strength, reaching 0.99907. The volume resistivity R_1 had a good linear relationship with the compressive strength, with correlation coefficients of 0.92 ($w_{DG} \leq 10\%$) and 0.98 ($w_{DG} > 10\%$).

References

Bai, R., Liu, S., Yin, L., Yan, C., Zhou, J., Lu, L., & Li, L. (2023). Freeze-thaw damage assessment of engineered cementitious composites using the electrochemical impedance spectroscopy method. *Materials & Design, 230*, 111965.

Gu, P., Xie, P., Fu, Y., & Beaudoin, J. J. (1994). AC impedance phenomena in hydrating cement systems: Frequency dispersion angle and pore size distribution. *Cement and Concrete Research, 24*(1), 86–88.

Hargis, C. W., Telesca, A., & Monteiro, P. J. (2014). Calcium sulfoaluminate (Ye'elimite) hydration in the presence of gypsum, calcite, and vaterite. *Cement and Concrete Research, 65*, 15–20.

Hu, X., Shi, C., Liu, X., Zhang, J., & De Schutter, G. (2019). A review on microstructural characterization of cement-based materials by AC impedance spectroscopy. *Cement and Concrete Composites, 100*, 1–14.

Macphee, D. E., Cormack, S. L., & Sinclair, D. C. (2000). AC impedance spectroscopy of pore reduced cements: Influence of contact resistance. *Journal of Materials Science, 35*, 4823–4826.

McCarter, W. J. (1994). A parametric study of the impedance characteristics of cement-aggregate systems during early hydration. *Cement and Concrete Research, 24*(6), 1097–1110.

Mota, R. M., Silva, A. S., Ramos, V. H. S., Rezende, J. C. T., & Jesus, E. D. (2020). Effects of storage temperature and time on false setting behavior of CPI-S Portland cement. *Ceramica, 66*, 321–329.

Song, G. (2000). Equivalent circuit model for AC electrochemical impedance spectroscopy of concrete. *Cement and Concrete Research, 30*(11), 1723–1730.

Sun, C., Zhang, J., Yan, C., Yin, L., Wang, X., & Liu, S. (2022). Hydration characteristics of low carbon cementitious materials with multiple solid wastes. *Construction and Building Materials, 322*, 126366.

Telesca, A., Marroccoli, M., Pace, M. L., Tomasulo, M., Valenti, G. L., & Monteiro, P. J. M. (2014). A hydration study of various calcium sulfoaluminate cements. *Cement and Concrete Composites, 53*, 224–232.

Trezza, M. A., & Lavat, A. E. (2001). Analysis of the system 3CaO· Al2O3–CaSO4· 2H2O–CaCO3–H2O by FT-IR spectroscopy. *Cement and Concrete Research, 31*(6), 869–872.

Wang, P., Li, N., & Xu, L. (2017). Hydration evolution and compressive strength of calcium sulphoaluminate cement constantly cured over the temperature range of 0 to 80 C. *Cement and Concrete Research, 100*, 203–213.

Yin, Y., Yin, J., Zhang, W., Tian, H., Hu, Z., Ruan, M., Xu, H., Liu, L., Yan, X., & Chen, D. (2018). FT-IR and micro-Raman spectroscopic characterization of minerals in high-calcium coal ashes. *Journal of the Energy Institute, 91*(3), 389–396.

Zhao, J., Yan, C., Liu, S., Zhang, J., Li, S., & Yan, Y. (2020). Effect of solid waste ceramic on uniaxial tensile properties and thin plate bending properties of polyvinyl alcohol engineered cementitious composite. *Journal of Cleaner Production, 268*, 122329.

Chapter 6
Mechanical Properties of Hardened Pastes of Seawater-Mixed Low-Carbon Cementitious Materials

6.1 Introduction

In this study, low-carbon cementitious materials (LCCM), with $C_4A_3\hat{S}$ and C_2S as the primary mineral components, were synthesized using coal gangue, carbide slag, and desulfurization gypsum as raw materials. A systematic investigation was conducted into the effects of various mixing water and curing conditions on the mechanical properties of LCCM, including fresh water mixing with standard curing, fresh water mixing with sea water curing, sea water mixing with standard curing, and sea water mixing with sea water curing. The variation patterns of free chloride ion content in LCCM under these different mixing and curing conditions were examined. The evolution of hydration products under varying mixing and curing conditions was characterized using thermogravimetry–differential scanning calorimetry (TG-DSC), Fourier transform infrared spectroscopy (FTIR), X-ray diffraction (XRD), and scanning electron microscopy (SEM). Nuclear magnetic resonance (NMR) technology was employed to analyze the pore structure, pore size distribution, and porosity characteristics of LCCM at different hydration ages. Based on grey theory, the relationships between the mass loss rate of hydration products and compressive strength, as well as the correlation between pore structure and compressive strength, were comprehensively evaluated.

6.2 Specimen Preparation and Research Methods

6.2.1 Raw Materials

The primary raw materials for preparing low-carbon cementitious materials include coal gangue, carbide slag, and desulfurization gypsum.

C. Yan et al., *Low-Carbon Cementitious Materials with 100% Solid Wastes*,
https://doi.org/10.1007/978-981-96-8742-8_6

Coal gangue, also known as coal-associated waste rock, is a mining solid waste produced during coal mining and washing processes. Its main components include oxides of silicon and aluminum. The coal gangue used in the experiment was sourced from a coal mine in Ordos. After crushing, grinding, and screening, it was processed into powder, as shown in Fig. 6.1a. Its XRD spectrum is shown in Fig. 6.2a.

Carbide Slag is an industrial solid waste produced during the production of acetylene. Its main component is calcium hydroxide. The carbide slag used in the experiment was sourced from a steel group in Baotou. It only needed to be screened, as shown in Fig. 6.1b. Its XRD spectrum is shown in Fig. 6.2b.

Desulfurization gypsum is a solid industrial waste generated during the recovery of sulfur dioxide from flue gas. Its main component is calcium sulfate dihydrate. The desulfurization gypsum used in the experiment was sourced from a power plant in Ordos. It only needed to be screened, as shown in Fig. 6.1c. Its XRD spectrum is shown in Fig. 6.2c (Table 6.1).

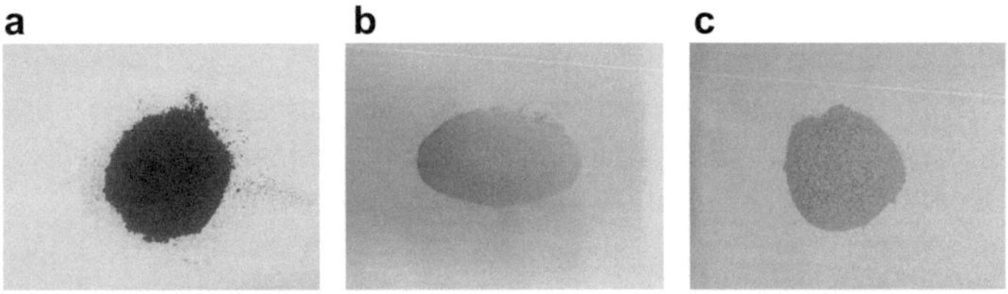

Fig. 6.1 **a** Coal gangue, **b** carbide slag, **c** desulfurization gypsum

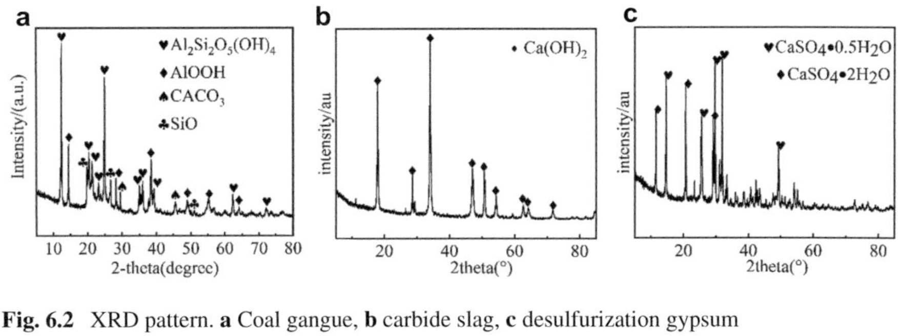

Fig. 6.2 XRD pattern. **a** Coal gangue, **b** carbide slag, **c** desulfurization gypsum

Table 6.1 Chemical composition of raw materials

Component	Al_2O_3	SiO_2	CaO	SO_3	Fe_2O_3	Na_2O	MgO	TiO_2	K_2O
Coal gangue	48.627	41.639	1.037	2.728	2.403	/	0.527	1.155	0.534
Carbide slag	0.463	10.09	84.75	1.047	/	1.092	1.102	0.114	0.313
Desulfurization gypsum	/	6.876	27.925	56.295	/	4.941	2.368	/	/

Table 6.2 The content of main ion mixed seawater

Main ions	Cl^{2-}	SO_4^{2-}	Mg^{2+}	K^{2+}
Testing equipment	Burette	Ion chromatograph	Plasma spectrometer	Plasma spectrometer
Percentage (mg/L)	1.65×10^4	2.32×10^3	1.00×10^3	330

The mixing water utilized natural seawater collected from the Bohai Sea coastal area (Laishan District, Yantai, China), with sampling coordinates approved by the local maritime administration. For small-scale usage (<100 L), seawater was directly collected from designated coastal zones, while larger quantities required formal authorization from the Maritime Safety Administration. Ionic chromatography revealed the seawater's predominant salt composition (Table 6.2), featuring chloride, sulfate, magnesium, and potassium as major ions. Control experiments employed municipal tap water, and distilled water was exclusively used for free chloride ion measurements to prevent interference.

6.2.2 Preparation Process

The coal gangue was ball milling and screening treatment to control its particle size less than 200 mesh. Carbide slag and desulfurization gypsum gypsum were screened to control their particle size less than 200 mesh.

Weigh the raw materials according to the ratio, and use a small mixer to mix well. After mixing evenly, 15% water was added to the mixture to make it easy to form with a certain humidity. The mixture was then pressed into a sample with a diameter of 50 mm and a thickness of 15 mm. Finally, the samples were put into an air drying oven and dried at 105 °C for 4 h.

The dried samples were calcined in a high temperature furnace. First, the temperature is raised to 900 °C at a heating rate of 10 °C/min. Then, the temperature was raised to 1280 °C at a heating rate of 5 °C/min, and then continued to calcination for 40 min. In order to prevent the decomposition of minerals after calcination, samples need to be taken out quickly after calcination and placed at room temperature for quench cooling.

After cooling, the sample was crushed, pulverized, and sieved to control its particle size to less than 200 mesh. The low-carbon cementitious material was obtained by adding the desulfurization gypsum with the mass fraction of 4% into the sample and mixing it evenly.

6.2.3 Test Methods

Compressive strength

The water-binder ratio was set to 0.47, and the cement-sand ratio was set to 1:3. Using seawater and fresh water as mixed water, the low-carbon cementitious material, mixied water, and sand were mixed in proportion and stirred evenly. Then, pour the evenly mixed mixture into a 40 mm × 40 mm × 40 mm mold and place it on a shaking table for vibration. After the sample was left standing for 4 h, the mold was removed. The types of mixed water and curing conditions were taken as study variables, which were fresh water mixed with standard curing (FN), fresh water mixed with seawater curing (FS), sea water mixed with standard curing (SN), and sea water mixed with seawater curing (SS). When the hydration age reaches 6 h, 1 d, 3 d, 7 d, 14 d, 28 d, and 90 d, the universal testing machine was used to test the compressive strength at the loading speed of 2400 N/s. The number of test samples in each group was three.

Free chloride ion content

Weigh 10 g of the sample and place it in a conical bottle, add 200 mL of distilled water and let it stand for 24 h. During the standing period, in order to ensure that the free chloride ions in the sample are fully dissolved in water, shake the conical bottle once every 4 h. The free chloride ion content was determined by using the chloride ion rapid analyzer. Before the measurement, the electrode activation solution was used to activate the lower electrode of the instrument, and the instrument was calibrated with standard NaCl solution with a concentration of 1.0×10^{-3} and 1.0×10^{-4}, and then calibrated again when the test exceeded 1 h. Each sample was tested three times, and the average value was taken as the final measurement result.

Thermogravimetric analysis

The endothermic and exthermic behavior of samples in different temperature ranges was measured by a comprehensive thermal analyzer. The particle size of the tested sample is controlled under 200 mesh. The heating rate was set at 10 °C/min, the nitrogen flow rate was set at 50 mL/min, and the data collection frequency was set at 1 °C/time.

Fourier transform infrared spectroscopy

The functional group composition of samples was characterized by Fourier-transform infrared (FTIR) spectroscopy. The particle size of the test sample should be controlled under 200 mesh and homogenously mixed with KBr at a mass ratio of 100:1. FTIR spectra were acquired in the range of 340–4700 cm^{-1} with a resolution of 2 cm^{-1}, accumulating 40 scans per measurement to ensure signal-to-noise ratio.

X-ray diffractometer

The phase composition of samples was analyzed using an X-ray diffractometer (XRD). The particle size of the test sample should be controlled under 200 mesh.

The XRD measurements were performed with a scanning rate of 0.02°/s over a 2θ range of 5°–90°.

Scanning electron microscope

The microstructural morphology of both clinker and hydration products was examined using scanning electron microscopy (SEM). The samples were trimmed to dimensions below 1 cm and sputter-coated with a thin gold layer to enhance surface conductivity and improve image resolution.

Nuclear magnetic resonance spectroscopy

NMR technology was used to analyze the pore structure of the specimens. The tests were conducted using a MesoMR23-060H-I NMR imaging analyzer (Suzhou Niuman Analytical Instrument Corporation, China). Prior to testing, the specimens were subjected to vacuum saturation with water. To prevent the loss of surface moisture during testing, which could affect the results, the vacuum-saturated specimens were sealed with plastic film and then placed into the testing coil for pore structure analysis using the NMR analyzer.

6.3 Free Chloride Ions of Hardened Pastes of Seawater-Mixed LCCM

Figure 6.3a and b show the variation patterns of free chloride ion content in LCCM and ordinary Portland cement with the curing age of specimens. As can be seen from the figures, the development trends of free chloride ion content in specimens under different curing conditions vary. The main differences are reflected in the significant fluctuation of free chloride ion content during the early hydration stage, while the values tend to stabilize during the later hydration period.

The content of free chloride in the samples of SS group decreased first, and then gradually tended to be stable. However, the content of free chloride ions increased slightly in the late curing period. This shows that in the initial stage of hydration,

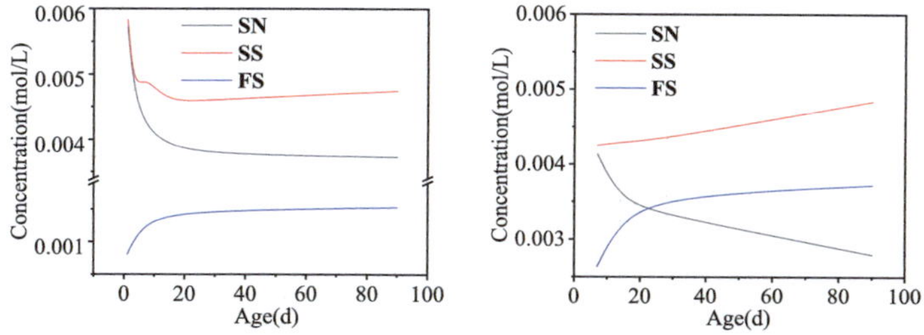

(a) LCCM (b) Portland cement

Fig. 6.3 Free chloride ions content

the chloride ions in seawater react with the hydration products, resulting in a small decrease in the content of free chloride ions. With the progress of hydration, the hydration products reacting with chloride ions are gradually consumed, and the change range of free chloride ion content in the late system begins to decrease.

The content of free chloride in the samples of FS group increased first and then tended to be stable. In the early stage of hydration, under the condition of seawater curing, chloride ions in seawater will invade the inside of the sample, resulting in an increase in the content of free chloride ions. With the progress of hydration, the hydration products continue to develop, and the porosity of the specimen gradually decreases, making it difficult for chloride ions to invade, which makes the content of free chloride ions in the system tend to be flat.

The content of free chloride ions in the samples of SN group showed a continuous decreasing trend. During the hydration process, some chloride ions in seawater react with the hydration products, resulting in a significant decrease in the content of free chloride ions. At the same time, some chloride ions are affected by internal and external potential difference in the later stage of hydration, resulting in precipitation phenomenon. Since there are few connected pores in the hardened slurry in the late hydration period, the precipitation phenomenon is not obvious, so the content of free chloride ions decreases slowly.

The change range of free chloride ion content in the samples of the Portland cement group and the LCCM group is basically similar, showing a trend of greater change in the early stage and less change in the late stage, but its development trend is slightly different.

Compared with the SS group, the free chloride ion content in the OPC-SS group always showed a trend of continuous increase. This indicates that although chloride ions in seawater continue to react with the hydration products in the system, external chloride ions continue to invade due to the slow strength development of Portland cement in the early stage, the internal porosity of the specimens is high and there are many connected pores. In addition, the compactness of Portland cement is lower than that of LCCM, so the external intrusion of chloride ions has a more significant impact on the overall free chloride ion content, resulting in a continuous increase in the phenomenon.

Compared with the FS group, the free chloride ion content in the OPC-FS group was higher on the whole, and the early chloride ion concentration increased more. This indicates that the OPC-FS samples have higher porosity, and chloride ions are more likely to invade the inside of the samples, especially in the early stage of curing, and the rapid invasion rate of chloride ions is much higher than the rate of binding. However, with the progress of hydration, the compactness of the specimen gradually increased, which made the rise of free chloride content in the later period tend to be gentle.

Compared with the SN group, the free chloride ion content in OPC-SN group was lower on the whole. This indicates that Portland cement has better ability to bind chloride ions than LCCM. However, due to the large number of connected pores in the OPC-SN group, the chloride ion precipitation is larger in the curing process, and the specimen is more susceptible to damage by chloride ion erosion.

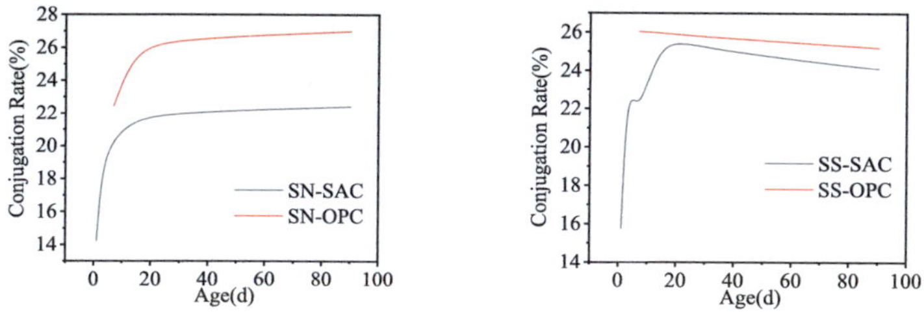

(a) Sea water mixed with standard curing (b) Sea water mixed with seawater curing

Fig. 6.4 Chloride ion binding rate

The relationship between the initial chloride ions concentration and the free chloride ions concentration is expressed by the chloride ion binding rate, as shown in Eq. 6.1.

$$\beta = \frac{C_i - C_w}{C_i} \times 100\% \tag{6.1}$$

where: C_i is the initial chloride ions concentration. C_w is the the free chloride ions concentration. β is the chloride ion binding rate.

The chloride ion binding rate of the samples is shown in Fig. 6.4. It can be seen that when seawater is introduced into the LCCM, its binding effect on chloride ions in seawater is mainly concentrated in the early stage of hydration. This is because LCCM have the characteristics of rapid early hydration, and seawater can accelerate the formation of AFt, thus combining a certain proportion of chloride ions in the hydration process. However, in the later stage of hydration, due to the low content of key components in the hydration products that can effectively bind chloride ions, such as C-S-H gel and AFm, the binding rate of chloride ions decreases significantly. In contrast, when seawater is introduced into Portland cement, the chloride ion binding rate is still faster in the early stage of hydration. In the later stage of hydration, due to the high content of C-S-H gel and AFm in the hydration products, the binding reaction of chloride ions can still maintain a high rate, and finally show a higher chloride ion binding rate than LCCM.

6.4 Hydration Process of Seawater-Mixed LCCM

6.4.1 Hydration Products

The XRD pattern of the LCCM mixed with fresh water under standard curing conditions (FN) are shown in Fig. 6.5a. Because part of the standard sand is mixed in the sampling process, there is a high SiO_2 diffraction peak, which is truncated. It can be seen that the diffraction peak of AFt as the main hydration product is observed at the positions of diffraction angles of 9.171° and 17.422°. The diffraction peak is observed for unhydrated $C_4A_3\hat{S}$ at a diffraction angle of 23.772°. The diffraction peaks of C_2S were observed at the angles of 32.528° and 39.115°. The diffraction peak of $CaSO_4$ was observed at the position of 25.829°. The diffraction peak of $CaCO_3$ is observed at the position of 50.281°. The diffraction peak of AFm was observed at the angle of 35.274°.

After 1 day of hydration, a relatively significant AFt diffraction peak was observed on the XRD pattern, which was mainly generated by the hydration reaction of $C_4A_3\hat{S}$. Because of the amorphous structure of AH_3, its diffraction peak cannot be observed in the XRD pattern. At the age of 3 d and 7 d, the diffraction peak of AFt is further enhanced. At the same time, the diffraction peak of $C_4A_3\hat{S}$ gradually weakened and disappeared, indicating that in the early stage of hydration, $C_4A_3\hat{S}$ has basically completely reacted. At the age of 28 d, the diffraction peak of AFt did not change significantly compared with that at the age of 7 d, while the diffraction peak of C_2S became sharper and the diffraction intensity decreased, indicating that C_2S continued to participate in the reaction slowly during the hydration process. In addition, the

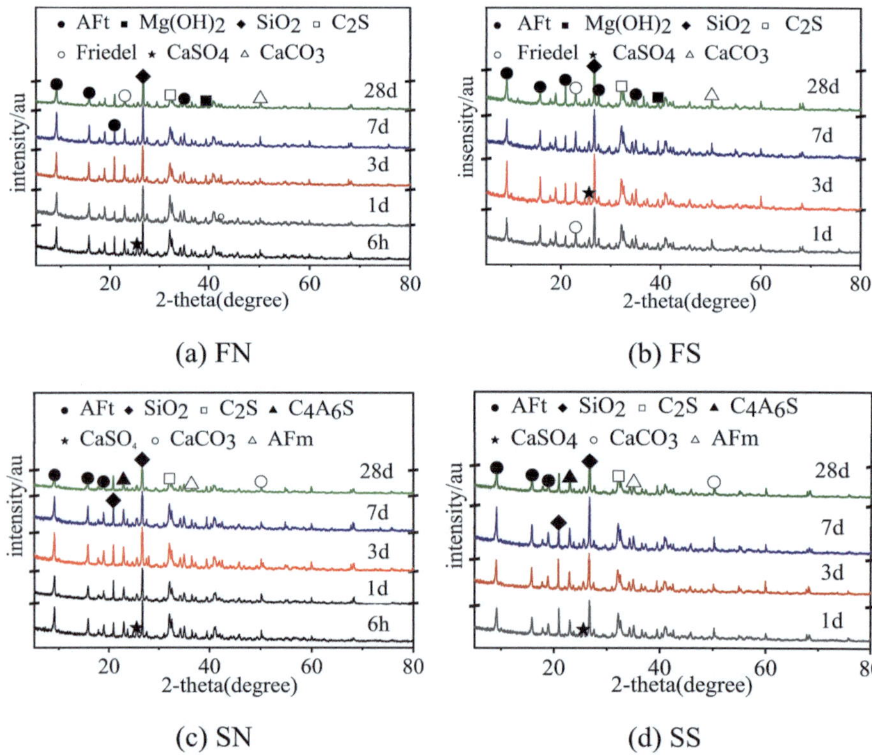

(a) FN (b) FS

(c) SN (d) SS

Fig. 6.5 XRD pattern of LCCM mixed with seawater

diffraction peak of AFm appeared after 1 d of hydration, and was slightly enhanced with the extension of hydration age.

As shown in Fig. 6.5b, the observed diffraction peak of the specimen (FS) mixed with fresh water and preserved in seawater is basically the same as that of the specimen in the FN group. At the first day of hydration, a large number of AFt diffraction peaks were detected, and the diffraction peaks of unhydrated phases such as $C_4A_3\hat{S}$, C_2S were also observed. Although the specimen was preserved in seawater at the early stage of hydration, a certain degree of erosion might occur, but due to the small amount of seawater intrusion, significant diffraction peaks of erosion products could not be detected. This indicates that the LCCM has a certain ability to resist seawater erosion in the early stage of hydration.

As depicted in Fig. 6.5c and d, a substantial formation of AFt is evident in the two sample groups (SN and SS) mixed with seawater at 1 d. This observation suggests that seawater mixing significantly enhances the precipitation of AFt as a hydration product in LCCM. Simultaneously, the presence of Friedel's salt was detected at the diffraction angle of 22.014°, indicating that chloride ions in seawater reacted with AFm in the hydration products. Notably, among the SN group specimens, the Friedel's salt diffraction peak observed at 6 h is weaker and more dispersed compared to that observed at 1 d. Conversely, the Friedel's salt diffraction peak is clearly discernible at 1 d and remains consistently visible as hydration progresses. This implies that Friedel's salt primarily forms during the hydration period from 6 h to 1 d and does not decompose during subsequent hydration stages. Furthermore, the diffraction peak of brucite ($Mg(OH)_2$) is observable in the XRD pattern. However, no diffraction peaks for $Ca(OH)_2$ or $CaCl_2$ were detected. This can be attributed to the fact that the LCCM inherently produces minimal $Ca(OH)_2$ during hydration, which rapidly reacts with other hydration products and thus does not interact with salt ions in seawater to form corresponding compounds. Additionally, the XRD pattern after 3 d reveals that the diffraction peak of brucite increases with extended hydration time.

Figure 6.6a illustrates the XRD pattern of Portland cement (OPC-FN) with a hydration period of 28 d under fresh water mixed with standard curing. As depicted in the Fig. 6.6a, the primary hydration products of OPC-FN comprise AFt, $Ca(OH)_2$, and AFm, while the incomplete hydration of C_2S is evident. Since C-S-H gel lacks a crystalline structure, its diffraction peak does not appear on the XRD pattern. At 28 d, the presence of C_2S was still evident, suggesting that the hydration process proceeded slowly and did not fully engage in the reaction. Moreover, the diffraction peak intensity of AFt was lower compared to that of LCCM, indicating that AFt is not a primary hydration product of Portland cement. Simultaneously, the diffraction peak intensity of AFm was higher than that of LCCM, implying that AFm may possess a stronger ability to bind chloride ions

As illustrated in Fig. 6.6b–d, the XRD diffraction patterns of OPC-SN, OPC-SS, and OPC-FS samples at 28 d revealed the presence of Friedel salt diffraction peaks, while AFm diffraction peaks were nearly absent. This observation suggests that seawater reacts with AFm in the OPC system. Furthermore, the XRD patterns of all three sample groups exhibited diffraction peaks corresponding to $Ca(OH)_2$ and $CaCl_2$, while the intensity of $Ca(OH)_2$ diffraction peaks was notably reduced,

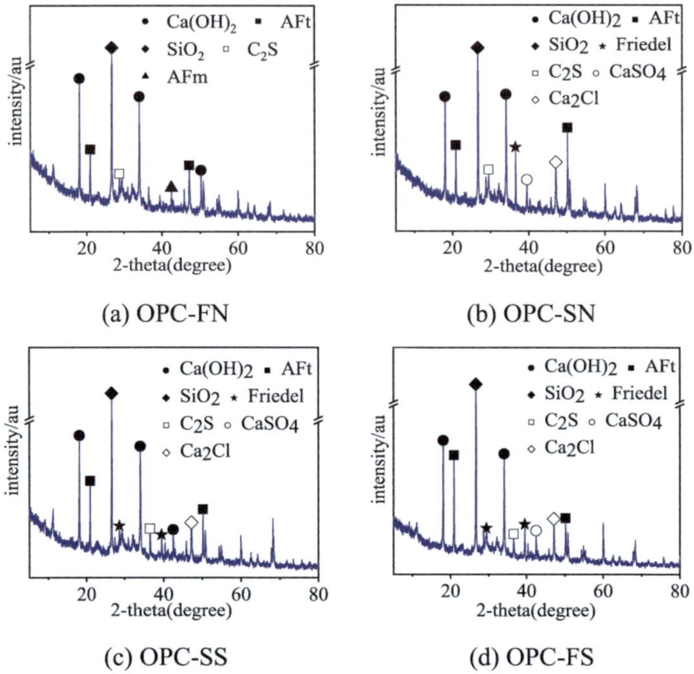

Fig. 6.6 XRD patterns of Portland cement

indicating that seawater ions also reacted with $Ca(OH)_2$. Notably, the diffraction intensity of AFt in these three groups was higher than that in the OPC-FN group, with an increased number of diffraction peaks. Specifically, the AFt diffraction intensity in the OPC-SS and OPC-SN groups was higher than that in the OPC-FS group. These findings demonstrate that the incorporation of seawater, particularly through mixed with seawater, promotes the precipitation of AFt in the hydration products of Portland cement.

6.4.2 Thermogravimetric Analysis

Figure 6.7 illustrates the TG-DSC curve for fresh water mixied with standard curing of the group FN. At 1d, a distinct dehydration endothermic peak of AFt is observed within the temperature range of 120–160 °C, indicating significant AFt formation during the early hydration phase. In the temperature range of 160–220 °C, the dehydration endothermic peak of AFm is evident and remains stable throughout the hydration period from 1 to 28 d. Additionally, the endothermic peak corresponding to AH_3 is detected in the range of 220–280 °C. Within the temperature range of 600–800 °C, the endothermic peak associated with

$CaCO_3$ is identified. Furthermore, the characteristic peak of AFm-CO_3 is successfully recognized in the range of 800–900 °C. By comparing the TG curves at different hydration ages, it is revealed that the endothermic peak intensity of AFt at 3 d is significantly higher than that at the 1 d, suggesting ongoing reactions of $C_4A_3\hat{S}$ during the early hydration phase. Although AFt exhibits a slight increasing trend during the 3–28 d, the increment is relatively minor.

Figure 6.8 presents the TG-DSC curve for seawater mixed with standard curing of the the group SN. The results indicate that, the peak value of AFt in the SN group is higher at 1 d. However, as the hydration age increases, the variation amplitude of the AFt peak value becomes relatively small, and the changes in both the peak value and mass loss rate of AFt remain stable and closely aligned between the ages of 3 d and 28 d. Furthermore, compared with the FN group, the SN group exhibits a lower peak value of AFm, along with a relatively smaller mass loss rate.

Figure 6.9 presents the TG-DSC curve for seawater mixed with seawater curing of the group SS. It is evident that AFt exhibits a higher dehydration heat absorption peak at an age of 1 d, which is comparable to the SN group. However, the variation in the peak value of AFt dehydration absorption remains relatively small

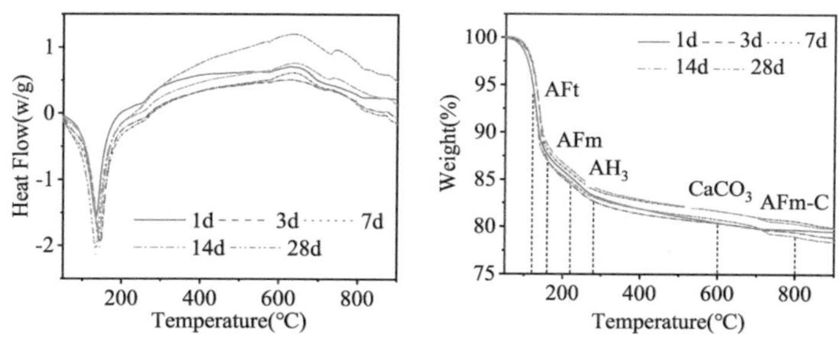

(a) DSC curve (b) TG curve

Fig. 6.7 TG-DSC curve of FN group

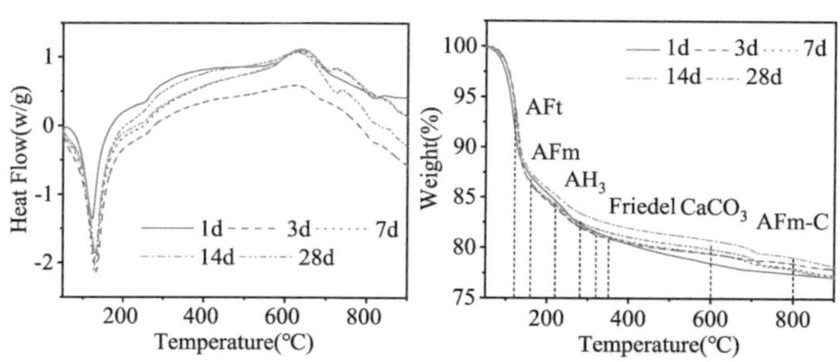

(a) DSC curve (b) TG curve

Fig. 6.8 TG-DSC curve of SN group

from 1 to 28 d. Additionally, it was observed that the endothermic peak of AH_3 was minimal, suggesting that salt ions in seawater reacted with AH_3, thereby facilitating the formation of AFt. Based on the TG-DSC curve of SS and SN, it can be concluded that seawater mixing promotes AFt precipitation to some extent while partially consuming the hydration products AH_3 and AFm. Furthermore, the dehydration heat absorption peak (320–350 °C) of Friedel's salt was detected in both SN and SS samples, albeit not prominently.

Figure 6.10 illustrates the TG-DSC curves for fresh water mixed with seawater curing of the group FS. The analysis revealed that the curves for the FS group were largely consistent with those of the FN group, indicating minimal differences in the mass loss rates of hydration products such as AFt and AH_3, as well as the absence of the dehydration heat absorption peak (320–350 °C) associated with Friedel's salt. These findings suggest that early-stage seawater curing exerts limited influence on the types and contents of hydration products. Furthermore, this result corroborates the excellent anti-seawater erosion properties of LCCM.

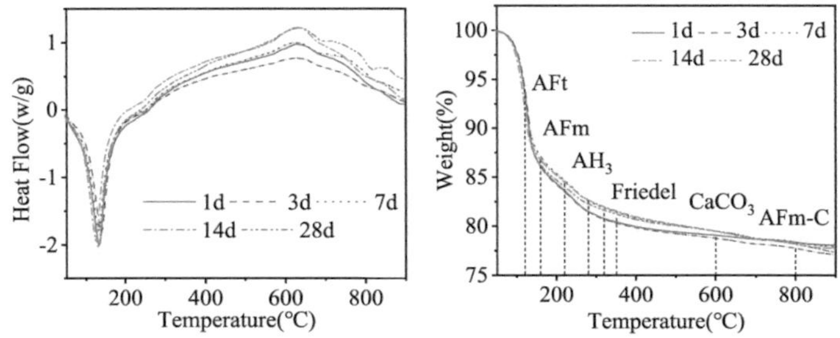

(a) DSC curve (b) TG curve

Fig. 6.9 TG-DSC curve of SS group

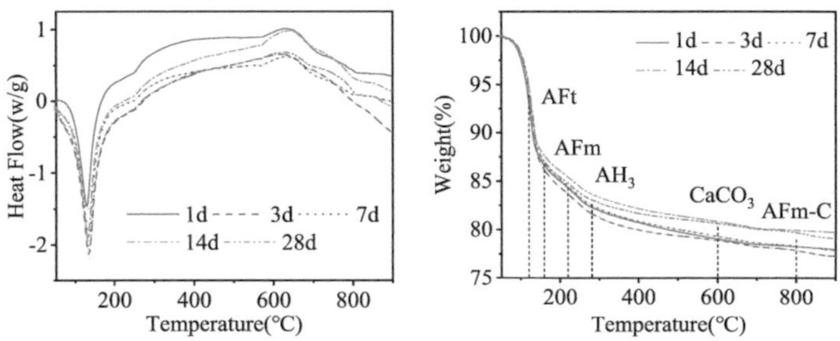

(a) DSC curve (b) TG curve

Fig. 6.10 TG-DSC curve of FS group

6.4.3 Molecular Structure of Hydration Products

Figure 6.11 presents the infrared spectra of LCCM, with spectral curves recorded within the wavenumber range of 400 to 4000 cm^{-1}. The hydration ages encompass 6 h, 1 d, 3 d, 7 d, 28 d, and 90 d. Through the analysis of peak intensity variations in specimens cured under fresh water and seawater, the presence of hydration products were further substantiated.

Higher absorption peaks (or lower transmission peaks) indicate a higher concentration of the corresponding molecules (Omrani et al., 2023). The characteristic peaks at wavenumbers of 535, 617, 1115, and 3431 cm^{-1} reflect the presence of AFt. Specifically, the absorption peak at 535 cm^{-1} is caused by the strong vibration of the Al-O bond in the AlO_4 group; the absorption peak at 617 cm^{-1} is due to the vibration of the S-O bond in the SO_4^{2-} group; the absorption peak at 1115 cm^{-1} is attributed to the asymmetric stretching vibration of the SO_4 group; and the absorption peak at 3431 cm^{-1} is caused by the stretching vibration of the O-H bond in the H_2O molecule (Horgnies et al., 2013). Additionally, the waveform changes in the ranges of 1600–1700 and 3200–3700 cm^{-1} respectively reflect the intensity characteristics of the O-H bond in the H_2O molecule and the bending vibration characteristics of the H-O-H bond (Haha et al., 2011; Lemougna et al., 2017; Bayat et al., 2018). The absorption peak at 3460 cm^{-1}, caused by the vibration of the O-H bond, further confirms the presence of AH_3. The changes at 617, 1115 cm^{-1}, and in the 1100–1200 cm^{-1} region not only reflect the stretching vibration characteristics of the S-O bond in the SO_4^{2-} group in AFt, but are also closely related to the presence of AFm, $CaSO_4$, and brucite. Meanwhile, the stretching vibration characteristics of the Si-O

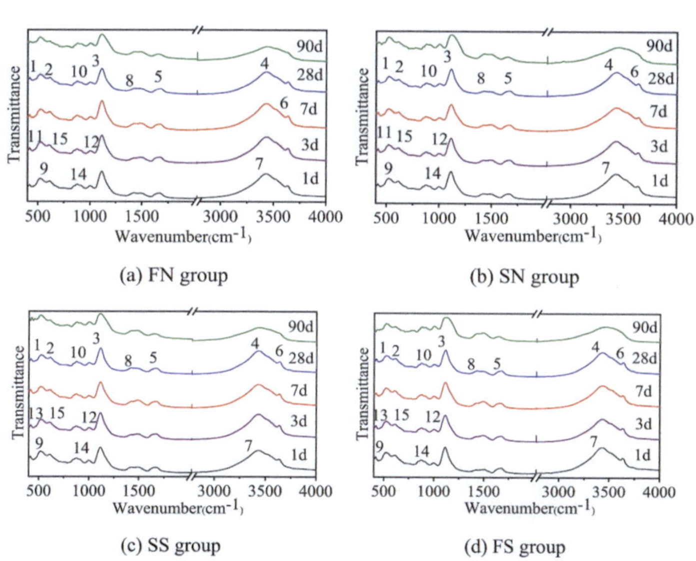

Fig. 6.11 FTIR spectrum of LCCM mixed with seawater

bond at 450 and 970 cm^{-1} further verify the existence of C-S-H gel (Tan et al., 2019). For the SN and SS group mixed with seawater, a minor peak caused by the Al-O lattice vibration in the $[Al(OH)_6]^{3-}$ group at 463 cm^{-1} was found in their infrared spectra, which confirms the presence of Friedel salt (Shao et al., 2013). Notably, this minor peak only appears in the specimens mixed with seawater, while other spectral bands or peaks that might be used to confirm the existence of Friedel salt cannot be clearly identified due to overlapping with those of other hydration products. Moreover, the relevant characteristic peaks of unhydrated clinker minerals were also observed in the figure, such as the absorption peak at 882 cm^{-1} caused by the asymmetric stretching vibration of the (AlO_4) group in $C_4A_3\hat{S}$, and the absorption peak at 617 cm^{-1} caused by the coupled vibration of the AlO_4 and SO_4 groups.

Furthermore, comparative analysis revealed that the AFt peak intensity in the infrared spectra of the SN and SS groups was notably more pronounced, suggesting that seawater exerts a certain promoting effect on the hydration reaction. The relatively weaker peak intensity of AH_3 at 3640 cm^{-1} in the SN and SS groups provides further evidence that salt ions in seawater may have undergone reactions with the hydration products.

6.4.4 Microstructural of Hydration Products

As shown in Fig. 6.12, the SEM images of the FN group are presented. A large number of needle-like and rod-like AFt of various lengths and thicknesses can be observed at 1 d, indicating that the $C_4A_3\hat{S}$ rapidly undergoes hydration reaction at the initial stage of hydration, generating a large amount of AFt, which leads to a disordered distribution and subtle differences in morphology. At the same time, a small amount of spherical AH_3 and unhydrated $C_4A_3\hat{S}$ can be observed, and there are many small pores inside the hardened paste. The content of $C_4A_3\hat{S}$ significantly decreases at 3 d, and AFt is more columnar, with more obvious interlacing. Meanwhile, a large amount of AH3 are mutually cemented, making the structure more compact, and the pores and cracks begin to gradually decrease. AFt is widely distributed and aggregated at 28 d, forming a uniform and thick columnar shape, and almost no unhydrated $C_4A_3\hat{S}$ can be observed. At the crack of the sample, AFt fully plays the cementing role, beginning to cover and fill the cracks. The aggregation of AFt becomes more compact, and its length and thickness gradually tend to be consistent at 90 d, forming a large area of uniform columnar AFt, and tightly wrapping the aggregates. Although the cracks have not been completely filled, the AFt covering the cracks significantly increases and is longer.

Figure 6.13 presents the SEM images of the SS group under different hydration age. A significant amount of AFt is observed at 1 d. Compared with the FN group, the AFt in the SS group exhibits a relatively regular and robust needle-like morphology, with additional spherical substances at the tips, suggesting that seawater influences the microstructure of AFt. Moreover, no spherical AH_3 was detected, indicating that AH_3 rapidly reacts upon the introduction of seawater, thereby further promoting AFt

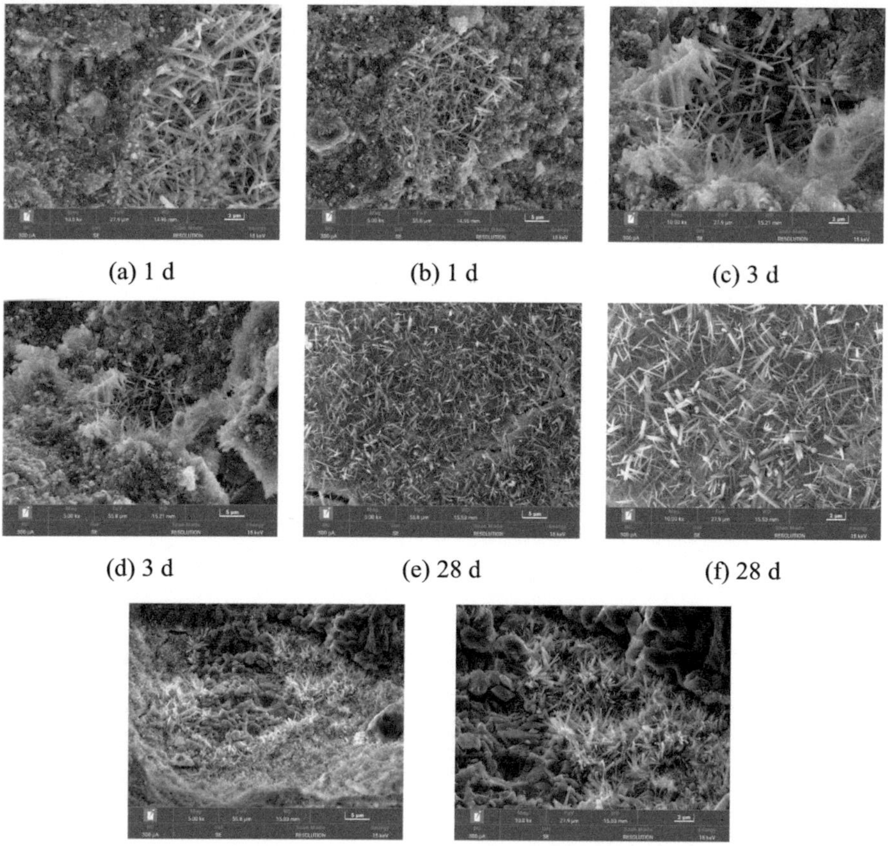

(a) 1 d (b) 1 d (c) 3 d

(d) 3 d (e) 28 d (f) 28 d

(g) 90 d (h) 90 d

Fig. 6.12 SEM images of FN group

formation. In comparison to the FN group, the AFt distribution in the SS group is more compact and the overall structure is more intricate. The AFt in the SS group demonstrates a uniform and widespread distribution, with reduced crystal growth space and a denser structure compared to the FN group at 28 d. When the age reached 90 d, the structural compactness further increased, and the AFt crystals changed from needle-like and rod-like to columnar.

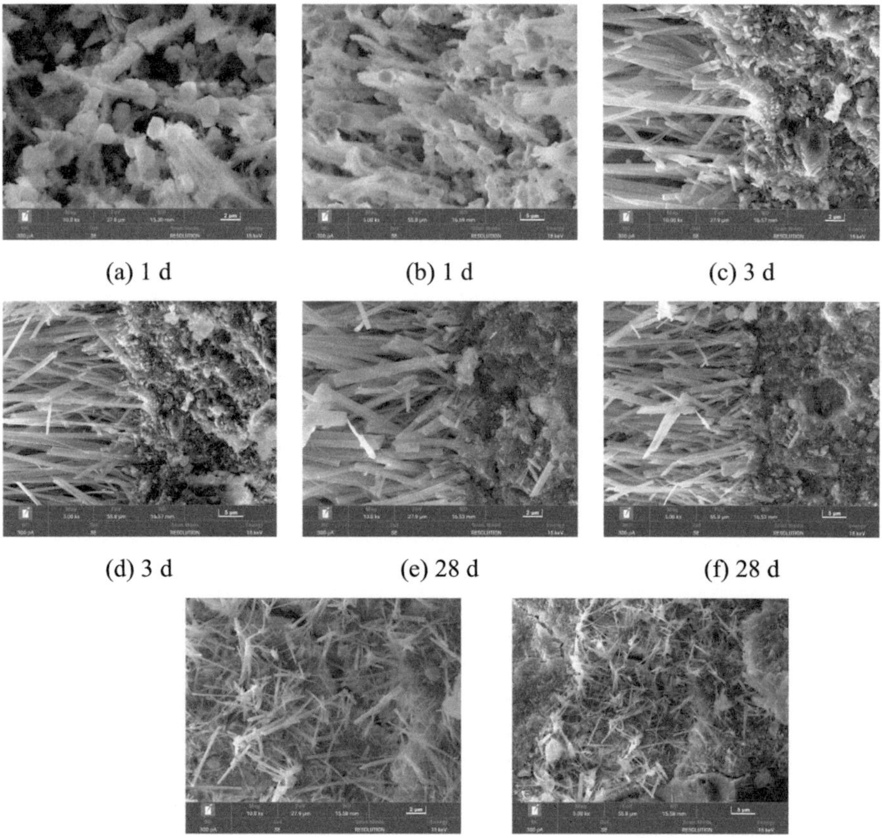

(a) 1 d (b) 1 d (c) 3 d

(d) 3 d (e) 28 d (f) 28 d

(g) 90 d (h) 90 d

Fig. 6.13 SEM images of SS group

6.5 Compressive Strength of LCCM Mixed with Seawater

6.5.1 Compressive Strength Under Different Mixing and Curing Conditions

The compressive strength results of LCCM are presented in Fig. 6.14. As illustrated in Fig. 6.14, with the increase in curing age, the compressive strength of specimens under various mixing and curing conditions exhibits a continuous upward trend without any occurrence of strength regression. Moreover, during the early hydration stage (1–3 d), the strength development of all groups is notably rapid. In contrast, during the later hydration stage (after 28 d), the growth rate of compressive strength gradually becomes more moderate.

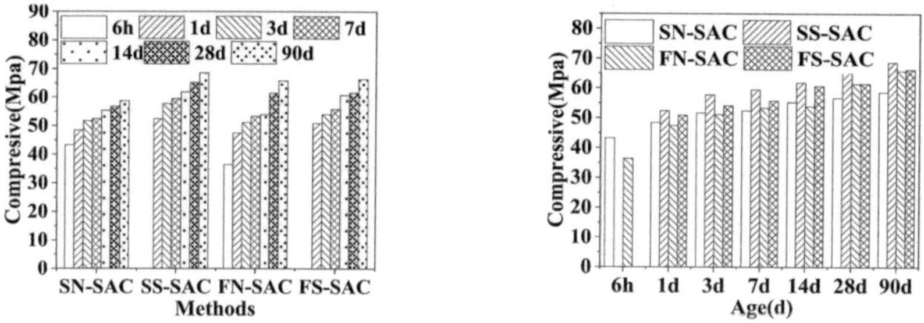

(a) Compressive strength under different mixing and curing conditions

(b) Compressive strength under different hydration age

Fig. 6.14 Compressive strength of LCCM mixed with seawater

The compressive strengths of the FN, FS, SN, and SS groups were recorded as 47.47, 50.94, 48.44, and 52.41 MPa at the age of 1 d, respectively. By the age of 3 d, these values had increased to 51.12, 54.07, 51.67, and 57.74 MPa, respectively. At this stage, the compressive strengths had already reached approximately 80% of the strength at 28 days and about 75% of the strength at 90 days, demonstrating a significant characteristic of high early-age strength. This phenomenon can be attributed to the substantial presence of $C_4A_3\hat{S}$ in LCCM, which reacts rapidly during the initial hydration phase to form AFt, thereby providing critical support for the development of early-age strength.

The compressive strength results of Portland cement are presented in Fig. 6.15. As illustrated in Fig. 6.15, the trend in the development of compressive strength for Portland cement under various mixing and curing conditions is largely consistent with that of LCCM. However, overall, the compressive strength of Portland cement at all ages is lower than that of LCCM.

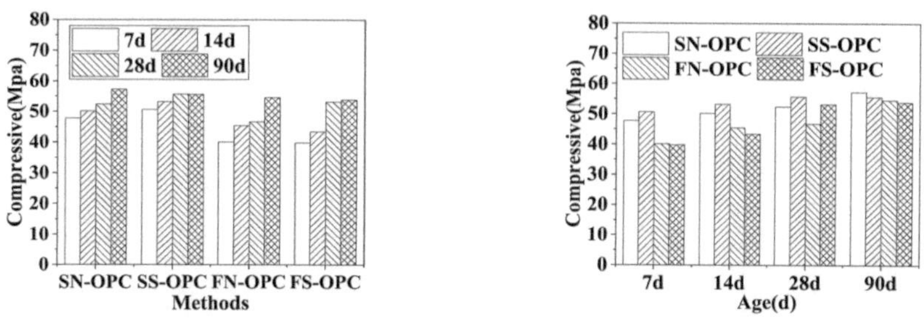

(a) Compressive strength under different mixing and curing conditions

(b) Compressive strength under different hydration age

Fig. 6.15 Compressive strength of Portland cement mixed with seawater

As shown in Fig. 6.15, at the age of 7 d, the compressive strength of the OPC-SN and OPC-SS specimens is higher than that of the OPC-FN specimens. However, after 7 d, the compressive strength of both OPC-SN and OPC-SS specimens becomes lower than that of OPC-FN specimens. Moreover, from 28 to 90 d, the growth in compressive strength is relatively limited. Additionally, the compressive strength of the OPC-FS is nearly identical to that of OPC-FN specimens during the 28 d. From 28 to 90 d, both the compressive strength and its growth rate are relatively low. At the age of 90 d, the compressive strength of OPC-FS is lower than that of OPC-FN. These findings suggest that seawater has a certain promoting effect on the early-age compressive strength of OPC but exerts an inhibitory influence on its later-stage strength development.

6.5.2 *Mechanism Analysis Chloride Ion Influence on Compressive Strength in Seawater*

The primary hydration products of LCCM comprise AFt, AH_3, AFm, and AFm-CO_3. Among these, AFt is abundantly present in specimens under various curing conditions and forms rapidly during the early hydration phase. It serves not only as a critical source of strength for cement paste but also as a key contributor to enhanced early-age strength. In comparison with Portland cement, LCCM demonstrate a more rapid rate of strength development and achieve higher strength levels, primarily attributed to the substantial presence of AFt.

Taking the compressive strength of samples under different curing conditions at different ages as the reference sequence:

$$Y_n = [y_n(1), y_n(2), ..., y_n(k)] \tag{6.2}$$

where: Y_n represents the compressive strength of LCCM at different curing ages. $y_n(k)$ represents the compressive strength of LCCM at the kth curing age.

Taking the mass loss rates of AFt (Z_1), AH_3 (Z_2), AFm (Z_3), AF-CO_3 (Z_4), and Friedel's salt (Z_5) as the comparison sequence, as shown in Eq. 6.3:

$$Z_n = [z_n(1), z_n(2), ..., z_n(k)] \tag{6.3}$$

where: Z_n is the mass loss rate of different hydration products. $z_n(k)$ is the mass loss rate of different hydration products at the kth curing age.

The original sequence of samples under different mixing and curing conditions is shown in Table 6.3.

The original sequence in Table 6.3 is subjected to mean normalization as follows:

Table 6.3 Original sequence of samples under different mixing and curing conditions

Number	Curing age	1 d	3 d	7 d	14 d	28 d
FN	Z1/%	7.618	8.261	8.6	9.071	9.342
	Z2/%	2.183	2.354	2.407	2.493	2.551
	Z3/%	1.932	1.85	1.777	1.853	1.666
	Z4/%	0.202	0.489	0.535	0.544	0.67
	Z5/%	\	\	\	\	\
	Y/MPa	47.47	51.12	53.3	53.92	61.4
FS	Z1/%	6.4	7.275	7.54	7.75	8.125
	Z2/%	2.288	2.43	2.423	2.462	2.506
	Z3/%	1.987	2.029	1.914	1.725	1.668
	Z4/%	0.205	0.256	0.383	0.531	0.567
	Z5/%	\	\	\	\	\
	Y/MPa	50.94	54.07	55.77	60.69	61.5
SN	Z1/%	8.117	8.581	8.723	8.813	8.927
	Z2/%	2.424	2.358	2.322	2.225	2.156
	Z3/%	2.309	1.935	1.919	1.814	1.74
	Z4/%	0.358	0.522	0.595	0.646	0.641
	Z5/%	0.311	0.325	0.33	0.345	0.348
	Y/MPa	48.44	51.67	52.38	55.24	56.72
SS	Z1/%	7.955	8.497	8.73	8.921	9.341
	Z2/%	2.441	2.5	2.383	2.178	2.187
	Z3/%	2.088	2.023	2.07	1.815	1.734
	Z4/%	0.34	0.468	0.412	0.522	0.515
	Z5/%	0.352	0.371	0.401	0.397	0.38
	Y/MPa	52.41	57.74	59.44	61.77	65.2

$$\begin{cases} z_n{\prime} = \dfrac{Z_n}{z_n(a)} = (z_n{\prime}(1), z_n{\prime}(2), \ldots, z_n{\prime}(k)) \\ y_m{\prime} = \dfrac{Y_m}{y_m(a)} = (y_m{\prime}(1), y_m{\prime}(2), \ldots, y_m{\prime}(k)) \end{cases} \tag{6.4}$$

where: $z'_{\mathrm{n}}(k)$ is the average value of $z_1(k), z_2(k), \ldots, z_{\mathrm{n}}(k)$. $y'_{\mathrm{m}}(k)$ is the average value of $y_1(k), y_2(k), \ldots, y_{\mathrm{n}}(k)$.

Calculate the difference between the reference sequence and the comparison sequence.

$$\begin{cases} \Delta_{mn}(k) = |y_m{\prime}(k) - x_n{\prime}(k)| \\ \Delta_{mn} = (\Delta_{mn}(1), \Delta_{mn}(2), \ldots, \Delta_{mn}(k)) \\ (m = 1, 2, \ldots, l; n = 1, 2, \ldots, q; k = 1, 2, \ldots., p) \end{cases} \tag{6.5}$$

Calculate the maximum difference value and the minimum difference value:

$$\begin{cases} M = \max\limits_{m} \max\limits_{n} \Delta_m(k) \\ m = \min\limits_{m} \min\limits_{n} \Delta_m(k) \end{cases} \tag{6.6}$$

Calculate the correlation coefficient:

$$\gamma_{nm} = \frac{m + \zeta M}{\Delta_n(k) + \zeta M} \tag{6.7}$$

where: γ_{nm} is the resolution coefficient, which is generally taken as 0.5.

Calculate the grey entropy correlation density:

$$P_h = \frac{\gamma[y_n(h), z_n(h)]}{\sum\limits_{k=1}^{n} \gamma[y_n(h), z_n(h)]} (h = 1, 2, ..., n) \tag{6.8}$$

Calculate the grey correlation entropy and the grey entropy correlation degree:

$$H = -\sum_{k=1}^{n} P_n \ln P_n \tag{6.9}$$

$$E = \frac{1}{n} H \tag{6.10}$$

The grey relational entropy and grey entropy correlation coefficient between the mass loss rate of hydration products and the compressive strength of LCCM were derived through mean normalization processing and difference analysis. The detailed results are presented in Table 6.4. These findings indicate that the mass loss rate of AFt exerts the most significant influence on the compressive strength of mortar specimens prepared with LCCM under various curing conditions. Specifically, the grey entropy correlation coefficients for the four specimen groups are 0.997, 0.999, 0.998, and 0.991, respectively. This phenomenon can be attributed to the fact that the primary mineral constituents of LCCM include $C_4A_3\hat{S}$, which reacts rapidly during the early hydration phase to produce a substantial volume of AFt. Additionally, given the relatively low content of gel-like substances in the hydration products, AFt serves as the predominant contributor to the compressive strength of LCCM.

Table 6.4 Grey entropy calculation results of hydration products and compressive strength

Number	Curing age	Sequence values					Grey relational entropy	Grey entropy correlation degree
		1 d	3d	7d	14d	28d		
FN	Z1/%	0.332	0.329	0.330	0.310	0.305	1.605	0.997
	Z2/%	0.326	0.325	0.333	0.322	0.299	1.605	0.997
	Z3/%	0.287	0.328	0.345	0.348	0.270	1.579	0.981
	Z4/%	0.240	0.354	0.337	0.335	0.299	1.566	0.973
	Z5/%	\	\	\	\	\	\	\
FS	Z1/%	0.315	0.322	0.319	0.321	0.332	1.608	0.999
	Z2/%	0.321	0.319	0.337	0.315	0.317	1.607	0.999
	Z3/%	0.307	0.319	0.355	0.314	0.297	1.592	0.989
	Z4/%	0.272	0.292	0.366	0.292	0.272	1.494	0.929
	Z5/%	\	\	\	\	\	\	\
SN	Z1/%	0.316	0.313	0.340	0.318	0.319	1.606	0.998
	Z2/%	0.293	0.331	0.349	0.326	0.294	1.593	0.990
	Z3/%	0.282	0.311	0.331	0.345	0.327	1.596	0.991
	Z4/%	0.250	0.343	0.365	0.297	0.287	1.542	0.958
	Z5/%	0.292	0.339	0.329	0.316	0.325	1.601	0.995
SS	Z1/%	0.279	0.335	0.342	0.323	0.316	1.595	0.991
	Z2/%	0.262	0.353	0.357	0.306	0.284	1.561	0.970
	Z3/%	0.258	0.365	0.340	0.310	0.272	1.546	0.960
	Z4/%	0.334	0.330	0.301	0.299	0.337	1.594	0.990
	Z5/%	0.287	0.344	0.326	0.343	0.291	1.590	0.988

6.6 Pore Structure of LCCM Mixed with Seawater

6.6.1 Porosity

Figure 6.16 shows the variation of porosity of LCCM with curing age under different curing conditions. By comparing the porosity of specimens under different curing conditions at the same age, it can be found that the porosity of the two groups of specimens mixed with seawater is significantly lower than that of the specimens mixed with fresh water. Specifically, the porosity difference between FN and SN specimens is 0.34%, and that between FS and SS specimens is 0.37% at 1 d; while the differences are only 0.06% and 0.11% at 90 d. This indicates that seawater has a certain optimizing effect on the porosity of specimens, but this effect is mainly reflected in the early stage of hydration, and gradually weakens and tends to disappear as hydration continues.

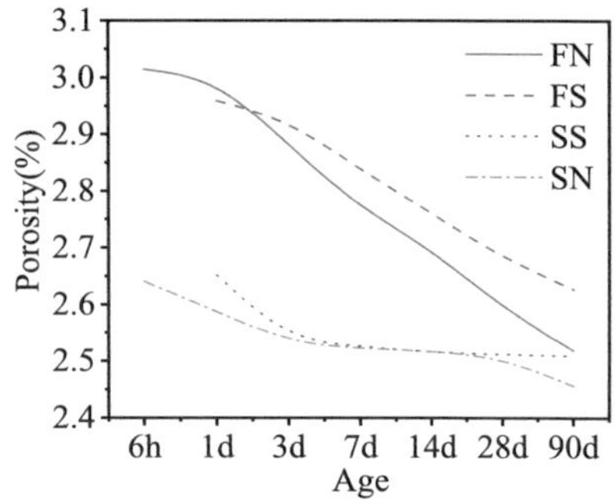

Fig. 6.16 Change law of porosity of LCCM mixed with seawater

Further analysis of the compressive strength variation of the specimens before the 3 d reveals a negative correlation between the porosity and the compressive strength. Among them, the SN and SS group, due to their lower porosity, exhibit higher compressive strength; while the FN and FS group specimens, with higher porosity, have relatively lower compressive strength. Additionally, in the early stage of hydration, the porosity reduction rate of the specimens mixed with seawater is similar to that of the specimens mixed with freshwater. However, as the age increases, the compressive strength growth rate of the SN group, which have the lowest porosity, begins to slow down, and the porosity reduction is relatively small. In contrast, the compressive strength of the other three groups continues to increase steadily, and their porosity reduction is relatively large. Nevertheless, the porosity reduction rate of the specimens mixed with seawater significantly decreases after 14 d, almost all falling below 1%.

The above results indicate that although seawater used as mixing water can accelerate the precipitation of AFt in the early stage of hydration, thereby effectively reducing the porosity of the specimens, its effect on the continuous development of porosity and the continuous growth of compressive strength is limited. For the SS group mixed with seawater and cured with seawater, due to the use of seawater for curing in the early stage of hydration, their porosity is slightly higher than that of the SN group. However, as hydration proceeds, the decrease in porosity is greater than that of the SN group, so the density of the specimens continuously increases, which also has a certain optimizing effect on the increase in compressive strength. As for the FS group mixed with fresh water but cured with seawater, their porosity is the highest among the four groups of specimens at any age, so their compressive strength is always relatively low.

6.6.2 T_2 Spectrum

Figure 6.17 illustrates the T2 spectrum distribution of LCCM mixed with seawater at the age range of 1–90 d. As observed from the figure, under varying curing conditions, the distribution curves of LCCM mixed with seawater exhibit a generally consistent trend, with three distinct peaks. Specifically, the first peak for the FN, FS, SN, and SS groups is located at 40, 42, 35, and 33 m/s, respectively. The second peak occurs at positions of 2.5, 2.9, 4.0, and 3.5 m/s, respectively. Due to the difficulty in accurately determining the position of the third peak, its specific location is not further analyzed. The first, second, and third peaks correspond to small pores, medium pores, and large pores, respectively. The significantly higher amplitude of the first peak compared to the other two indicates that the volume of small pores within the specimens is the largest.

By analyzing the area enclosed by each peak in the T2 spectrum, it can be observed that when the curing age extends from 14 to 90 d, the average areas of the first peak for the FN, FS, SN, and SS groups decrease by 5.29, 4.76, 1.49, and 1.52%, respectively; the average areas of the second peak decrease by 1.88, 1.73, 1.74, and 1.76%, respectively. The reduction in peak area for specimens mixed with seawater is smaller compared to those mixed with fresh water. After 14 days, the rate of decrease in the area of the second peak becomes significantly slower. This phenomenon suggests that as the curing age increases, the internal structure of the cement paste becomes increasingly stable, leading to minimal changes in the internal pore volume.

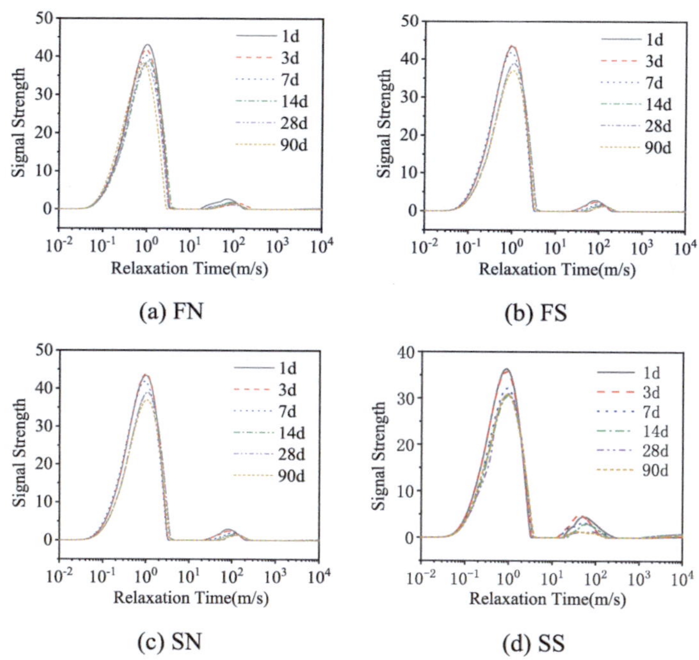

Fig. 6.17 T_2 spectrum of LCCM mixed with seawater

6.6.3 Pore Size Distribution

The relationship between pore size and relaxation time T_2 can be represented by Eq. 6.15.

$$\frac{1}{T_2}=\frac{1}{T_s}+\rho\left(\frac{S}{V}\right)+\frac{D(\gamma GT_E)}{12} \tag{6.11}$$

where: T_s represents the free relaxation time of the pore fluid, ms. ρ denotes the transverse surface relaxation strength of T_2, μm/ms. S is the surface area of the pores, cm^2. *V* is the volume of the fluid within the pores, cm^3. D represents the diffusion coefficient. γ is the gyromagnetic ratio, rad/(s·T). G represents the magnetic gradient, Gs/cm. T_E is the echo time, measured in milliseconds, ms.

When only one fluid exists in the internal pores of the concrete material and the volume relaxation time is less than the size relaxation time, T_s can be ignored. Meanwhile, if the magnetic field is uniform, the magnetic gradient is extremely small, and a short echo time is adopted, the diffusion time can also be disregarded. Therefore, Eq. 6.11 can be simplified to Eq. 6.12:

$$\frac{1}{T_2}=\rho\left(\frac{S}{V}\right) \tag{6.12}$$

According to Eq. 6.12, the relaxation time is directly proportional to the pore diameter, meaning that the larger the relaxation time, the larger the corresponding pore diameter. Here, the transverse surface relaxation intensity ρ varies depending on the sample characteristics. For concrete materials, ρ typically ranges from 0.003 to 0.01 μm/ms. In this study, it is assumed that $\rho = 0.05$ μm/ms, and further assuming that the pores are ideal spheres, the following formula can be derived:

$$\frac{S}{V}=\frac{3}{r} \tag{6.13}$$

Figure 6.18a–d respectively depict the pore size distribution of LCCM in the FN, FS, SN, and SS groups. It can be observed that the pore size distribution curves primarily exhibit two peaks, and their trends are consistent with those reflected by the T_2 spectra. Specifically, the area enclosed by the first peak is the largest, while the area enclosed by the second peak is relatively smaller. By comparison, the first peak for the FN and FS groups is higher than that of the SN and SS groups; however, during the early hydration stage, the situation for the second peak is reversed. As the hydration process progresses, the first peak of the FN and FS groups gradually decreases with increasing age but remains higher than the corresponding peak of the SN and SS groups. Additionally, the second peak of the SN and SS groups exhibits a significant decrease, and by the later stages of hydration, its value becomes largely consistent with the second peak of the FN and FS groups.

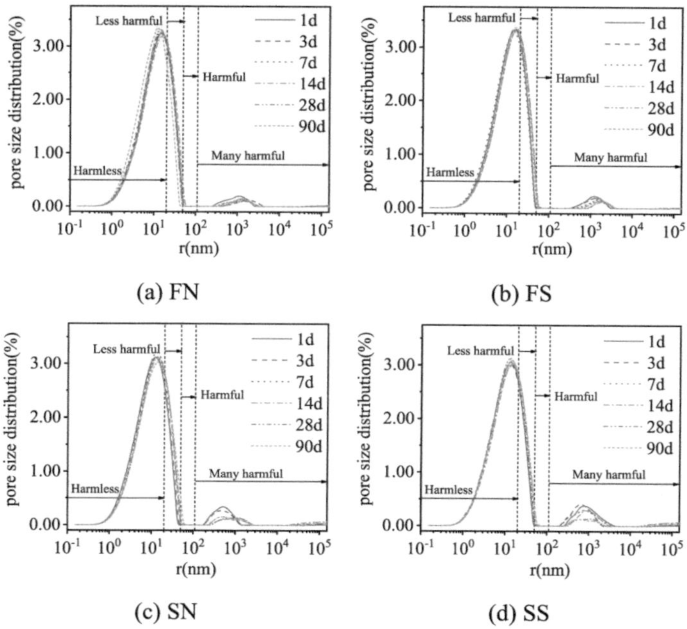

Fig. 6.18 The pore size distribution of LCCM under different mixing and curing conditions

The pore structure can be categorized into four types: harmless pores (pore diameter < 20 nm), slightly harmful pores (20 nm ≤ pore diameter < 50 nm), harmful pores (50 nm ≤ pore diameter < 200 nm), and highly harmful pores (pore diameter ≥ 200 nm). By utilizing Eqs. 6.15–6.17 to calculate the pore size distribution of the specimens, a detailed quantitative analysis was performed regarding the proportions of pore structures under various curing conditions. The specific results are presented in Table 6.5.

As shown by the data analysis in Table 6.5, the proportion of harmless pores in LCCM under different curing conditions exhibits an increasing trend with age. Specifically, within the age range of 1–90 d, the proportions of harmless pores in the FN, FS, SN, and SS groups increased by 1.9, 2.73, 1.365, and 2.701%, respectively, indicating that seawater curing positively promotes the growth of harmless pore proportions. For less harmful pores, the proportion changes for the FN, FS, SN, and SS groups during 1–90 d increased by −0.371, −0.407, 0.96, and 0.111%, respectively. It can be seen that seawater mixing has a certain promoting effect on the development of less harmful pores. Regarding harmful pores, the proportion changes for the FN, FS, SN, and SS groups increased by −0.264, −0.433, 0.487, and 1.104% at 1–90 d, respectively, further confirming that seawater mixing promotes the growth of harmful pores. Moreover, the trend of highly harmful pores indicates that the FN, FS, SN, and SS groups decreased by 1.289, 1.89, 3.262, and 3.988%, respectively. It is evident that the highly harmful pores of all four groups decreased to

Table 6.5 The proportion of pore structure of LCCM under different mixing and curing conditions

Curing age		1 d	3 d	7 d	14 d	28 d	90 d
FN	Harmless pores	65.544	66.015	66.005	66.640	66.996	67.467
	Less harmful pores	27.577	27.175	27.989	27.749	27.588	27.206
	Harmful pores	1.029	0.410	0.748	0.774	0.715	0.765
	Highly parmful pores	5.850	6.399	5.258	4.837	4.701	4.562
FS	Harmless pores	66.022	66.462	66.452	67.232	68.312	68.752
	Less harmful pores	26.131	25.882	26.67	26.325	26.268	25.724
	Harmful pores	1.895	1.317	1.633	1.47	1.415	1.462
	Highly harmful pores	5.952	6.339	5.245	4.973	4.005	4.062
SN	Harmless pores	62.011	62.231	62.226	62.616	63.156	63.376
	Less harmful pores	28.271	28.969	29.58	29.468	29.653	29.681
	Harmful pores	1.271	1.282	2.952	1.764	1.712	1.758
	Highly harmful pores	8.447	7.518	5.242	6.152	5.479	5.185
SS	Harmless pores	61.933	62.075	62.769	63.148	64.592	64.634
	Less harmful pores	27.961	28.034	27.764	27.954	27.947	28.072
	Harmful pores	0.85	0.737	1.774	1.738	1.823	1.954
	Highly harmful pores	9.352	9.175	7.726	7.182	5.651	5.364

varying degrees, with the two seawater-mixed groups (SN and SS) showing a more pronounced downward trend.

6.6.4 *Relationship Between Pore Structure and Compressive Strength*

The influence of harmless pores, less harmful pores, harmful pores, and highly harmful pores on the compressive strength of LCCM mixed with seawater was quantitatively analyzed by using the grey relational segment entropy analysis method. The original sequence is shown in Table 6.6.

Taking the values of harmless pores, less harmful pores, harmful pores, and highly harmful pores at different ages as influencing factors, the grey theory is used to analyze the influence of different pore structures on compressive strength. The values of compressive strength of samples under different mixing and curing conditions are taken as the reference sequence, as follows:

$$Y_n = [y_n(1), y_n(2), \ldots, y_n(k)] \tag{6.14}$$

where: Y_n represents the compressive strength values of LCCM mixed with seawater at different curing ages. $Y_n(k)$ is the compressive strength value at the kth curing age.

Table 6.6 Original sequence under different mixing and curing conditions

Number	Curing age	1 d	3 d	7 d	14 d	28 d	90 d
FN	X1/%	65.544	66.015	66.005	66.64	66.996	67.467
	X2/%	27.577	27.175	27.989	27.749	27.588	27.206
	X3/%	1.029	0.41	0.748	0.774	0.715	0.765
	X4/%	5.85	6.399	5.258	4.837	4.701	4.562
	Y/Mpa	47.47	51.12	53.3	53.92	61.4	65.74
FS	X1/%	66.022	66.462	66.452	67.232	68.312	68.752
	X2/%	26.131	25.882	26.67	26.325	26.268	25.724
	X3/%	1.895	1.317	1.633	1.47	1.415	1.462
	X4/%	5.952	6.339	5.245	4.973	4.005	4.062
	Y/Mpa	50.94	54.07	55.77	60.69	61.5	66.22
SN	X1/%	62.011	62.231	62.226	62.616	63.156	63.376
	X2/%	28.271	28.969	29.58	29.468	29.653	29.681
	X3/%	1.271	1.282	2.952	1.764	1.712	1.758
	X4/%	8.447	7.518	5.242	6.152	5.479	5.185
	Y/Mpa	48.44	51.67	52.38	55.24	56.72	58.63
SS	X1/%	61.933	62.075	62.769	63.148	64.592	64.634
	X2/%	27.961	28.034	27.764	27.954	27.947	28.072
	X3/%	0.85	0.737	1.774	1.738	1.823	1.954
	X4/%	9.352	9.175	7.726	7.182	5.651	5.364
	Y/Mpa	52.41	57.74	59.44	61.77	65.2	68.63

Taking the proportion of harmless pores (X_1), the proportion of less harmful pores (X_2), the proportion of harmful pores (X_3), and the proportion of highly harmful pores (X_4) as the comparison sequence, as follows:

$$X_n = [x_n(1), x_n(2), ..., x_n(k)] \tag{6.15}$$

where: X_n represents the proportion of pores with different sizes. $X_n(k)$ is the proportion of pores with different sizes at the kth curing age.

The original sequence in Table 6.6 is subjected to mean normalization as follows:

$$\begin{cases} x_n\prime = \dfrac{X_n}{x_n(a)} = (x_n\prime(1), x_n\prime(2), ..., x_n\prime(k)) \\ y_m\prime = \dfrac{Y_m}{y_m(a)} = (y_m\prime(1), y_m\prime(2), ..., y_m\prime(k)) \end{cases} \tag{6.16}$$

where: $x'_{\mathrm{n}}(k)$ represents the mean values of $x_1(k)$, $x_2(k)$, …, $x_{\mathrm{n}}(k)$. $y'_{\mathrm{m}}(k)$ represents the mean values of $y_1(k)$, $y_2(k)$, …, $y_{\mathrm{n}}(k)$.

Calculate the differences between the reference sequences and the comparative sequences:

$$\begin{cases} \Delta_{mn}(k) = |y_m\prime(k) - x_n\prime(k)| \\ \Delta_{mn} = (\Delta_{mn}(1), \Delta_{mn}(2), ..., \Delta_{mn}(k)) \\ (m = 1, 2, ..., l; n = 1, 2, ..., q; k = 1, 2, ..., p) \end{cases} \tag{6.17}$$

Calculate the maximum and minimum differences:

$$\begin{cases} M = \max_m \max_n \Delta_m(k) \\ m = \min_m \min_n \Delta_m(k) \end{cases} \tag{6.18}$$

Calculate the grey relational coefficients:

$$\gamma_{nm} = \frac{m + \zeta M}{\Delta_n(k) + \zeta M} \tag{6.19}$$

where: γ_{nm} is the distinguishing coefficient, generally taken as 0.5.

Calculate the grey entropy relational density:

$$P_h = \frac{\gamma[y_n(h), x_n(h)]}{\sum_{k=1}^{n} \gamma[y_n(h), x_n(h)]} (h = 1, 2, ..., n) \tag{6.20}$$

Calculate the grey relational entropy and grey entropy correlation degree:

$$H = -\sum_{k=1}^{n} P_n \ln P_n \tag{6.21}$$

$$E = \frac{1}{n} H \tag{6.22}$$

The grey relational entropy and grey entropy correlation degree of the pore structure and compressive strength are detailed in Table 6.7. As can be seen from the table, harmless pores have the greatest impact on the compressive strength of the material, with their grey entropy correlation degrees being 0.985, 0.988, 0.992 and 0.982. Less harmful pores, harmful pores, and highly harmful pores have an impact on the compressive strength in sequence. Through comparative analysis of the porosity and pore structure of FN, FS, SS, and SN groups, it was found that in the early stage of hydration, the two groups of specimens mixed with seawater (SS and SN) showed a higher proportion of harmless pores, indicating that seawater has a positive effect on promoting the development of pore structure in the early stage of hydration, thereby enhancing the compressive strength of these two groups of specimens in the

Table 6.7 Grey entropy calculation results of pore structure and compressive strength

Number	Curing age	P*ln(P)						Grey relational entropy	Grey entropy correlation degree
		1 d	3 d	7 d	14 d	28 d	90 d		
FN	X1/%	0.338	0.316	0.302	0.301	0.263	0.245	1.764	0.985
	X2/%	0.340	0.312	0.307	0.301	0.261	0.240	1.761	0.983
	X3/%	0.367	0.245	0.289	0.291	0.256	0.250	1.699	0.948
	X4/%	0.350	0.346	0.290	0.271	0.238	0.221	1.717	0.958
FS	X1/%	0.335	0.316	0.305	0.279	0.279	0.257	1.771	0.988
	X2/%	0.338	0.314	0.313	0.279	0.274	0.247	1.765	0.985
	X3/%	0.363	0.277	0.306	0.268	0.259	0.248	1.721	0.961
	X4/%	0.353	0.352	0.291	0.259	0.227	0.215	1.697	0.947
SN	X1/%	0.329	0.319	0.294	0.283	0.283	0.269	1.778	0.992
	X2/%	0.325	0.320	0.297	0.285	0.284	0.269	1.780	0.993
	X3/%	0.324	0.315	0.183	0.297	0.308	0.319	1.746	0.974
	X4/%	0.353	0.328	0.270	0.275	0.264	0.251	1.741	0.972
SS	X1/%	0.351	0.292	0.303	0.280	0.272	0.262	1.759	0.982
	X2/%	0.352	0.294	0.302	0.280	0.269	0.259	1.756	0.980
	X3/%	0.364	0.296	0.254	0.275	0.272	0.264	1.723	0.962
	X4/%	0.363	0.310	0.295	0.269	0.244	0.234	1.715	0.957

early stage of hydration. However, as hydration continued, the growth rate of harmless pores in these two groups of specimens gradually decreased in the later stage, resulting in relatively lower growth in compressive strength in the later stage. Furthermore, research indicates that seawater mixing can effectively reduce the porosity of specimens, while seawater curing increases the porosity of specimens. For instance, the porosity of specimens in the SS group was lower than that of the SN group at all ages, while the porosity of specimens in the FS group was higher than that of the FN group at all ages. Notably, the two groups of specimens mixed with seawater (SS and SN) exhibited lower porosity at the initial stage of curing, but throughout the curing process, the rate of porosity reduction was lower than that of the groups mixed with fresh water (FN and FS).

6.7 Conclusions

In this chapter, an investigation was conducted into the compressive strength, hydration products, microstructure, free chloride ion concentration, and pore structure of LCCM mixed with seawater. Based on the analysis, the following conclusions were drawn:

(1) In LCCM mixed with seawater, the concentration of free chloride ions decreases rapidly during the early stages of hydration but gradually slows down with increasing curing age. This behavior is attributed to the rapid reaction of AFm with chloride ions during the initial hydration phase, which diminishes as hydration progresses. Conversely, LCCM cured with seawater exhibit a significant increase in free chloride ion concentration during the early stages of hydration. However, this increase weakens over time as the cement structure becomes denser and the effects of seawater corrosion diminish.
(2) During the early stages of hydration, the precipitation of AFt is more pronounced in LCCM mixed with seawater. The sulfate ions present in seawater accelerate the formation of AFt. Additionally, Friedel's salt was observed in the hydration products of specimens mixed with seawater, indicating a limited capacity to bind chloride ions.
(3) In the early stages of hydration, disordered and irregular AFt crystals can be observed in specimens mixed with seawater. This phenomenon gradually weakens with curing age due to the uneven distribution caused by the accelerated precipitation of AFt during the early stages. As the accelerating effect of seawater diminishes, the distribution of AFt becomes more regular in the later stages.
(4) The overall trend in the development of compressive strength for LCCM mixed with seawater is similar to that of materials mixed with freshwater: rapid strength gain in the early stages followed by slower development in the later stages. However, specimens mixed with seawater exhibit higher early compressive strength, although their growth rate diminishes with curing age, resulting in late-stage compressive strength comparable to that of freshwater-mixed specimens.
(5) LCCM mixed with seawater exhibit lower porosity in the early stages compared to those mixed with freshwater. However, the rate of porosity reduction is slower, leading to similar porosity levels under both mixing conditions in the later stages. The peak area of T_2 spectra for LCCM mixed with seawater is smaller and persists throughout the later stages of hydration. The pore size distribution curve aligns closely with the trend of the T_2 spectrum curve.
(6) Grey entropy analysis was employed to model the relationship between the mass loss rate of hydration products and compressive strength, as well as the relationship between pore structure and compressive strength. The results indicate that the mass loss rate of AFt and the proportion of harmless pores significantly influence compressive strength.

References

Bayat, A., Hassani, A., & Yousefi, A. A. (2018). Effects of red mud on the properties of fresh and hardened alkali-activated slag paste and mortar. *Construction and Building Materials, 167*, 775–790.

Haha, M. B., Le Saout, G., Winnefeld, F., & Lothenbach, B. (2011). Influence of activator type on hydration kinetics, hydrate assemblage and microstructural development of alkali activated blast-furnace slags. *Cement and Concrete Research, 41*(3), 301–310.

Horgnies, M., Chen, J. J., & Bouillon, C. (2013). Overview about the use of Fourier transform infrared spectroscopy to study cementitious materials. *WIT Transactions on Engineering Sciences, 77*, 251–262.

Lemougna, P. N., Wang, K. T., Tang, Q., & Cui, X. M. (2017). Study on the development of inorganic polymers from red mud and slag system: Application in mortar and lightweight materials. *Construction and Building Materials, 156*, 486–495.

Omrani, I. A., Maniukiewicz, W., Świątkowski, M., Colombel, M. C., Bednarska, D., & Koniorczyk, M. (2023). The effects of sample preparation on the pore structure and phase composition of the sulfated cement pastes. *Construction and Building Materials, 386*, 131592.

Shao, Y., Zhou, M., Wang, W., & Hou, H. (2013). Identification of chromate binding mechanisms in Friedel's salt. *Construction and Building Materials, 48*, 942–947.

Tan, H., Deng, X., He, X., Zhang, J., Zhang, X., Su, Y., & Yang, J. (2019). Compressive strength and hydration process of wet-grinded granulated blast-furnace slag activated by sodium sulfate and sodium carbonate. *Cement and Concrete Composites, 97*, 387–398.

Chapter 7
Mechanical Properties of Grouting Materials Prepared by Low-Carbon Cementitious Materials

7.1 Introduction

In this study, high-performance grouting materials were prepared using LCCM as the cementitious material, fly ash as the admixture, and steel slag as the fine aggregate. The effects of steel slag content and fly ash content on the fluidity, vertical expansion rate, and compressive strength of the grouting materials were systematically investigated. Through microscopic testing methods such as X-ray diffraction (XRD), the formation and evolution laws of AFt at different hydration ages were analyzed, and the regulatory effects of steel slag and fly ash on the content of AFt were revealed. Further, the bonding performance tests of grouting steel sleeve were carried out, and the variation characteristics of the bond stress-displacement curve and load–displacement curve under different anchorage lengths were studied. The development law of the average bond stress with the hydration age and the failure mode of the specimens were clarified. In addition, the response laws of the circumferential and longitudinal characteristic points on the surface of the sleeve to the load were compared and analyzed.

7.2 Specimen Preparation and Test Methods

7.2.1 Raw Materials

Coal gangue

The coal gangue used in this experiment is sourced from the Zhungeer mining area in Ordos, Inner Mongolia Autonomous Region. The supplier is Inner Mongolia Jiutai Energy Group, and it is used for the calcination of cement clinker, as shown in Fig. 7.1. The composition of the material is listed in Table 7.1.

C. Yan et al., *Low-Carbon Cementitious Materials with 100% Solid Wastes*,
https://doi.org/10.1007/978-981-96-8742-8_7

Fig. 7.1 Coal gangue

Table 7.1 Chemical composition of raw materials

Composition	Al_2O_3	SiO_2	CaO	Fe_2O_3	SO_3	P_2O_5	TiO_2	MgO	Na_2O
Coal gangue	48.914	43.118	2.198	0.47	1.09	0.05	1.246	1.138	0
Carbide slag	0.668	10.2	84.39	0	1.03	0	0.112	1.355	1.026
Desulfurization gypsum	0	6.876	27.925	0	56.295	0.584	0	2.368	4.941

Carbide slag

Carbide slag is the waste residue produced after the hydrolysis of calcium carbide to obtain acetylene. The carbide slag used in this experiment is sourced from Baotou, Inner Mongolia Autonomous Region, and is used as a raw material for the calcination of cement clinker, as shown in Fig. 7.2. The composition of the material is listed in Table 7.1.

Desulfurization gypsum

Desulfurization gypsum is a byproduct of the limestone-gypsum flue gas desulfurization process. The desulfurization gypsum used in this experiment is sourced from Gongyi, Henan Province, and is used both as a raw material for the calcination of cement clinker and as a mineral admixture, as shown in Fig. 7.3. The composition of the material is listed in Table 7.1.

Fly ash

Fly ash is used as a mineral admixture for grouting materials prepared by LCCM. It is sourced from the Jinqiao Thermal Power Plant in Hohhot, Inner Mongolia, as shown in Fig. 7.4. Fly ash not only enhances the flowability of the grouting materials but also

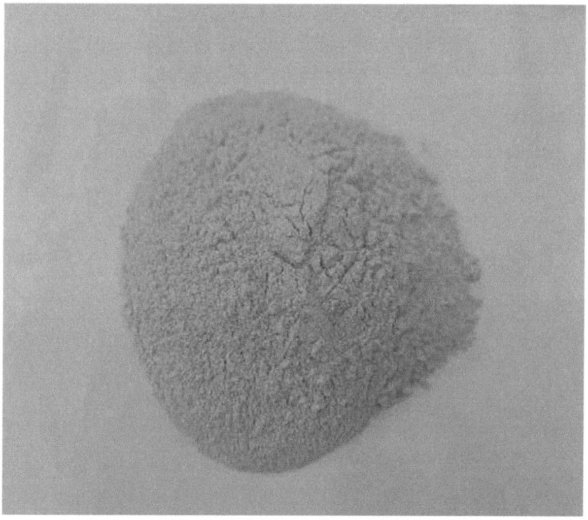

Fig. 7.2 Carbide slag

Fig. 7.3 Desulfurization gypsum

acts as an activator during the later stages of hydration, facilitating the development of strength. The primary chemical composition of the fly ash is listed in Table 7.2.

Aggregate

Steel slag is a solid by-product generated during the steel smelting process. Characterized by high hardness, significant specific gravity, and a rough, irregular surface texture, steel slag exhibits excellent frictional properties, making it a suitable complete substitute for river sand as fine aggregate. In this study, steel slag with a particle size ranging from 0.15 mm < d < 4.75 mm was utilized as the fine aggregate for low-carbon cementitious grouting materials, as illustrated in Fig. 7.5.

Fig. 7.4 Fly ash

Table 7.2 Main chemical composition of fly ash (%)

Composition	CaO	SiO_2	Al_2O_3	MgO	Fe_2O_3	Na_2O	MgO	SO_3	L.O.I
Fly ash	19.9	41.1	20.2	2.32	8.85	2.69	2.32	1.69	5.42

Note L.O.I stands for loss on ignition

Fig. 7.5 Steel slag

Admixtures

The retarder employed in this experiment is tartaric acid retarder, sourced from Shanxi Feke New Material Technology Co., Ltd., as illustrated in Fig. 7.6. The recommended dosage of the retarder ranges from 0.1 to 0.3%.

The high-efficiency water-reducing agent selected for this experiment is the FK-A powdered polycarboxylate high-performance water reducer, supplied by Shanxi

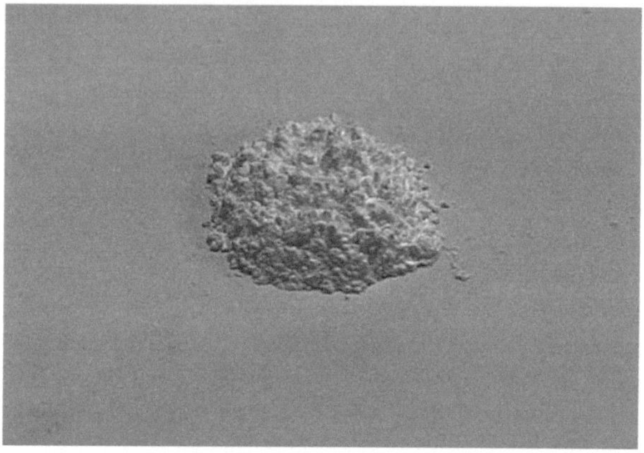

Fig. 7.6 Tartaric acid retarder

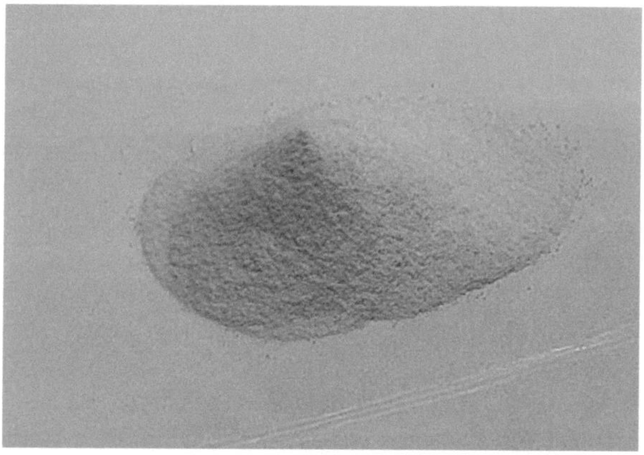

Fig. 7.7 Polycarboxylic acid water reducer

Feke New Material Technology Co., Ltd., as illustrated in Fig. 7.7. The recommended dosage of the water reducer ranges between 0.1 and 1.5%.

Water

The mixing water used in this experiment is ordinary tap water.

7.2.2 Preparation Process

The coal gangue was first crushed using a jaw crusher, followed by grinding in a cement test mill for 30 min. The ground material was then screened through a 200-mesh square hole sieve, and the resulting coal gangue was stored in a sealed bag for subsequent use. Similarly, carbide slag and desulfurization gypsum were also screened through a 200-mesh square hole sieve and subsequently stored in sealed bags.

The preliminary experimental results indicate that when the proportions of coal gangue, carbide slag, and desulfurization gypsum are 0.46:0.34:0.2. The LCCM hardened paste exhibits optimal mechanical properties, characterized by high early strength, stable hydration development, and no strength regression in the later stages. The compressive strengths of the LCCM hardened paste at 1 day, 3 days, and 28 days are 68.5, 77.63, and 83.4 MPa, respectively.

The coal gangue, carbide slag, and desulfurization gypsum were weighed and mixed according to the mix ratio. The blended materials were thoroughly mixed in a blender, and 12% water was added to ensure complete homogenization. The mixture was then pressed into a round shape using a powder forming hydraulic press and dried at 105 °C in a blast drying oven for 24 h. Following drying, the samples were calcined in a high-temperature electric furnace. The calcination process involved heating from 25 to 900 °C at a rate of 10 °C/min, followed by further heating to 1280 °C at a rate of 5 °C/min. After maintaining the temperature at 1280 °C for 40 min, the calcined samples were rapidly removed and cooled. Finally, the cooled samples were ground in a cement test mill for 30 min and sieved through an ultrasonic vibrating screen with a 200-mesh sieve to produce cement clinker.

Appropriate amounts of desulfurization gypsum can have a retarding effect on grouting materials. For this purpose, desulfurization gypsum was added to LCCM as an admixture at 3, 5, 7, 9, and 11% by weight, and the compressive strength and expansion performance of the hardened LCCM paste at different dosages were studied. The results show that when the dosage of desulfurization gypsum is 7, 9, and 11%, the hardened LCCM paste exhibits excessive expansion, causing the samples to crack. As the hydration age increases, this phenomenon becomes more pronounced, reducing the compressive strength of the hardened paste from over 60 MPa at 1 day to below 20 MPa. This may be due to the excessive addition of desulfurization gypsum, which leads to an excessively high concentration of Ca^{2+}. Under the action of retarders and water reducers, Ca^{2+} is encapsulated and precipitated, thereby causing excessive expansion of the test blocks and loss of strength. Considering the expansion performance and mechanical properties of the hardened paste comprehensively, the optimal dosage of desulfurization gypsum is determined to be 3%.

7.2.3 *Test Methods*

Molding and curing of grouting materials containing LCCM

(1) The water-binder ratio was set at 0.26. According to 0.14 and 0.6% of the powder mass, the retarder and water reducer were weighed respectively and dissolved in the mixing water, then thoroughly stirred and mixed evenly.
(2) LCCM and fly ash were accurately weighed according to the mix proportion and placed in a planetary mixer for uniform mixing.
(3) The prepared mixture of retarder, water reducer and water was then poured into the mixer. The mixture was stirred at low speed for 30 s, left to stand for 30 s,

and then stirred at high speed for 60 s. Then, the steel slag was added to the mixture and stirred at high speed for 30 s.

(4) The mixture was then layered into molds of 40 × 40 × 160 mm^3 and vibrated to ensure density. Afterwards, the surface of the specimens was covered with polyethylene film. The specimens were left to stand at room temperature for 3 h until completely solidified, then demolded and transferred to a curing box for curing at 1 day, 3 days and 28 days respectively. The curing conditions were a temperature of 20 ± 1 °C and a relative humidity of 99%.

Flowability test

The flowability of the grouting material was tested using a truncated cone-shaped circular mold with an upper inner diameter of 70 ± 0.5 mm, a lower inner diameter of 100 ± 0.5 mm, and a height of 60 ± 0.5 mm. The testing steps for flowability of grouting material are as follows:

(1) Preparatory work: Place the glass plate on a level test bench. Wipe the mixer bowl, mixer blades, glass plate, and the inner walls of the truncated cone mold with a damp cloth to ensure they are free of visible water. Place the truncated cone mold in the center of the glass plate and cover it with a damp cloth to prevent moisture loss.
(2) Preparation of grouting material: The preparation steps for the grouting material follow steps (1)–(3) in molding and curing of grouting materials containing LCCM.
(3) Initial flowability measurement: Pour the grouting material into the truncated cone mold until the material level touches the edge of the mold. Use a spatula to level the top surface, then slowly lift the mold to allow the grouting material to flow freely on the glass plate until it stops. Measure the flow diameter along two perpendicular directions using a ruler and take the average value as the flowability of the grouting material. The measurement should be accurate to 1 mm, and the entire process of mixing and measuring should be completed within 5 min. After measuring the initial value, quickly transfer the grouting material from the glass plate back into the mixer bowl and cover it with a damp cloth.
(4) Flowability measurement at 10 min: Ten minutes after the initial flowability measurement, stir the grouting material in the mixer bowl slowly for 60 s, then quickly for another 60 s. Measure the flowability again following the same steps as in (3) to obtain the 10-min retained value, and record the data.

Vertical expansion rate test

A vertical expansion rate tester was used to test the expansibility of the slurry. The experimental device is shown in Fig. 7.8. The testing steps for vertical expansion rate of grouting material are as follows:

Fig. 7.8 Vertical expansion rate experimental device

(1) Place a glass plate measuring 140 × 80 × 5 mm horizontally at the center of the top surface of the cubic mold, with the long edge of the glass plate approximately 1 cm from the mold edge.
(2) The preparation steps for the grouting material follow steps (1) to (3) in molding and curing of grouting materials containing LCCM.
(3) Gently press the glass plate and slowly pour the grouting material into the mold from one side, allowing it to overflow from the other side and rise above the mold edge. Use a spatula to smooth the excess grouting material at an angle, and cover the exposed grouting material on both sides of the glass plate with a moist cotton cloth to maintain humidity. Then, vertically position and secure a dial gauge at the center of the glass plate.
(4) Read the initial dial gauge reading h_0 within 30 s of securing the gauge. Record the readings every 5 min until the value stabilizes for an extended period. Additionally, record the readings at 3 h and 24 h, respectively, and calculate the vertical expansion rate using Eq. 7.1.

$$\varepsilon_1 = \frac{h_t - h_0}{h} \times 100 \tag{7.1}$$

where: ε_1 is the vertical expansion rate, accurate to 0.01%. h_0 is the initial reading of the specimen, mm. h_t is the reading of the specimen at age t, mm. h is the reference height of the specimen, which is 100 mm.

Compressive strength test

The mechanical properties of the grouting materials were tested at hydration ages of 1 day, 3 days and 28 days. The size of the specimen is 40 mm × 40 mm × 160 mm. First, the flexural strength of the specimen was tested. Then, the compressive strength of the specimen was tested at a loading rate of 2400 N/s. Test results remove the maximum and minimum values of the 6 data, and take the average value as the test result.

Uniaxial tensile test of steel bar sleeves with grouting materials test

First, both ends of the steel bar sleeve should be sealed using rubber plugs. Subsequently, the pre-marked steel bars should be inserted through the reserved holes in the rubber plugs according to the specified anchorage depth. The prepared grouting material then be poured into the grouting device and injected into the grouting sleeve via the grouting hole until the grouting material continuously overflows from the overflow hole. Finally, the specimen should remain undisturbed for 3 h and be cured for the specified aging period in preparation for subsequent testing.

The rebars used in the experiment are HRB400 hot-rolled ribbed rebars produced by Hengshui Anda Machinery Equipment Co., Ltd. The diameter of the rebars selected for this experiment is 12 mm. The mechanical properties of the rebars are shown in Table 7.3.

The grouted steel sleeves used in this experiment were produced by Hengshui Anda Machinery Equipment Co., Ltd. The diameter of the rebar matches the dimensions of the grouted sleeves. The specifications and parameters of the grout sleeves are shown in Table 7.4. The uneven surface of the sleeves effectively increases the mechanical interlock between the grouting material and the steel sleeves, preventing the rebar from detaching during the uniaxial tension test. The schematic diagram of the grout sleeve structure is shown in Fig. 7.9, and the picture of grout sleeve is shown in Fig. 7.10.

X-ray diffraction (XRD) test

The XRD was performed using a Rigaku Ultima IV X-ray diffractometer with a Cu target. The operating voltage was 40 kV, and the current was 40 mA. The scanning angle range was 5–85°, with a step size of 0.02°, and a scanning speed of 8°/min. Samples weighing more than 1 g and sieved through a 200-mesh sieve were pressed into square shapes and placed in the instrument for testing.

Table 7.3 Mechanical properties of rebars

Diameter/mm	Tensile strength/MPa	Yield strength/MPa	Maximum force/KN
12	580	405.8	65.71

Table 7.4 Parameters of grout sleeve

Size	Material	Sleeve material	L (mm)	Weight (kg)	L_z (mm)
φ12	45#	φ38.5 × 2.5	247	0.619	123.5
$D_1{}^{+0.2}+0$	A_p (mm)	A_z (mm)	Set screw	L_p (mm)	L_1 (mm)
φ28.5	27.5	27.5	M5 × 35	123.5	192

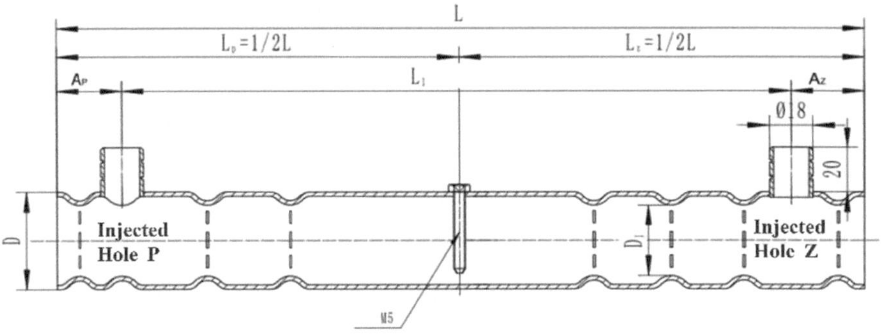

Fig. 7.9 Schematic diagram of the grout sleeve structure

Fig. 7.10 Picture of grout sleeve

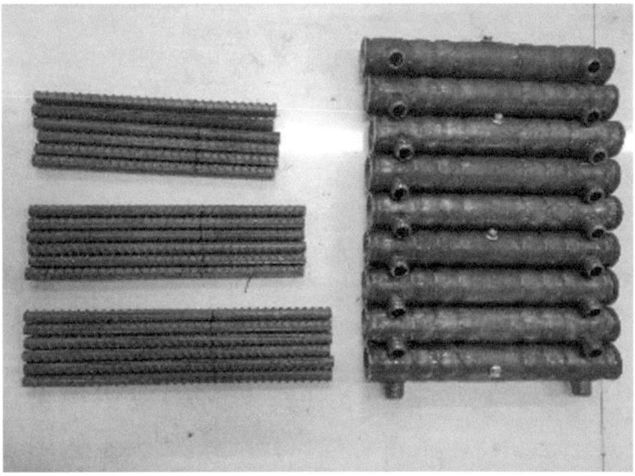

7.3 Flowability of Grouting Materials Prepared by LCCM

7.3.1 Effect of Steel Slag Content on Flowability

Steel slag with a particle size of less than 4.75 mm is selected as the fine aggregate. The content of the aggregate significantly affects the flowability of the grouting material. Before preparing the specimens, the steel slag is sieved to remove small particles and dust, making it suitable for use in the specimens. In grouting materials, the aggregate serves as the framework. Steel slag is more cost-effective compared to river sand and other aggregates, and it helps to manage solid waste. The effect of steel slag content on the flowability of the waste material and the grouting material is illustrated in Fig. 7.11.

Based on Fig. 7.11, it can be observed that as the steel slag content in the grouting material increases, the flowability of the grouting material shows a decreasing trend.

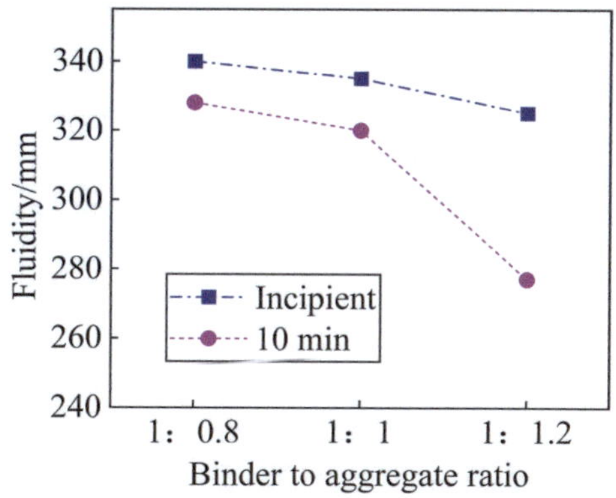

Fig. 7.11 Effect of ratio of binder to aggregate on fluidity of grouting material

Both the initial flowability and the 10-min flowability decrease with the increase in steel slag content.

When the ratio of binder to aggregate is 1:1.2, the 10-min flowability of the grouting material decreases the most compared to the initial flowability, with a reduction of 48 mm, a decrease of 14.77%. When the ratio of binder to aggregate is 1:1, the 10-min flowability decreases by 15 mm, a reduction of 4.48%. When the ratio of binder to aggregate is 1:0.8, the flowability decreases by only 12 mm, a reduction of 3.53%. This indicates that when steel slag is used as the fine aggregate, the loss of flowability increases with the content of the aggregate. This is because the rough and uneven surface of the steel slag increases friction, which hinders the flow of the grouting material. As the steel slag content increases, the decrease in flowability becomes more pronounced.

The initial flowability at a ratio of binder to aggregate of 1:1 decreases by 5 mm, a reduction of 1.47%, compared to a ratio of 1:0.8. The initial flowability at a ratio of 1:1.2 decreases by 2.99% compared to a ratio of 1:1. The 10-min flowability at a ratio of 1:1 decreases by 8 mm, a reduction of 2.44%, compared to a ratio of 1:0.8. The 10-min flowability at a ratio of 1:1.2 decreases by 43 mm, a reduction of 13.44%, compared to a ratio of 1:1. This demonstrates that the increase in steel slag content has a relatively minor impact on the initial flowability but a significant effect on the 10-min flowability of the grouting material. The higher the steel slag content, the greater the decrease in 10-min flowability. This is because steel slag has certain water absorption properties; the more steel slag is added, the more water is absorbed, indirectly reducing the water-cement ratio, thereby increasing the loss of 10-min flowability.

7.3.2 Effect of Fly Ash Content on Flowability

When fly ash is used as a mineral admixture in grouting materials, its particle shape and size can significantly enhance the flowability of the grouting materials,

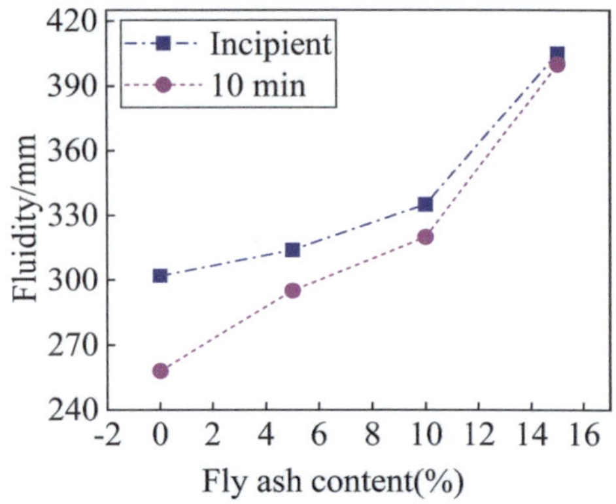

Fig. 7.12 Effect of fly ash content on the fluidity of grouting material

acting as a lubricant within the mix. Fly ash particles are fine, smooth, and spherical, with low aggregation and numerous individual beads. These spherical particles provide a rolling effect, reducing friction between the aggregate and cement, and improving flowability. Additionally, fly ash absorbs onto the cement surface, dispersing cement particles and preventing the formation of flocculated structures. The impact of different fly ash contents on the initial flowability and the 10-min flowability of grouting materials is illustrated in Fig. 7.12.

Based on Fig. 7.12, it is evident that the flowability of grouting materials increases with the addition of fly ash. The initial flowability becomes more uniform as the fly ash content increases. When the fly ash content is 5%, the flowability increases by 12 mm, which is a 3.97% increase. At a fly ash content of 10%, the flowability increases by 21 mm, or 6.69%. When the fly ash content reaches 15%, the flowability increases by 70 mm, or 20.9%. This demonstrates that the higher the fly ash content, the more pronounced the improvement in initial flowability, with a significant and exponential growth rate. This is due to the smooth, spherical nature of the fly ash particles, which act as lubricants, reducing internal friction.

From the 10-min flowability graph, it can be seen that the 10-min flowability of the grouting materials also increases with the fly ash content, but at different rates. When the fly ash content is 5%, the flowability increases by 37 mm compared to 258 mm without fly ash. At 10% fly ash content, the flowability increase is 25 mm. At 15% fly ash content, the flowability increases by 80 mm. This indicates that the fly ash content between 5 and 10% has a relatively moderate impact on the growth rate of flowability, while at over 10% fly ash content, the improvement becomes more significant. However, at 15% fly ash content, there is some settling of the steel slag aggregate, leading to a separation of the aggregate and paste. This is because excessive fly ash significantly increases the flowability, causing the dense steel slag to settle.

When the fly ash content ranges from 0 to 5%, its impact on improving flowability is most noticeable. Without adding fly ash, the 10-min flowability loss is 44 mm. With 5% fly ash, the 10-min flowability loss is 19 mm. At 10% fly ash content, the 10-min flowability loss is 15 mm. At 15% fly ash content, the 10-min flowability loss is only

5 mm. This shows that as the fly ash content increases, the flowability loss gradually decreases. The most significant reduction is seen when the fly ash content increases from 0 to 5%. This is because more fly ash particles adsorb onto the cement surface, dispersing the surrounding cement particles, preventing the formation of flocculated structures, and hindering the aggregation of cement particles, which delays the hydration reaction.

7.4 Vertical Expansion Rate of Grouting Materials Prepared by LCCM

7.4.1 Effect of Steel Slag Content on Vertical Expansion Rate

As one of the three main characteristics of grouting materials, the vertical expansion rate is crucial for its applications. Small expansibility can make grouting materials play an essential role when used as repair materials and anchoring grouts. Excessive expansion can result in the grouting materials losing their required high strength, while insufficient expansion or even shrinkage means the grouting materials cannot perform their intended function. The aggregate content has a specific impact on the vertical expansion rate of grouting materials. This section mainly explores the influence of steel slag content as an aggregate on the vertical expansion rate of grouting materials. The variation curves of vertical expansion rate with different steel slag contents are shown in Fig. 7.13.

Based on Fig. 7.13, it is clear that different steel slag contents significantly impact the vertical expansion rate of grouting materials. As the steel slag content increases, the vertical expansion rate gradually rises. The cement-based material used is LCCM, which has certain expansibility and a fast hydration rate. The expansion occurs quickly within 3 h, with a rapid development phase lasting about an hour. When the expansion reaches 3 h, the vertical expansion rate remains stable.

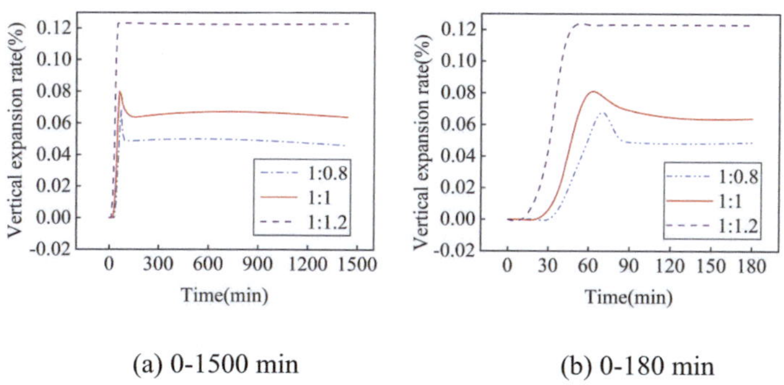

Fig. 7.13 Vertical expansion rate of grouting material with different aggregate to binder ratio

When the ratio of binder to aggregate is 1:1, the maximum vertical expansion rate of 0.08% is 0.012% higher than the 0.068% rate at a ratio of 1:0.8. At a ratio of 1:1.2, the vertical expansion rate of 0.123% is 0.043% higher than the 0.08% rate at a ratio of 1:1. Furthermore, when the ratio of binder to aggregate is 1:1.2, the vertical expansion rate remains constant at its maximum value until 24 h. In contrast, at ratios of 1:1 and 1:0.8, there is some degree of shrinkage after reaching the maximum expansion rate, but the shrinkage remains within standard requirements. As the steel slag content increases, the shrinkage after reaching the maximum expansion rate decreases.

At a ratio of binder to aggregate of 1:0.8, the maximum vertical expansion rate within 3 h is 0.068%, while at 24 h, it is 0.046%, with a difference of 0.022%. At a ratio of 1:1, the maximum vertical expansion rate within 3 h is 0.08%, and at 24 h, it is 0.064%, with a difference of 0.016%. At a ratio of 1:1.2, the vertical expansion rate shows an increasing trend with no reduction phase within 24 h, maintaining its maximum value after rapid growth. The difference between the maximum value within 3 h and the 24-h vertical expansion rate is 0. This indicates that as the steel slag content increases, the difference between the maximum vertical expansion rate within 3 h and at 24 h decreases.

The reasons for these phenomena are that steel slag, as the aggregate in l grouting materials, acts as the framework. The individual volume of steel slag does not change, and increasing the steel slag content increases the proportion of the non-changing part in the paste. Therefore, the more steel slag is added, the greater the expansion and the lesser the shrinkage after expansion, or it may even remain unchanged.

The vertical expansion rate is related to the hydration rate. The greater the steel slag content, the larger the vertical expansion rate, and the faster it reaches the maximum vertical expansion rate, indicating a faster hydration rate.

7.4.2 Effect of Fly Ash Content on Vertical Expansion Rate

This experiment continuously observes the vertical expansion rate measurements under different fly ash contents, thereby plotting the curve of vertical expansion rate over time. The expansion of grouting materials is accompanied by the hydration process, and to a certain extent, the expansion rate curve can reflect the hydration rate. The impact of different fly ash contents on the vertical expansion rate of grouting materials is illustrated in Fig. 7.14.

Based on Fig. 7.14, it can be seen that the vertical expansion rate of grouting materials gradually decreases with the increase in fly ash content. When the fly ash content is 0, 5, 10, and 15%, the grouting materials exhibit some degree of expansion. After adding fly ash, the vertical expansion rate undergoes a rapid growth phase until it reaches its maximum value, followed by a quick slight shrinkage phase until 24 h. In contrast, the grouting material without fly ash only shows an increase in vertical expansion rate within 24 h. The early stage shows rapid growth, with slow and slight growth from 3 to 24 h.

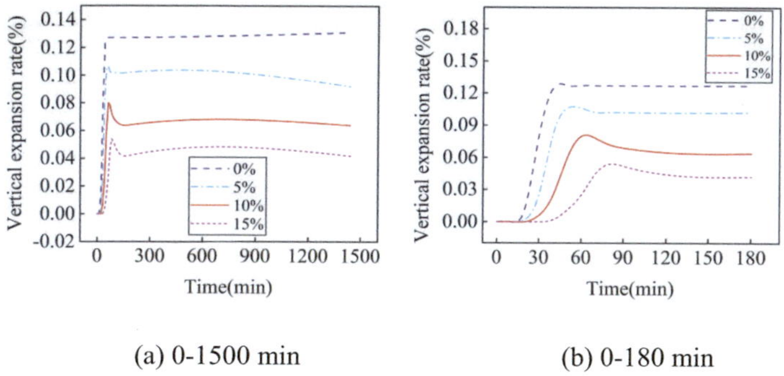

Fig. 7.14 Vertical expansion rate of grouting material with different fly ash content

When the fly ash content is 5%, the maximum vertical expansion rate is 0.106%, which is 0.025% lower than the 0.131% at 0% fly ash content. When the fly ash content is 10%, the maximum vertical expansion rate is 0.08%, which is 0.026% lower than at 5% fly ash content. When the fly ash content is 15%, the maximum vertical expansion rate is 0.054%, which is 0.026% lower than at 10% fly ash content. This indicates that as the fly ash content increases uniformly, the vertical expansion rate decreases uniformly, with similar differences. This is because the spherical particles of fly ash make the internal structure of the grouting material denser, reducing the number of air bubbles.

When the fly ash content is 0%, there is no shrinkage phase. The maximum vertical expansion rate during the early rapid growth phase is 0.127%, and at 24 h, it is 0.131%, with a difference of −0.004%. When the fly ash content is 5%, the maximum vertical expansion rate within 3 h is 0.106%, and at 24 h, it is 0.092%, with a difference of +0.014%. When the fly ash content is 10%, the maximum vertical expansion rate within 3 h is 0.08%, and at 24 h, it is 0.064%, with a difference of 0.016%. When the fly ash content is 15%, the maximum vertical expansion rate within 3 h is 0.054%, and at 24 h, it is 0.042%, with a difference of 0.01%. This shows that the difference between the maximum vertical expansion rate within 3 h and at 24 h increases with the increase in fly ash content. This could be due to the release of a large amount of heat during the setting process of the grouting material, and the greater the fly ash content, the more significant the effect of thermal expansion and contraction on the test block.

Additionally, Fig. 7.14 shows that as the fly ash content increases, the time for the vertical expansion rate curve to reach its maximum value gradually delays. It is evident that the slope of the expansion phase curve decreases with the increase in fly ash content. Since the expansion of the grouting materials in this study is provided by the formation and expansion of AFt crystals during hydration, the hydration time-vertical expansion rate curve can reflect the hydration rate. As the fly ash content increases, the hydration rate of the grouting material gradually slows down. After the expansion curve peak, the hydration rate gradually decreases. At 3 h of hydration, the vertical expansion rate curve has already leveled off, and the hydration rate further slows down, indicating that the fastest hydration rate period for the grouting material is within the first 3 h.

7.5 Compressive Strength of Grouting Materials Prepared by LCCM

7.5.1 Effect of Steel Slag Content on Compressive Strength

Compressive strength is a fundamental property of grouting materials. After curing the specimens with different steel slag content ratios of grouting materials for 1 day, 3 days, and 28 days, their compressive strength is measured. The impact of steel slag content on the compressive strength of grouting materials is analyzed, as shown in Fig. 7.15.

Based on Fig. 7.15, it is evident that the compressive strength of grouting materials at each age increases with the addition of steel slag. When the grouting material is cured for 1 day, the compressive strength increases from 56.4 to 59 MPa as the binder to aggregate ratio changes from 1:0.8 to 1:1.2. The compressive strength at a ratio of 1:1 increases by 4.08% compared to a ratio of 1:0.8, and by approximately 0.5% when the ratio changes from 1:1 to 1:1.2. Thus, the 1-day compressive strength of grouting materials increases with the addition of steel slag, but the growth rate diminishes.

After curing for 3 days, the compressive strength increases from 63.95 MPa to 73.8 MPa as the binder to aggregate ratio changes from 1:0.8 to 1:1.2. The compressive strength at a ratio of 1:1 increases by 3.28% compared to a ratio of 1:0.8, and by 11.73% when the ratio changes from 1:1 to 1:1.2. This indicates that for the 3-day curing period, the increase in compressive strength is more substantial with higher steel slag content, and the growth rate accelerates with increasing steel slag content.

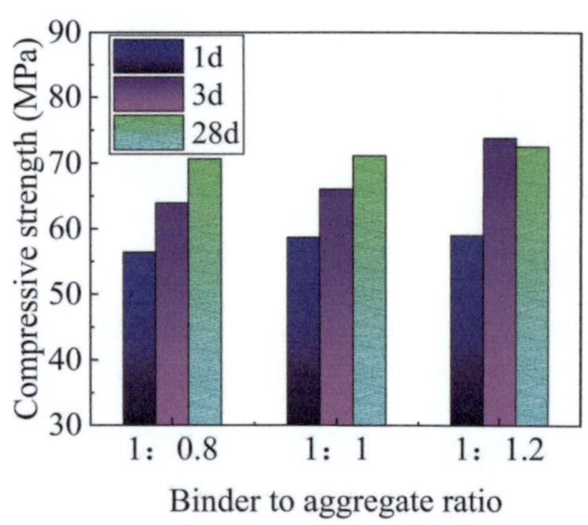

Fig. 7.15 Compressive strength of grouting materials with different binder to aggregate ratio

After curing for 28 days, the compressive strength increases slightly from 70.65 MPa to 72.5 MPa as the binder to aggregate ratio changes from 1:0.8 to 1:1.2. The impact of steel slag content on the 28-day compressive strength is minimal.

In conclusion, increasing the binder to aggregate ratio from 1:0.8 to 1:1 significantly enhances the compressive strength of grouting materials with the addition of steel slag. However, the enhancement effect weakens when the ratio changes from 1:1 to 1:1.2. This is because as the steel slag content increases, the cementitious material surrounding it decreases, reducing the amount of cementitious material per unit volume. Since the strength of grouting materials mainly depends on the cementitious material, the enhancement effect diminishes with higher steel slag content.

When the binder to aggregate ratio is 1:0.8, the compressive strength of cementitious grouting material increases by 7.88 and 6.7 MPa with curing age, resulting in growth rates of 13.38 and 10.4%, respectively, as shown in Fig. 7.16. The 1-day compressive strength is 79.83% of the 28-day strength, and the 3-day compressive strength is 90.51% of the 28-day strength.

When the binder to aggregate ratio is 1:1, the compressive strength increases by 7.35 and 5.05 MPa with curing age, resulting in growth rates of 12.52 and 7.65%, respectively. The 1-day compressive strength is 80.13% of the 28-day strength, and the 3-day compressive strength is 94.09% of the 28-day strength.

When the binder to aggregate ratio is 1:1.2, the compressive strength increases by 14.8 MPa and decreases by 1.3 MPa with curing age, resulting in growth rates of 25.08 and −1.76%, respectively. The 1-day compressive strength is 81.38% of the 28-day strength, and the 3-day compressive strength is slightly higher than the 28-day strength, showing a minor reduction at 28 days.

In summary, the influence of steel slag content on the compressive strength of grouting materials for different curing ages is in the order of 3 days > 1 day > 28 days. The impact is greatest for the 3-day curing age. When the binder to aggregate ratio is 1:1.2, a reduction in compressive strength is observed at 28 days. This indicates that excessive steel slag can cause a decrease in the 28-day compressive strength due to its water absorption properties, which indirectly reduce the water-cement ratio,

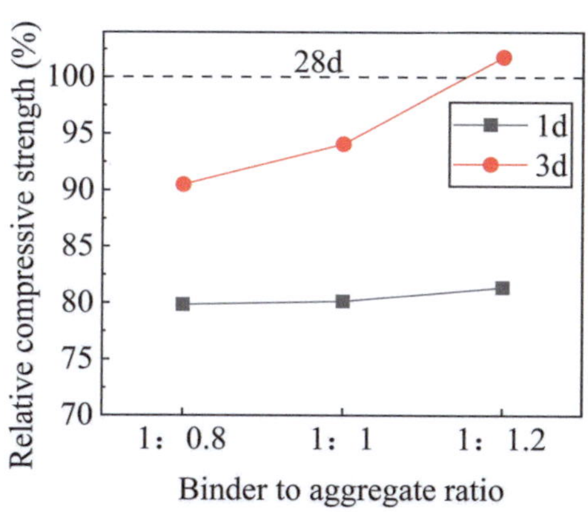

Fig. 7.16 Relative compressive strength of grouting matrial with different binder to aggregate ratio

accelerating hydration and increasing early strength. However, excessive steel slag may hinder hydration, and its expansibility could create micro-cracks, leading to strength reduction.

7.5.2 *Effect of Fly Ash Content on Compressive Strength*

Based on the measured compressive strength values from the experiments, an analysis chart depicting the influence of fly ash content on the compressive strength of grouting materials at different curing ages is created, as shown in Fig. 7.17.

Based on Fig. 7.17, it can be seen that the compressive strength of grouting materials initially decreases and then increases with the addition of fly ash, affecting both early and late compressive strengths differently. The lowest compressive strength at 1 day of curing occurs with 10% fly ash content at 56.25 MPa, while the lowest strengths at 3 days and 28 days of curing are observed with 5% fly ash content, with strengths of 63.32 and 67.4 MPa, respectively.

For the 1-day curing period, the compressive strength with 5% fly ash content is 2.3 MPa lower than without fly ash, a decrease of 3.6%. With 10% fly ash content, the compressive strength decreases by 5.35 MPa compared to 5% fly ash, a decrease of 8.69%. However, when the fly ash content reaches 15%, the compressive strength significantly increases by 15.41 MPa, a growth of 27.4%. This indicates that as the fly ash content increases, the 1-day compressive strength of the grouting material initially decreases, but the decrease rate gradually increases, followed by a sharp increase in strength. Low fly ash content inhibits the 1-day compressive strength of the grouting material, whereas sufficient fly ash content can promote its development.

The reason for this is that fly ash replaces part of the LCCM, reducing the amount of hydration products. Low fly ash content has a lesser effect on promoting the grouting material's compressive strength and does not make it dense enough, leaving small, dense air bubbles within the test block.

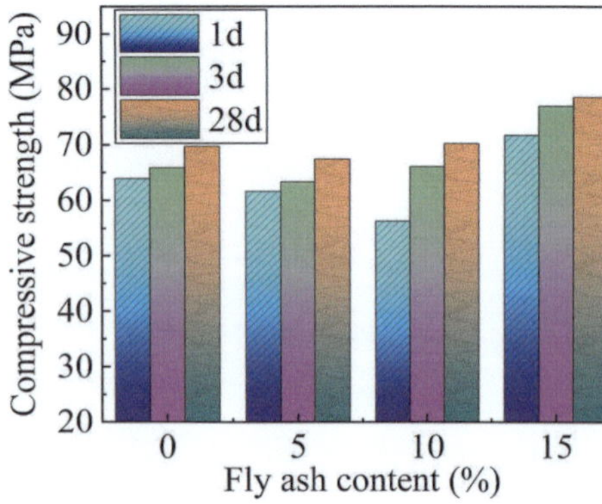

Fig. 7.17 Compressive strength of grouting materials with different fly ash content

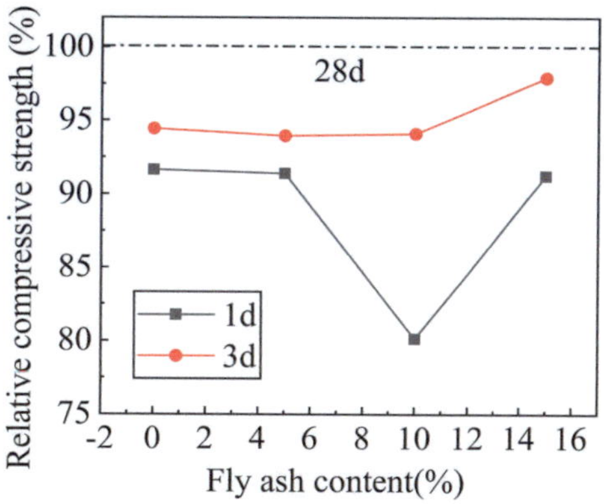

Fig. 7.18 Relative compressive strength of grouting matrial with different fly ash content

As illustrated in Fig. 7.18, when the fly ash content is 0%, the compressive strength at 1 day is 91.64% of that at 28 days, and the strength at 3 days is 94.44% of that at 28 days. When the fly ash dosage increases to 5%, the compressive strengths at 1 day and 3 days are 91.39 and 93.95% of that at 28 days, respectively. At a dosage of 10%, the corresponding values decrease to 80.13 and 94.09%, respectively. When the dosage reaches 15%, the compressive strength at 1 day becomes 91.23% of that at 3 days, while the strength at 3 days is 97.90% of that at 28 days. These results indicate that low fly ash content has minimal impact on early-age compressive strength, whereas high fly ash content significantly suppresses early-age strength development but enhances late-age strength. This phenomenon can be attributed to the hydrophobic and lubricating properties of the spherical structure of fly ash. At low dosages, these effects are limited; however, at high dosages, the hydrophobicity inhibits early hydration reactions, while the lubricity improves the compactness of the matrix, thereby enhancing the late-age strength.

7.6 Content of Ettringite of Grouting Materials Prepared by LCCM

7.6.1 Effect of Steel Slag Content on Ettringite Content

This experiment conducts XRD tests on grouting materials with different steel slag contents, cured for 1 day, 3 days, and 28 days. By observing the height differences of diffraction peaks in the spectra, the relative content of various hydration products can be reflected. The XRD patterns for binder to aggregate ratios of 1:0.8 (Fig. 7.19), 1:1 (Fig. 7.20), and 1:1.2 (Fig. 7.21) are shown below.

Observing Figs. 7.19, 7.20 and 7.21, which depict the XRD spectra of grouting materials with varying steel slag contents cured for 1 day, 3 days, and 28 days, it

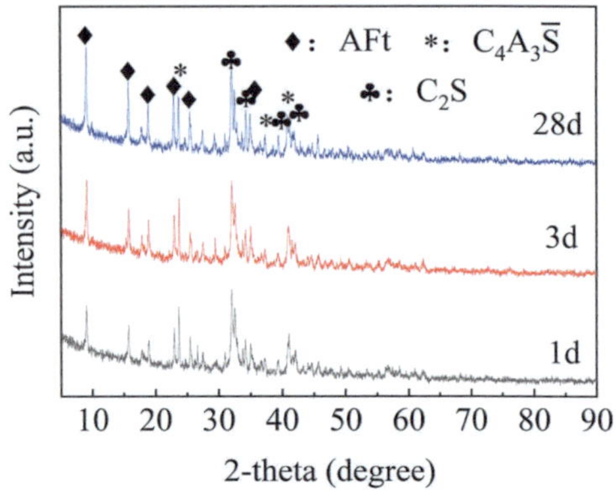

Fig. 7.19 XRD pattern of the sample with a binder to aggregate ratio of 1:0.8

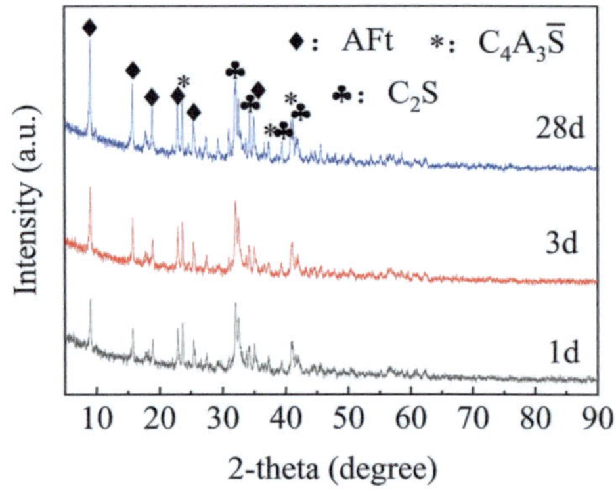

Fig. 7.20 XRD pattern of the sample with a binder to aggregate ratio of 1:1

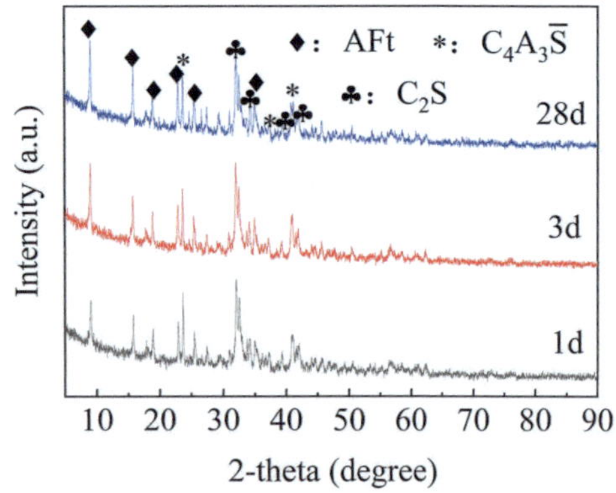

Fig. 7.21 XRD pattern of the sample with a binder to aggregate ratio of 1:1.2

is evident that the types of phases in the grouting materials remain largely consistent over time. No new crystalline substances are detected, and AFt continues to be the predominant hydration product. The diffraction peaks of $C_4A_3\overline{S}$ decrease progressively with increasing curing age, while the diffraction peaks of AFt increase correspondingly. This suggests that as the curing age advances, an increasing amount of $C_4A_3\overline{S}$ is transformed into AFt.

Observing Figs. 7.19, 7.20 and 7.21, which present the XRD spectra of grouting materials with different steel slag contents cured for 1 day, 3 days, and 28 days, it is clear that the types of phases in the grouting materials do not change significantly over time. No new crystalline substances are formed, and AFt remains the primary hydration product. The diffraction peaks of $C_4A_3\overline{S}$ decrease continuously with increasing curing age, while the diffraction peaks of AFt increase. This indicates that more $C_4A_3\overline{S}$ is converted to AFt as the curing age progresses. However, at 28 days of hydration, there is still some unhydrated $C_4A_3\overline{S}$ in the grouting materials. The diffraction peaks of C_2S do not show significant changes with curing age, suggesting that only a small amount of C_2S participates in the hydration reaction to form C-S-H. Given that some unhydrated $C_4A_3\overline{S}$ and a large amount of C_2S still exist in the grouting materials after 28 days of hydration, these materials have significant potential for further mechanical strength development.

Observing Figs. 7.19, 7.20 and 7.21, it is found that the diffraction peaks of AFt in the first three days compared to the AFt diffraction peaks at 28 days show that as the steel slag content increases, the height difference of the AFt diffraction peaks gradually decreases, while the diffraction peaks of unhydrated $C_4A_3\overline{S}$ show an increasing trend. This is because steel slag, as an aggregate, has some water absorption properties. When the steel slag content increases, it indirectly reduces the water-cement ratio of the grouting material, preventing sufficient hydration of $C_4A_3\overline{S}$. Consequently, the amount of hydration products in the grouting materials decreases.

7.6.2 *Effect of Fly Ash Content on Ettringite Content*

Figure 7.22 shows the XRD patterns of grouting materials with fly ash contents of 0, 5, 10, and 15% at different curing ages. Based on Fig. 7.22a–d, it can be seen that the diffraction peak angles in the XRD patterns for the four different fly ash contents are the same. This indicates that changes in fly ash content do not affect the types of hydration products in grouting materials. However, the heights of the diffraction peaks vary with different fly ash contents, indicating that fly ash content affects the quantity of hydration products. As the curing age increases, the diffraction peaks of AFt in the XRD patterns for each fly ash content also increase. This suggests that the amount of AFt increases with the curing age for different fly ash contents, providing a foundation for the sustained growth of compressive strength in grouting materials. It is also observed that the diffraction peaks of AFt are prominent at 1 day of hydration across all fly ash contents, explaining why the grouting materials exhibit early and high strength.

Comparing Fig. 7.22a–d, it is found that as the fly ash content increases, the difference in peak heights between the 28-day and 3-day XRD patterns initially increases and then decreases, with the maximum difference observed at 10% fly

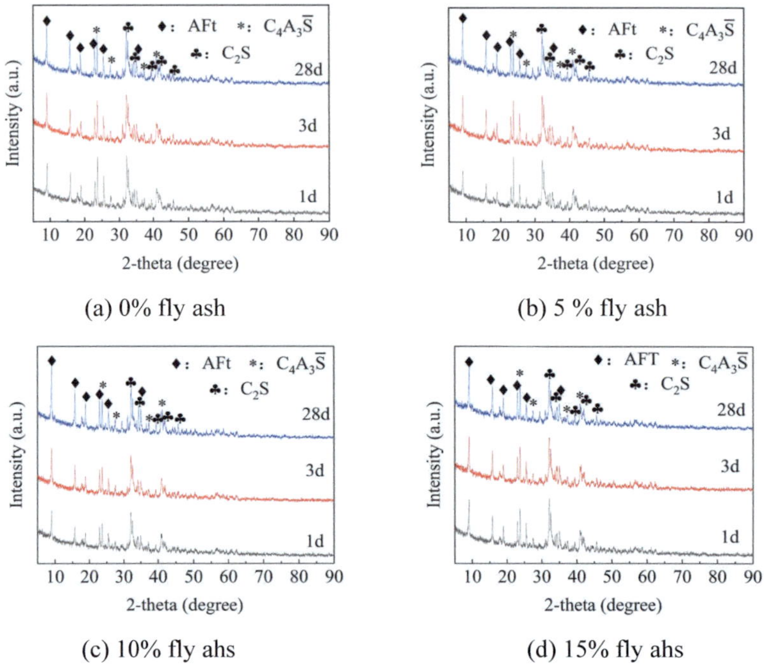

Fig. 7.22 XRD patterns of grouting materials with different fly ash content at different ages

ash content. This indicates that at this fly ash content, the AFt crystals in grouting materials grow uniformly with curing age, promoting the healthy development of AFt crystals. The grouting materials continue to produce a significant amount of AFt from 3 to 28 days.

Comparing Fig. 7.23a–c, it is observed that as the fly ash content increases, the height of the AFt diffraction peak at a 1-day curing age shows a slight upward trend, while the peak height at a 3-day curing age remains almost unchanged. However, the peak height at a 28-day curing age shows a downward trend. This indicates that the addition of fly ash has a slight promoting effect on the formation of hydration products in grouting materials from 1 to 3 days. From 3 to 28 days, the addition of fly ash inhibits the formation of hydration products.

This inhibition is due to the partial replacement of cement clinker by fly ash, reducing the amount of $C_4A_3\overline{S}$. Therefore, fly ash addition promotes hydration for less than 3 days, but after 3 days, insufficient $C_4A_3\overline{S}$ results in reduced AFt formation. Although the amount of AFt is relatively low with 15% fly ash content, the structure is dense, leading to high strength.

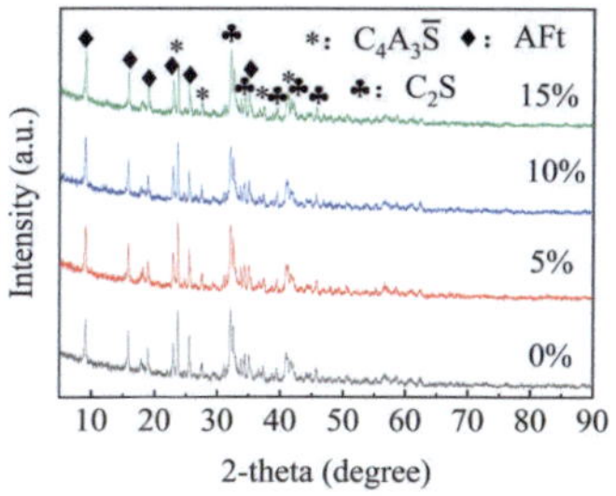

(a) Curing age of 1 d

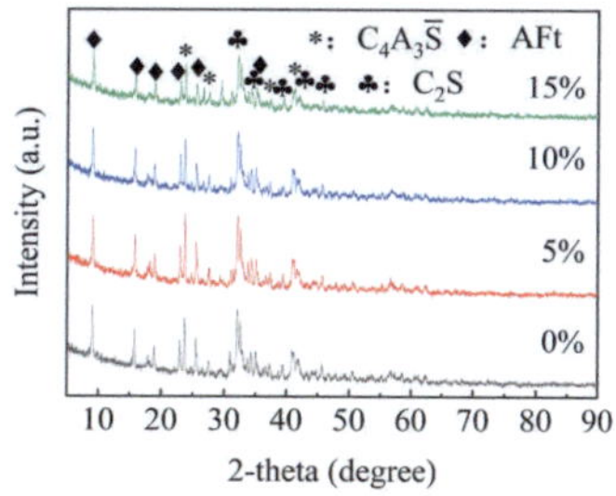

(b) Curing age of 3 d

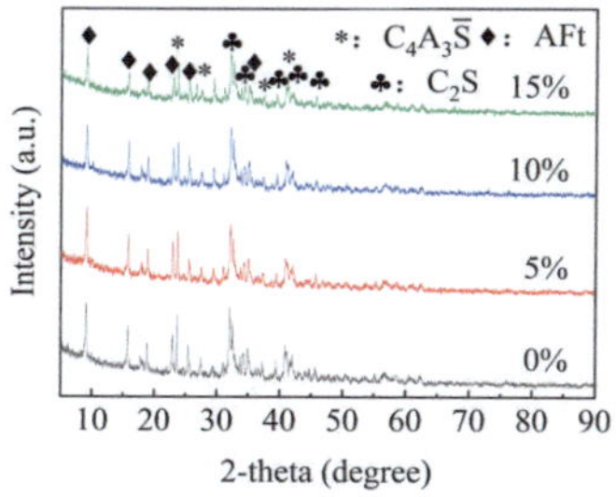

(c) Curing age of 28 d

Fig. 7.23 XRD patterns of grouting materials with different fly ash contents at different ages

7.7 Uniaxial Tension Test of Reinforcement Sleeve with Grouting Material

7.7.1 Bond Stress Analysis

This experiment analyzes the tensile performance of grouted rebar sleeve specimens by examining the influence of anchorage length and curing age. The study investigates the load–displacement curve variations for grouted sleeves made with grouting material cured for 1 day, 3 days, and 28 days. It explores the anchorage effects of the grouting material on grouted rebar sleeves at rebar anchorage depths of 7 D, 8.5 D, and 10 D. Strain measurements at selected test points for each group are analyzed to evaluate the sleeve deformation during uniaxial tension tests and to assess failure patterns. Table 7.5 presents the characteristic force values and failure states for different rebar anchorage depths at various curing ages.

Based on the tensile load P obtained from the uniaxial tension test, the horizontal bond stress between the grouting material and the anchored rebar can be calculated using Eq. 7.2.

$$\tau = \frac{p}{\pi d l_{\mathrm{a}}} \tag{7.2}$$

Table 7.5 Failure patterns of the samples with different anchorage length at different ages

Anchorage length (D)	Curing age (d)	Maximum tensile force (MPa)	Yield strength (MPa)	Failure mode
7	1	47.25	38.53	Relative slip occurs between the rebar and the grouting sleeve
	3	55.88	31.93	Relative slip occurs between the rebar and the grouting sleeve
	28	66.90	46.63	Broken at the tensile position of the grout rebar sleeve
8.5	1	61.7	56.34	Relative slip occurs between the rebar and the grouting sleeve
	3	66.57	44.30	Broken at the tensile position of the grout rebar sleeve
	28	66.42	46.66	Broken at the tensile position of the grout rebar sleeve
10	1	65.71	45.90	Broken at the fixed position of the grout rebar sleeve
	3	67.05	42.93	Broken at the fixed position of the grout rebar sleeve
	28	66.75	47.03	Broken at the fixed position of the grout rebar sleeve

where: τ represents the average bond stress. p represents the tensile load. d represents the diameter of the rebar. l_a represents the anchorage length.

When p takes the maximum tensile load, the calculated bond stress is also the maximum bond stress, denoted as τ_u.

The average bond stress between the grouting material and the anchored rebar can be calculated using Eq. 7.3:

$$\tau_u = k\sqrt{f_c} \tag{7.3}$$

where: k represents a constant coefficient. f_c represents the compressive strength of the grouting material.

The bond between the anchored reinforcing bars and the inner reinforcing bars in the grouting steel sleeve primarily originates from the adsorption and cementation effect caused by the contact between the cementitious material and the ribbed surface of the reinforcing bars during the hydration and hardening process. This bonding strength progressively increases as the hardening strength of the grouting material improves. Furthermore, the grouting material exhibits expansion and contraction characteristics. Upon injection into the sleeve, it initially expands during the setting phase, reaches a peak, and subsequently undergoes slight contraction, thereby tightly encapsulating the reinforcing bars. When the reinforcing bars are subjected to tensile forces, frictional forces are generated at the interface between the grouting material and the anchored reinforcing bars. Simultaneously, the ribs on the anchored reinforcing bars engage with the grouting material to produce mechanical interlocking forces during tensioning, effectively preventing the reinforcing bars from being pulled out of the sleeve. During loading, the primary bonding mechanisms between the anchored reinforcing bars and the grouting material are mechanical interlocking and frictional forces, while adhesive forces are relatively minor and can be easily overcome.

When tensile loads are applied to the reinforcing bars at both ends of the grouted steel sleeve, shear stress (referred to as bonding anchorage stress) develops at the interface between the grouting material and the anchored reinforcing bars. This shear stress is distributed along the direction of reinforcement tension and varies across different locations. Specifically, the shear stress progressively diminishes from the loaded end toward the center of the grouted steel sleeve. At the head of the anchored reinforcing bars within the sleeve (near both sides), the shear stress attains its minimum value, with a consistent pattern observed on both segments of the reinforcing bars.

Figure 7.24a–c illustrate the average bond stress-displacement curves for specimens at various curing ages under differing reinforcement anchorage depth conditions

As depicted in Fig. 7.24, the bond force–displacement curve of the specimen with a steel bar pulled apart within the grouted sleeve undergoes four distinct stages: elasticity, yielding, strain hardening, and necking.

Elastic stage: The tensile load is relatively low, and the grouted steel sleeve demonstrates high stiffness. As tensile displacement increases gradually, the load grows linearly, and the bonding stress follows a linearly increasing trend. During this phase, the bonding force between the grouting material and the surface of the anchoring rebar diminishes, while the squeezing interaction between the ribs of the ribbed rebar and the surrounding grout intensifies. This may result in the formation of micro-cracks or localized deformations.

Yield stage: Upon entering the yield stage, as tensile displacement continues to increase, the average bond stress fluctuates around a constant level, forming a

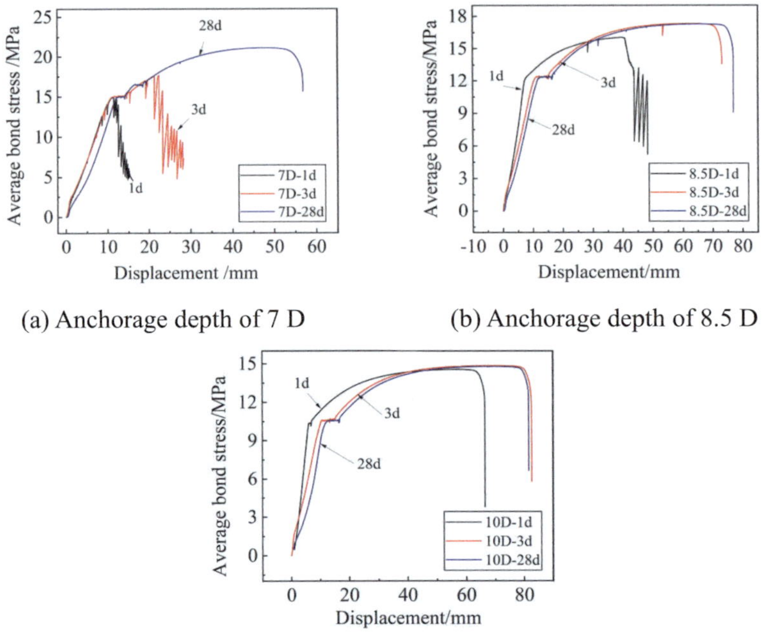

(a) Anchorage depth of 7 D

(b) Anchorage depth of 8.5 D

(c) Anchorage depth of 10 D

Fig. 7.24 Aaverage bond stress of samples with different anchorage depth

pronounced yield plateau. This stage signifies the onset of plastic deformation in the specimen.

Strengthening stage: During the strengthening phase, as tensile displacement further increases, the squeezing interaction between the grouting material within the sleeve and the transverse ribs of the reinforcing bars becomes more pronounced. Due to the rise in shear forces, additional micro-cracks may develop within the grouting material, leading to a gradual reduction in mechanical interlocking. At this stage, the bond stress primarily depends on mechanical interlocking and frictional forces, exceeding the ultimate load capacity of the reinforcing bars. Specimens with varying curing ages or anchorage depths may exhibit differing failure modes. For specimens with shorter curing periods, the lower strength of the grouting material prevents it from fully resisting the compressive forces induced by the tensile load on the reinforcing bars, causing slight fragmentation at the sleeve edges and a notable increase in internal cracking.

Necking stage: During the necking stage, as tensile displacement continues to increase, the reinforcing bars reach their ultimate load capacity and subsequently fracture.

As depicted in Fig. 7.24, The average bond force–displacement curve for specimens with reinforcing bars pulled out from grouting sleeves can be categorized into five distinct stages.

In the elastic stage, the relatively short curing period and shallow anchorage depth of the reinforcing bars result in lower mechanical strength of the grouting material within the grouting sleeves, leading to the formation of numerous cracks during this phase. As tensile displacement increases gradually, so does the applied load, causing a rapid decline in adhesion between the grouting material and the reinforcing bars. The insufficient strength of the grouting material between the reinforcing bar ribs to resist shear and compressive stresses leads to a gradual reduction in mechanical interlocking force, with bond force primarily relying on friction.

During the yielding stage, the behavior of these specimens is largely similar to that of others, though some specimens exhibit no clear yield point. In the later part of the elastic stage, the extrusion interaction between the steel bar ribs and the grouting material intensifies. With increasing shear stress, the mechanical interlocking force diminishes until it is entirely lost. At this point, the bond force between the grouting material and the steel bar depends mainly on friction, which is lower than the ultimate load capacity of the steel bar. During this phase, the grouting material develops extensive cracking, its expansion effect diminishes, and spalling of the grout begins.

In the descending stage, the load decreases abruptly. In the residual stage, the confining stress exerted by the steel sleeve on the grouting material and reinforcing bars plays a critical role, compensating for the stress loss caused by the shearing of transverse ribs at the bonding interface. Consequently, the connection retains good ductility. The bonding stress exhibits a uniform and cyclic downward trend, with residual bonding strength contributing significantly.

The symbols in the Eq. 7.4 correspond to the explanations provided in Eqs. 7.2–7.3. The bond stress and K values of the grouted sleeves made from grouting material are calculated using the above Eqs. 7.2–7.4, and the results are shown in Table 7.6.

$$K = \frac{p}{\pi d l_a \sqrt{f_c}} \tag{7.4}$$

The analysis of the K values presented in Table 7.6 demonstrates that the performance of grouted steel sleeves is significantly influenced by the anchorage depth of the reinforcing bars. Specifically, when the anchorage depth is 7 D, a K value below 2.5222 results in slippage of the reinforcing bar within the sleeve filled with low-carbon cementitious grout, thereby failing to satisfy the anchorage requirements. Conversely, at an anchorage depth of 8.5 D, if the K value is less than 2.14, the test concludes with the bar being pulled apart, indicating compliance with the anchorage requirements. Furthermore, at an anchorage depth of 10 D, a K value below 1.938 also leads to the bar being pulled apart during testing, confirming adherence to the anchorage criteria. In conclusion, as the anchorage depth increases, the K value exhibits a clear downward trend.

Table 7.6 Bond stress of samples with diffrernt anchorage depth

Anchorage length (D)	Curing age (d)	Peak load (MPa)	Failure mode	τ_u (MPa)	*K*
7	1	47.25	Rebar was pulled out	14.92834	1.990446
	3	55.88	Rebar was pulled out	17.65494	2.171579
	28	66.90	Rebar was broken	21.13664	2.522272
8.5	1	61.7	Rebar was pulled out	16.05366	2.140488
	3	66.57	Rebar was broken	17.32078	2.130477
	28	66.42	Rebar was broken	17.28175	2.062262
10	1	65.71	Rebar was broken	14.53247	1.937662
	3	67.05	Rebar was broken	14.82882	1.823963
	28	66.75	Rebar was broken	14.76247	1.761632

7.7.2 Effect of Anchorage Length on Bond Stress

The load–displacement curves of the grouting steel sleeve specimens with different anchorage lengths are shown in Fig. 7.25, and the failure morphology of samples is shown in Fig. 7.26.

1. With an anchorage depth of 10 D, the grouted steel sleeve undergoes four distinct stages: elastic, yield, strengthening, and necking, with a clearly defined yield point. The failure mode concludes with the reinforcing bar being fractured, and no relative slippage occurs between the reinforcing bar and the sleeve.

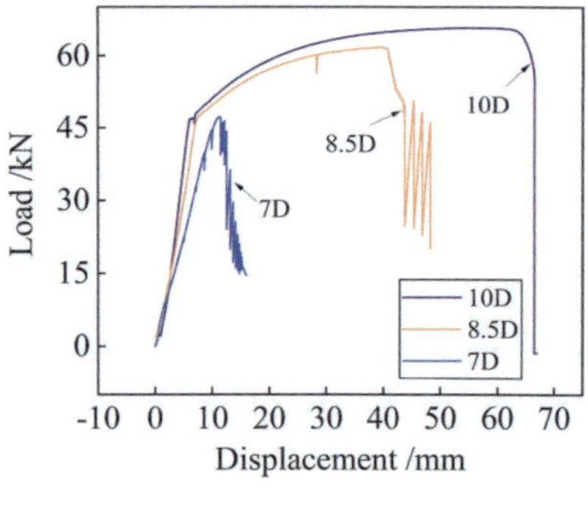

Fig. 7.25 Load–displacement curve of samples with different anchorage lengths at cuing age of 1 d

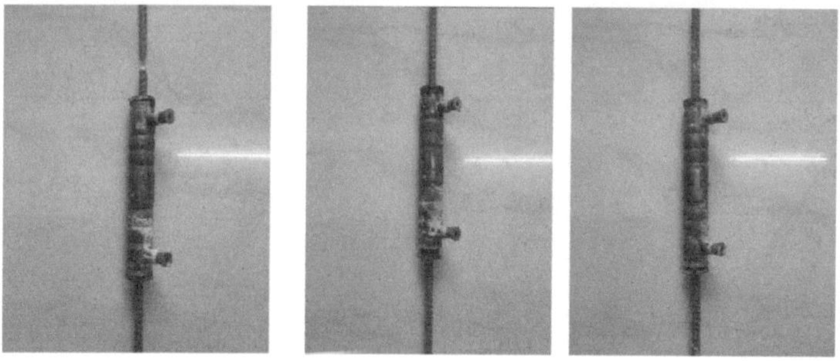

Fig. 7.26 Failure morphology of samples with different anchorage length at curing age of 1 d

2. With an anchorage depth of 8.5 D, the grouted reinforcing bar sleeve progresses through three stages: elastic, yield, and strengthening, without entering the necking stage and exhibiting an indistinct yield point. During the tensile process, when the maximum tensile force applied to the specimen has not yet reached the ultimate load of the reinforcing bar, relative slippage occurs between the reinforcing bar and the sleeve, resulting in a sudden drop in load and indicating specimen failure. Subsequently, the load exhibits a fluctuating and uniform decrease due to the frictional forces generated during the relative slippage between the reinforcing bar and the grout within the sleeve.
3. With an anchorage depth of 7 D, the grouted reinforcing bar sleeve demonstrates only a relatively pronounced elastic stage, with its maximum tensile force approaching the yield strength but failing to exhibit a yield stage. Damage occurs as the sleeve approaches the yield stage. Prior to reaching the maximum tensile strength, the load–displacement curve shows a phenomenon of initially decreasing and then increasing, which results from minor damage between the reinforcing bar and the grouting material inside the sleeve, causing cracking of the grouting material between the ribs of the reinforcing bar. At this point, although the grouting material inside the specimen has sustained initial damage and lost part of its mechanical interlocking force, it can still withstand a certain level of tensile force. After the tensile strength reaches its peak, the load exhibits a uniform and cyclic decreasing trend, indicating that the grouted reinforcing bar sleeve specimen has failed and significant slippage has occurred between the reinforcing bar and the sleeve. Despite this, during the cyclic reduction phase of the load, the specimen continues to bear some tensile force due to the residual mechanical interlocking force and frictional force between the ribs of the reinforcing bar and the damaged grouting material. As shown in the figure, as displacement increases, the ascending phase in each cycle gradually diminishes, reflecting the progressive intensification of damage until the complete failure of the grouted reinforcing bar sleeve specimen.

It is evident that, under a curing age of 1 day, the grouted steel sleeve can only satisfy the usage requirements when the anchorage depth is 10 D. In contrast, at anchorage depths of 7 D and 8.5 D, the grouted steel sleeve fails to meet the specified

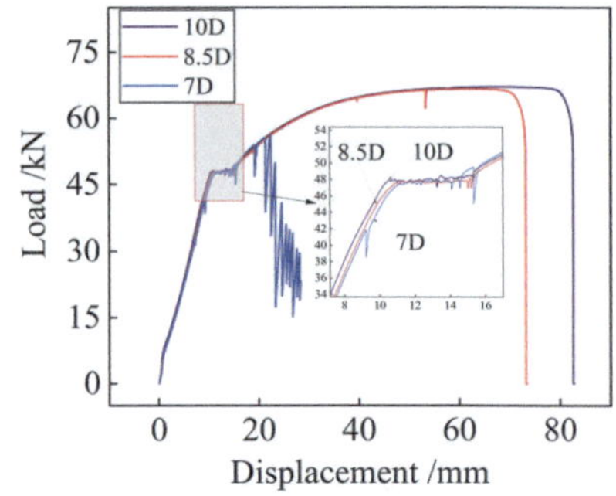

Fig. 7.27 Load–displacement curve of samples with different anchorage lengths at cuing age of 3 d

requirements. As the anchorage depth decreases, the tensile performance of the sleeve progressively diminishes. Furthermore, an increased anchorage depth enhances both the mechanical interlocking force and frictional resistance.

Figure 7.27 illustrates the load–displacement curves of the grouted steel sleeve specimens under varying steel bar anchorage depths after a curing period of 3 days. It is evident that, when the steel bar anchorage depth is 10 D, its load–displacement curve aligns with that observed after 1 day of curing. The curve progresses through four distinct stages: elastic, yield, strengthening, and necking, concluding with the fracture of the steel bar, which indicates a pronounced yield stage. When the steel bar anchorage depth is 8.5 D, the load–displacement curve also traverses these four stages, exhibiting a clear yield stage. However, in contrast to the specimen with an anchorage depth of 10 D, the specimen with an anchorage depth of 8.5 D experiences a brittle fracture during the test.

When the anchorage depth of the reinforcing bar is 7 D, the grouted reinforcing bar sleeve experiences three stages: elastic, yield and strengthening. However, before the maximum tensile force of the reinforcing bar is reached in the strengthening stage, the specimen fails, but still shows obvious yield point characteristics. In the elastic stage, the sudden drop of the load–displacement curve is due to the damage of the reinforcing bar at the fixed end by the fixture, but this situation has no significant impact on the test results. During the unidirectional tensile process, the curves of the elastic stage and the yield stage of the specimen are basically the same as those of the specimens with anchorage lengths of 10 D and 8.5 D; however, in the strengthening stage, relative slippage failure occurs between the reinforcing bar and the grouted sleeve, indicating that an anchorage depth of 7 D cannot provide sufficient anchorage force. At this time, the mechanical interlocking force and friction force provided by the specimen are insufficient, resulting in a uniform and cyclic decreasing trend of the load–displacement curve at the time of failure. Subsequently, after a significant drop in load, the curve enters a state of uniform and cyclic decrease again. This phenomenon may be caused by the large-scale slippage of the reinforcing bar ribs. Specifically, small-scale slippage occurs when the reinforcing bar ribs are located in the low-carbon cementitious grout section, while large-scale slippage occurs when the ribs slip to the next rib position, thus forming this type of failure mode. The failure mode of the grouted reinforcing bar sleeve is shown in Fig. 7.28.

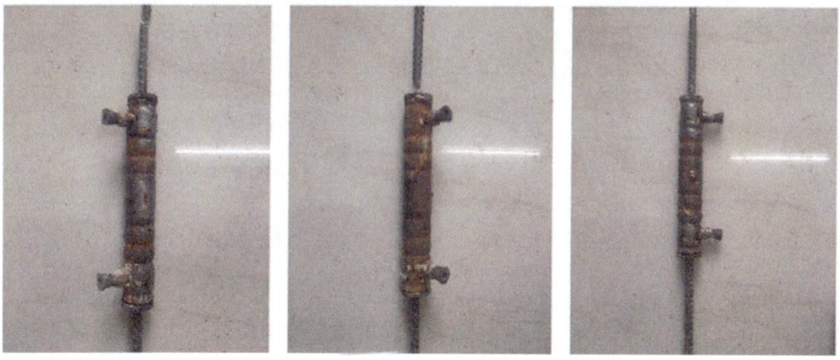

Fig. 7.28 Failure morphology of samples with different anchorage length at curing age of 3 d

As illustrated in Fig. 7.29, when the grouted steel sleeve is cured for 28 days, specimens with varying steel bar anchorage depths all conclude with the steel bar being pulled apart during the unidirectional tensile test. From the figure, it is evident that specimens with three distinct anchorage depths undergo four stages: elasticity, yield, strengthening, and necking. The differences in failure displacements among the specimens can be attributed to variations in the lengths of the steel bars. Furthermore, the load–displacement curves of the three specimens align closely during the elastic stage, while those of specimens with embedded depths of 8.5 D and 10 D also exhibit close agreement during the yield and strengthening stages. However, when the embedded depth is 7 D, its load–displacement curve demonstrates two distinct yield stages. This phenomenon may arise from slight mechanical property discrepancies between the two steel bars connected by the grouted sleeve. During the tensile process, one steel bar reaches its yield strength earlier and transitions into the strengthening phase, followed by the second steel bar entering the yield stage, thereby causing the occurrence of dual yield phenomena. Subsequently, the specimens progress through the strengthening and necking stages before ultimately failing via tensile fracture. Based on these observations, it can be concluded that after 28 days of curing, the grouting material adequately satisfies the anchoring requirements of the grouted sleeve.

Fig. 7.29 Load–displacement curve of samples with different anchorage lengths at cuing age of 28 d

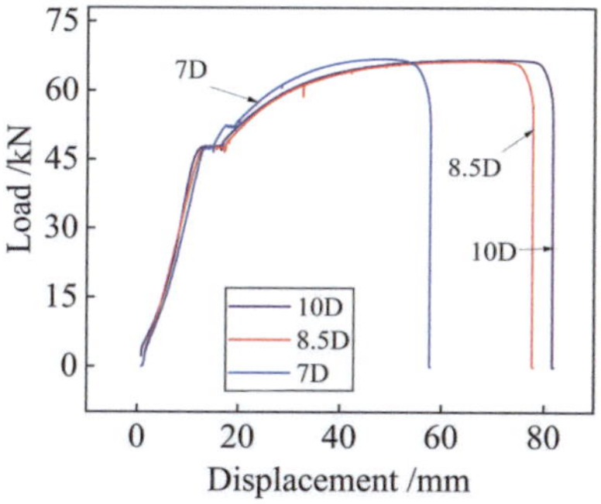

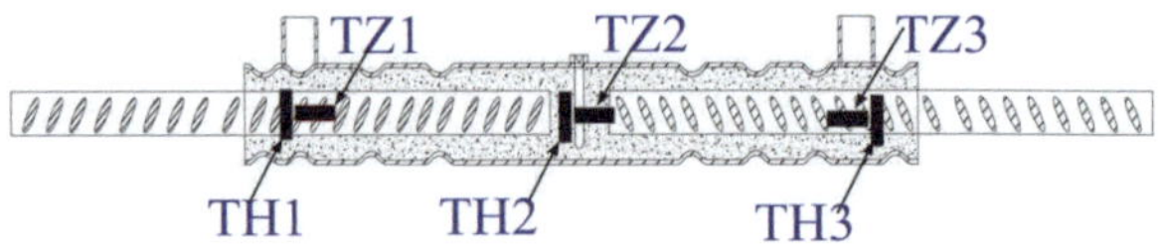

Fig. 7.30 Paste position of strain gauge

7.7.3 Strain Analysis of Grouted Rebar Sleeves

In this experiment, three points on the surface of the grouted steel sleeve were selected for measurement. The curves of circumferential strain and longitudinal strain at each point as functions of load were subsequently obtained. Based on these measurements, the load-strain variation characteristics of the grouted steel sleeve under different curing ages and varying steel bar anchorage depths were analyzed. It should be noted that positive strain indicates tensile deformation, while negative strain indicates compressive deformation. The specific locations for strain gauge attachment are illustrated in Fig. 7.30.

The load-strain curves of the grouting steel sleeve at the curing age of 1 day with different anchorage lengths (10 D, 8.5 D and 7 D) are shown in Fig. 7.31.

As illustrated in Fig. 7.31a, when the reinforcing bar is embedded at a depth of 10 D, the stress–strain variation curves of each point on the grouted steel sleeve exhibit notable differences. Specifically, the strain changes at points TH2 and TZ2 become

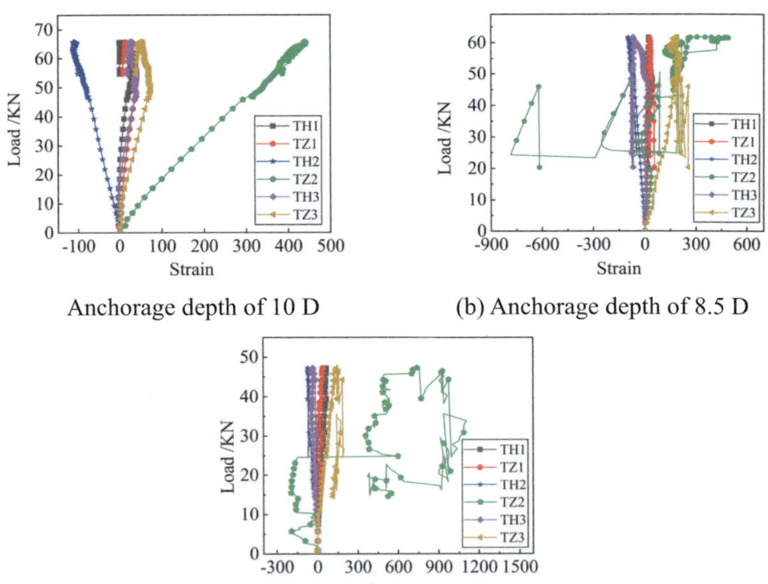

Anchorage depth of 10 D

(b) Anchorage depth of 8.5 D

(c) Anchorage depth of 7 D

Fig. 7.31 Load-strain curve of samples with different anchorage depth at the curing age of 1 d increasingly pronounced as the load progressively increases, whereas the strains at other points remain relatively small. This phenomenon can be attributed to the fact that points TH2 and TZ2 are located in the central region of the grouted steel sleeve, where no reinforcing bars are present, and this area is solely filled with grouting material. During uniaxial tensile loading, this region demonstrates substantial longitudinal deformation. Additionally, due to the Poisson effect, the circumferential strain in this region reaches its peak value. Consequently, the circumferential and longitudinal strains at points TH2 and TZ2 are significantly greater than those at other measurement points.

During the stretching process, the longitudinal length of the steel bar sleeve elongates, while the circumferential diameter at this point shows a tightening trend, which reflects the interaction between the steel bar sleeve and the low-carbon cementitious grout. For the four measuring points TZ1, TH1, TZ3, and TH3, the strain changes are relatively small as the load increases, and all are positive strains. Among them, TZ3 and TH3 are located at the stretching end, while TZ1 and TH1 are at the fixed end. As shown in the figure, the order of strain values is TZ3 > TZ1 and TH3 > TH1, indicating that the strain at the stretching end is greater than that at the fixed end. Additionally, only the strain value at point TH2 shows a negative strain as the load increases, which is due to the expansion effect of the internal grout. Specifically, during the stretching process, the grout in the middle of the grouted steel bar sleeve is subjected to the circumferential inward pressure of the sleeve. The Poisson effect and bonding effect exceed the expansion effect of the grout, resulting in circumferential tightening and slight deformation. In contrast, the strain values at points TH1 and TH3 both show positive strains as the load increases, because during the stretching process, the Poisson effect and bonding effect at both ends are less than the expansion effect of the internal grouting material, leading to circumferential expansion. Eventually, the steel bar breaks, and the strain values at all points return to zero.

Figure 7.31b illustrates the load-strain curve of the grouted steel sleeve when the anchorage depth of the steel bar is 8.5 D. As can be observed from the figure, the load-strain curve at point TZ2 exhibits the largest variation range. Specifically, during the elastic stage, the longitudinal strain at the midpoint of the steel sleeve undergoes continuous changes. At the onset of the tensile test, the load-strain curve at point TZ3 initially demonstrates positive strain, subsequently decreases to zero, transitions into negative strain, and ultimately reverts to positive strain. This behavior arises because the length of the steel bar anchored within the sleeve is 8.5 D, resulting in a relatively long distance between the end of the steel bar and the midpoint of the sleeve. During the initial tension phase, the tensile force acting on the sleeve is minimal, and no relative displacement occurs between the steel bar inside the sleeve and the grouting material. Consequently, points TZ1, TZ2, and TZ3 are all subjected to longitudinal tension, with their strains progressively increasing as the load increases. As the load continues to rise, the region covered by the steel bar inside the sleeve continues to develop rapidly along the initial large strain trend, whereas the region not covered by the steel bar exhibits a distinct change pattern

The deformation in the region covered by reinforcing bars is smaller than that in the region without reinforcing bar coverage. At this point, the stress distribution in the middle section of the sleeve undergoes a noticeable change. During the tensile process, the ribbed reinforcing bars within the covered section interact with the grouting material. Under the combined influence of mechanical interlocking and frictional forces, the grouting material surrounding the reinforcing bars experiences compression, leading to a circumferential outward expansion effect. Since this expansion effect exceeds the Poisson effect induced during the tensile process, points TH1 and TH3 exhibit positive strain. Conversely, point TH2 demonstrates negative strain; however, its strain magnitude is reduced compared to when the anchorage depth of the reinforcing bars is 10 D. This reduction occurs because the outward displacement of the reinforcing bars results in a shallower anchorage depth, thereby increasing the area of low-carbon cementitious grouting material in the middle section that is not covered by reinforcing bars and enhancing the expansion effect of the grouting material in this uncovered region. Nevertheless, despite the increased expansion effect, it remains weaker than the Poisson effect generated under unidirectional tensile loading. Consequently, the strain at point TH2 continues to exhibit negative strain.

As the load gradually increases, relative slippage occurs between the reinforcing bars and the grouting sleeves. Following failure, the ribs of the reinforcing bars slide toward the positions of the adjacent ribs, causing the strain in the sleeve to progressively diminish to zero. During the subsequent pullout process, the reinforcing bars transfer compressive forces from both ends toward the middle through the grouting material, inducing a negative strain change in the central region of the sleeve. Specifically, point TZ3, located near the tensile end, exhibits positive strain during the tensile process. The strain variation pattern after the elastic stage is similar to that observed at point TZ2, further validating the strain characteristics of point TZ2. Regarding points TH1, TH2, and TH3, no significant differences are noted in their strain behaviors: points TH1 and TH3 exhibit positive circumferential strain, whereas point TH2 demonstrates negative strain. Their force characteristics align with those observed when the anchorage depth is 10 D; however, there are notable distinctions. The strain amplitude at point TH3 is greater, while the strain variation at point TH2 is less pronounced. These observations are consistent with the trends exhibited in the load–displacement curve.

Figure 7.31c presents the load-strain curve of the grouted steel sleeve when the anchorage depth of the steel bar is 7 D. Notably, the strain at point TZ2 exhibits the most significant variation with increasing load. In comparison to the scenario where the anchorage depth is 8.5 D, this group of grouted steel sleeves features a shallower anchorage depth for the steel bar, as well as a greater distance from the middle stop nail. The circumferential strain at TH1 and TH3 has decreased due to the shortened anchorage length of the steel bars. This reduction in anchorage length weakens the expansion effect of the grouting material while enhancing the Poisson effect during the tensile process, thereby diminishing the overall circumferential expansion. In the middle region without steel bar coverage, the increased area results in a greater volume of grouting material contributing to the expansion effect during tension, thus amplifying the expansion in this region. However, the increase in the Poisson effect remains relatively small. Consequently, under the combined influence of these factors, although the circumferential strain remains negative, its magnitude has further decreased. The strain at point TZ2 is negative strain, which is caused by the combined effect of the expansion of the grouting material in the middle and the Poisson effect generated during the tensile process. As the anchorage length of the reinforcing bars shortens, the area subjected to compression in the middle increases. The longitudinal strain gauge is located at the junction or within the area of circumferential tightening where the compressive expansion effect and the circumferential tightening effect interact, thus showing negative strain. As the load gradually increases, the strain in this area turns positive until relative slippage occurs between the reinforcing bars within the grouting sleeve, leading to the ultimate failure of the specimen. The strain at TH1 and the load-strain curve at TH3 exhibit a relatively symmetrical relationship. Specifically, TH1, located near the fixed end, demonstrates positive strain, whereas TH3, situated closer to the tensile end, exhibits negative strain. This phenomenon arises because the anchorage-end reinforcing bars are shorter and positioned nearer to the sleeve edge. During the unidirectional tensile test, the deformation at the tensile end is more pronounced. The tensile force applied exceeds the inward compressive force exerted by the sleeve's outer wall on the low-carbon cementitious grouting material, leading to greater longitudinal deformation compared to circumferential deformation. Consequently, this results in negative circumferential strain at the tensile end.

The load-strain curves of the grouting steel sleeve at the curing age of 3 days with different anchorage lengths (10 D, 8.5 D, and 7 D) are shown in Fig. 7.32.

As shown in Fig. 7.32a–b, when the anchorage depths of the reinforcing bars are 10 D and 8.5 D, the maximum variations in circumferential strain and longitudinal strain occur at points TH2 and TZ2. At the age of 3 days, both specimens with these two anchorage depths failed due to tensile fracture of the reinforcing bars. The strain variation trends for the specimens with the two anchorage depths exhibit similar characteristics.

For the specimen with an anchorage depth of 10 D, the distance between the end of the reinforcing bar and the stop nail is relatively short, and the length of the central section without reinforcing bars is minimal. Under unidirectional tensile loading, the middle portion experiences the least deformation-induced compressive force, resulting in the greatest circumferential strain change.

When the anchorage depth is 8.5 D, the strain variation trend is similar to that observed at 10 D. However, the strain values are relatively lower, and the strain distribution across all test points is more uniform. As the load increases, the strain at point TZ2 becomes the largest due to the transfer of tensile force from the reinforcing bar through the grouting material to the external sleeve, causing maximum deformation of the sleeve. With further increases in tensile strength, the sleeve near the stopper elongates while simultaneously contracting radially, thereby generating negative strain.

Additionally, the load-strain curves of points TH1 and TH3 exhibit a largely symmetrical trend. The negative strain at point TH3 occurs because the strain gauge is positioned at the front of the end groove. During tensile loading, the rear portion of the

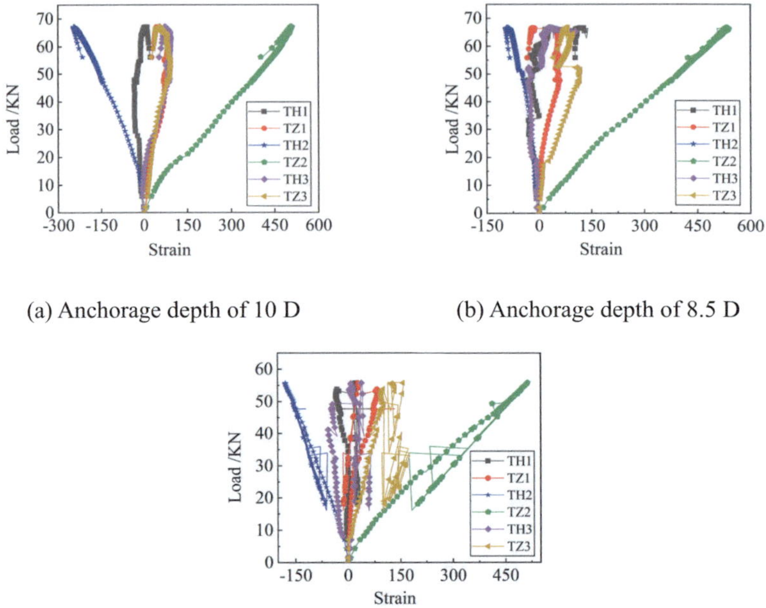

(a) Anchorage depth of 10 D

(b) Anchorage depth of 8.5 D

(c) Anchorage depth of 7 D

Fig. 7.32 Load-strain curve of samples with different anchorage depth at the curing age of 3 d

groove generates positive strain due to the expansion of the grouting material, whereas the front portion generates negative strain due to tensile effects. Consequently, the circumferential strain value at point TH3, which corresponds to the tensile end, is slightly greater than that at point TH1, which represents the fixed end. Meanwhile, the load-strain curves of points TZ1 and TZ3 largely overlap, indicating consistent deformation behavior of the sleeve at both the fixed end and the unidirectional tensile end.

In summary, the grouted steel sleeve achieves reliable anchorage of the two sections of reinforcing bars at an anchorage depth corresponding to 3 days of curing and demonstrates excellent mechanical performance during the unidirectional tensile test.

Figure 7.32c presents the load-strain curves for a reinforcing bar anchorage depth of 7.5 D. From the elastic stage to the yield stage, the circumferential strains at points TH1, TH2, and TH3 are all negative, with the trend following TH2 > TH3 > TH1. Simultaneously, the longitudinal strains at points TZ1, TZ2, and TZ3 are positive, exhibiting a magnitude relationship of TZ2 > TZ3 > TZ1. Furthermore, the circumferential load-strain curve at point TH1 is symmetrically aligned with the load-strain curve at point TZ1, while the circumferential load-strain curve at point TH3 is also symmetrically aligned with the longitudinal load-strain curve at point TZ3. Notably, the circumferential and longitudinal strains at point TH3 are slightly greater than those at point TH1. When the anchorage depth of the reinforcing bars is 7.5 D, the forces acting on each test point are uniformly distributed, and the deformations are consistently coordinated. After specimen failure, the strain changes at points TH2 and TZ2 remain relatively small, whereas the strain changes at points TH3 and TZ3 are more pronounced. Based on these observations, it can be concluded that after relative slip occurs in the reinforcing bars, the tensile end experiences the greatest strain influence, while the fixed end exhibits the least strain influence

As demonstrated above, the load-strain curves of the grouted steel sleeve with anchorage depths of 10 D and 8.5 D exhibit minimal differences, and the circumferential and longitudinal strains at each measurement point are largely consistent. This suggests that once the compressive strength of the grouting material satisfies the anchorage requirements, further increasing the anchorage depth of the reinforcing bar within the steel sleeve has a limited effect on the anchorage performance of the grouted steel sleeve.

The load-strain curves of the grouting steel sleeve at the curing age of 28 days with different anchorage lengths (10 D, 8.5 D, and 7 D) are shown in Fig. 7.33.

Figure 7.33a presents the load-strain curve of the grouted steel sleeve with a steel bar anchorage depth of 10 D. As observed from the figure, the circumferential strain at point TH2 and the longitudinal strain at point TZ2 exhibit the largest

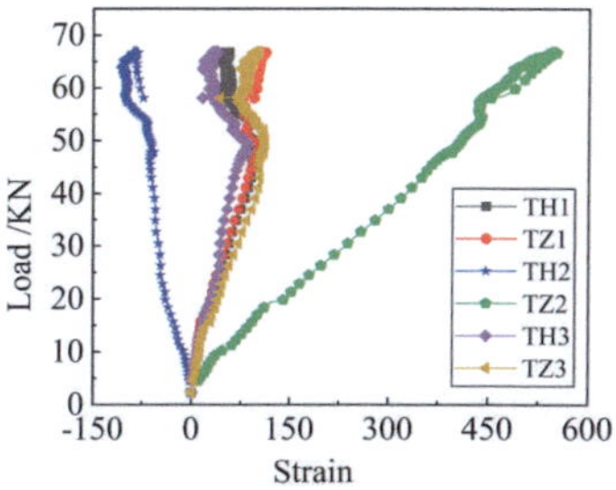

(a) Anchorage depth of 10 D

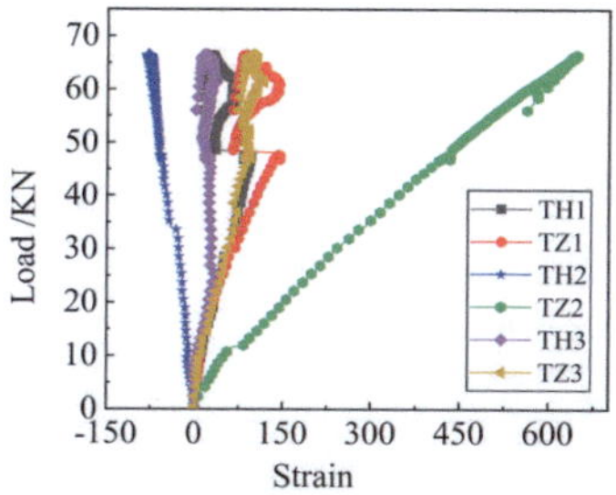

(b) Anchorage depth of 8.5 D

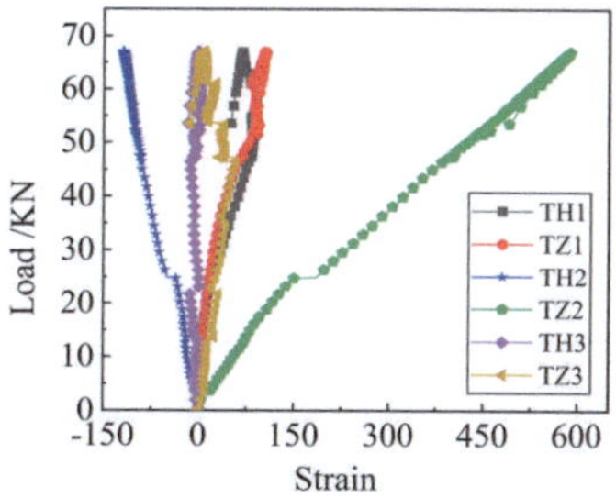

(c) Anchorage depth of 7 D

Fig. 7.33 Load-strain curve of samples with different anchorage depth at the curing age of 28 d variation range. Furthermore, the stress–strain curves at points TH1 and TZ1 nearly completely coincide, whereas noticeable differences exist in the stress–strain curves at points TH3 and TZ3. Based on these observations, it can be concluded that under unidirectional tensile loading, the strain variation at the tensile end of the sleeve is more pronounced. Additionally, upon entering the yield stage, the stress–strain curves at all measurement points demonstrate a certain degree of synchronization.

Figure 7.33b presents the load-strain curve of the grouted steel sleeve with an anchorage depth of 8.5 D. Among the test points, the longitudinal strain at point TZ2 exhibits the largest variation, followed by the circumferential strain at point TH2, while the strain variations at the remaining points are relatively minor. When the anchorage depth is reduced to 8.5D, the strain variation at point TH3 decreases compared to that at an anchorage depth of 10D and fluctuates within a range near zero. Furthermore, the strain values at points TH1, TZ1, and TZ3 nearly completely coincide.

As illustrated in Fig. 7.33c, when the anchorage depth of the reinforcing bar is 7 D, the stress–strain curve of the grouted steel sleeve exhibits a variation trend that is largely consistent with those observed at anchorage depths of 10 D and 8.5 D. However, there are notable differences: the strain at point TH3 transitions to negative values, and the strain curves at points TH1, TZ1, and TZ3 demonstrate a higher degree of overlap. Furthermore, the circumferential strain value at point TH2 is significantly greater than that of the corresponding points at anchorage depths of 10 D and 8.5 D.

7.8 Conclusions

The effects of steel slag and fly ash content on the fluidity, expansion rate, and compressive strength of grouting material prepared using LCCM were systematically investigated. Additionally, the influence of steel slag and fly ash dosage on the ettringite content in the LCCM-based grouting material was analyzed through microscopic testing methods. The impact of anchorage length on the bonding performance of grouted steel sleeves was studied by examining the load–displacement curves, bond stress-displacement curves, failure modes, and circumferential and longitudinal strains at characteristic points. The main conclusions drawn from this study are as follows:

(1) As the steel slag content increases, the fluidity of the grouting material decreases. Additionally, the vertical expansion rate of the grouting material rises with the increase in steel slag content. At various hydration ages, the compressive strength of the grouting material demonstrates a certain degree of enhancement. However, when the binder-to-aggregate ratio is 1:1.2, the compressive strength of the grouting material at 28 days exhibits a regression phenomenon. Based on comprehensive performance analysis, the optimal proportion of steel slag is determined to be 1:1.

(2) The increase in fly ash content can improve the fluidity of grouting materials. However, when the fly ash content reaches 15%, the excessive fluidity will cause the slurry to experience segregation. The increase in fly ash content will inhibit the early compressive strength of grouting materials, but it is beneficial to the development of later strength. When the fly ash content is 10%, the compressive strengths of the grouting materials at 1 day, 3 days and 28 days are 56.25, 66.05 and 70.2 MPa respectively, and there is no phenomenon of compressive strength regression.
(3) With the increase in reinforcing bar anchorage depth, the curing age required for the grouted steel sleeve to meet service requirements gradually decreases. The deeper the anchorage depth of the reinforcing bars, the lower the mechanical property requirements for the grouting material become. Under the same curing age, increasing the anchorage depth of the reinforcing bars enables compliance with anchoring service requirements. A larger anchorage area effectively compensates for the insufficient early-age compressive strength of the grouting material while enhancing its mechanical interlock and bond strength with the reinforcing bars and sleeve. Research findings indicate that the grouting material can satisfy anchoring service requirements after only 1 day of curing.
(4) During the unidirectional tensile process, the influence of different curing ages and reinforcement anchorage depths on strain changes shows significant differences. The variation range of longitudinal strain is greater than that of circumferential strain, and the position with the greatest strain value change is the midpoint of the sleeve. In contrast, the longitudinal and circumferential strain changes at both ends of the grouted steel sleeve are relatively small and are distributed more uniformly. The strain change at the midpoint of the sleeve is greatly affected by the reinforcement anchorage depth, and its deformation is caused by the combined effect of Poisson's effect and the expansion of low-carbon cementitious grout. When the expansion effect is dominant, positive strain is produced, while when the Poisson effect is dominant, negative strain is produced. At the initial stage of tensile, the circumferential and longitudinal strains at the midpoint of the sleeve will alternate between positive and negative.

Chapter 8
Natural Curing Properties of Concrete Containing Low-Carbon Cementitious Materials

8.1 Introduction

Low-carbon cementitious materials (LCCM) served as the primary binder, river sand was used as fine aggregate, and crushed stone acted as coarse aggregate for concrete preparation. Concrete made with sulphoaluminate cement was used as the control sample. A systematic investigation was conducted into the early hydration characteristics, pore structure, and mechanical properties of the developed concrete containing LCCM under natural curing. To characterize the types, quantities, and micro-morphologies of early hydration products, X-ray diffraction (XRD), thermogravimetric-differential scanning calorimetry (TG-DSC), chemically bound water analysis, Fourier transform infrared spectroscopy (FTIR), and scanning electron microscopy (SEM) were employed. Additionally, nuclear magnetic resonance (NMR) was utilized to analyze the pore structure of the concrete containing LCCM, focusing on porosity and pore size distribution. Regarding mechanical performance, the compressive strength and failure mode of early-age concrete were evaluated, and the relationships between pore structure and compressive strength, as well as the influence mechanisms of hydration products on compressive strength, were thoroughly examined.

8.2 Experimental Overview

8.2.1 Materials

Coal gangue was procured from the Inner Mongolia Jiutai Energy Group. The coal gangue was crushed using a jaw crusher to achieve particle sizes of approximately 1 cm. Subsequently, the crushed coal gangue was ground in a ball mill for 40 min to produce fine coal gangue powder. The resulting powder was dried and sieved

C. Yan et al., *Low-Carbon Cementitious Materials with 100% Solid Wastes*,
https://doi.org/10.1007/978-981-96-8742-8_8

through a 200-mesh sieve before being stored in bags for subsequent use. The physical diagram of coal gangue in bulk and powder is shown in Figs. 8.1 and 8.2. The XRD pattern of coal gangue is shown in Fig. 8.3. The chemical composition of coal gangue is presented in Table 8.1. The primary chemical constituents of coal gangue are SiO_2 and Al_2O_3, with respective contents of 46.32 and 45.34%.

Fig. 8.1 Bulk coal gangue

Fig. 8.2 Powder coal gangue

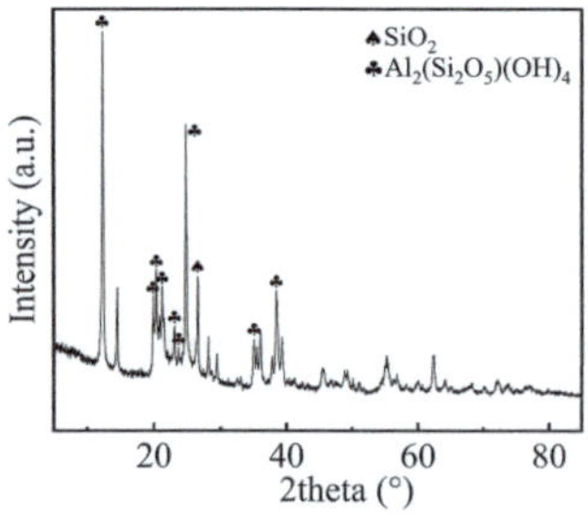

Fig. 8.3 XRD pattern of coal gangue

Table 8.1 Chemical composition of coal gangue (wt.%)

Component	SiO_2	Al_2O_3	CaO	MgO	Fe_2O_3	Na_2O	TiO_2	SO_3	Ignition loss
Proportion	46.32	45.34	2.23	0.36	0.87	0.02	1.20	1.67	28.20

Desulfurization gypsum was sourced from Gongyi City, Henan Province. To ensure its particle size is controlled to less than 200 mesh, the material was subjected to sieving. The physical diagram of the desulfurization gypsum is illustrated in Fig. 8.4. The chemical composition of the desulfurization gypsum, as determined by XRF analysis, is presented in Table 8.2. According to Table 8.2, the primary chemical components of the desulfurization gypsum are SO_3 and CaO, with respective contents of 54.9 and 41.4%. The XRD pattern of the desulfurization gypsum, as shown in Fig. 8.5, confirms that its main mineral components are $CaSO_4 \cdot 0.5H_2O$) and $CaSO_4 \cdot 2H_2O$.

Carbide slag, sourced from Gongyi City, Henan Province, was sieved through a 200-mesh sieve and subsequently stored in bags for further use. The physical diagram of the carbide slag is illustrated in Fig. 8.6, while its chemical composition is detailed in Table 8.3. As indicated in Table 8.3, the primary chemical constituent of the carbide slag is CaO, with a concentration of 94.75%. The XRD pattern of the carbide slag, presented in Fig. 8.7, confirms that its main component is $Ca(OH)_2$.

Sulphoaluminate cement, sourced from Weifang, Shandong Province, is illustrated in Fig. 8.8. Its XRD pattern is presented in Fig. 8.9. The chemical composition of the sulphoaluminate cement is detailed in Table 8.4. As indicated in Table 8.4, the

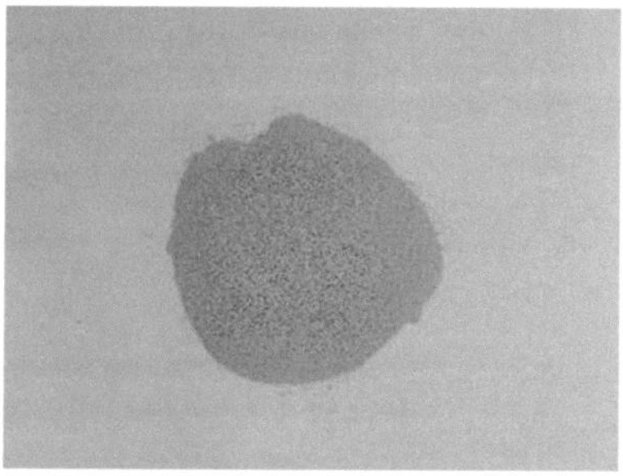

Fig. 8.4 Desulfurized gypsum

Table 8.2 Chemical composition of desulfurized gypsum (wt.%)

Component	SO_3	CaO	SiO_2	Al_2O_3	MgO	Fe_2O_3	Na_2O	K_2O	Ignition loss
Proportion	54.9	41.4	1.19	0.51	0.41	0.30	0.04	0.07	12.04

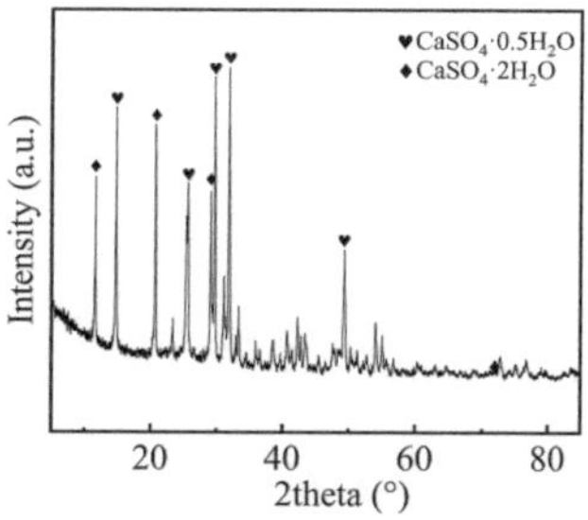

Fig. 8.5 XRD pattern of desulfurized gypsum

Fig. 8.6 Carbide slag

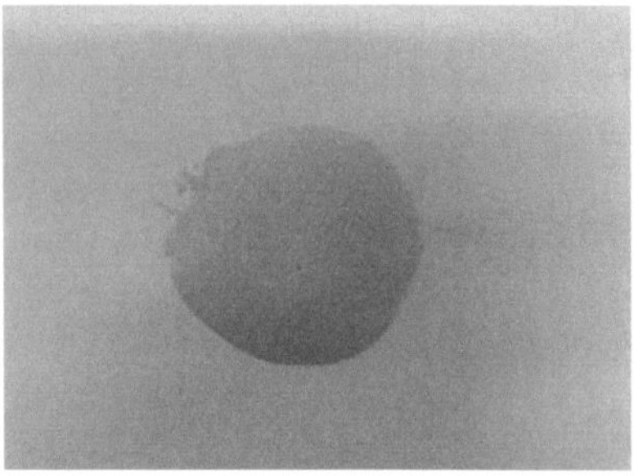

Table 8.3 Chemical composition of carbide slag (wt.%)

Component	CaO	SiO_2	Al_2O_3	SO_3	Fe_2O_3	MgO	Na_2O	TiO_2	Ignition loss
Proportion	94.75	3.04	1.09	0.47	0.28	0.07	0.03	0.04	22.03

Fig. 8.7 XRD pattern of carbide slag

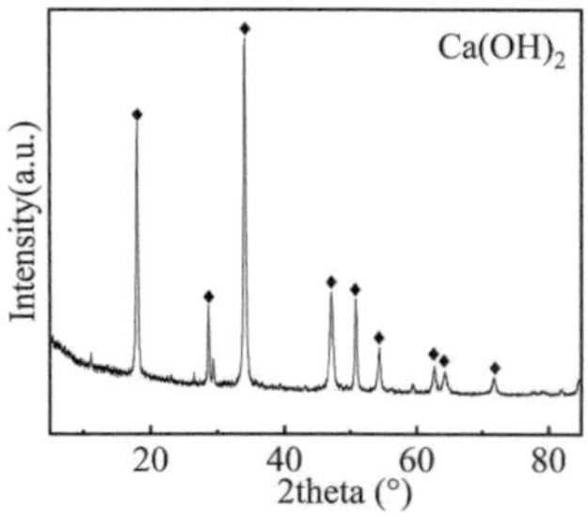

Fig. 8.8 Sulphoaluminate cement

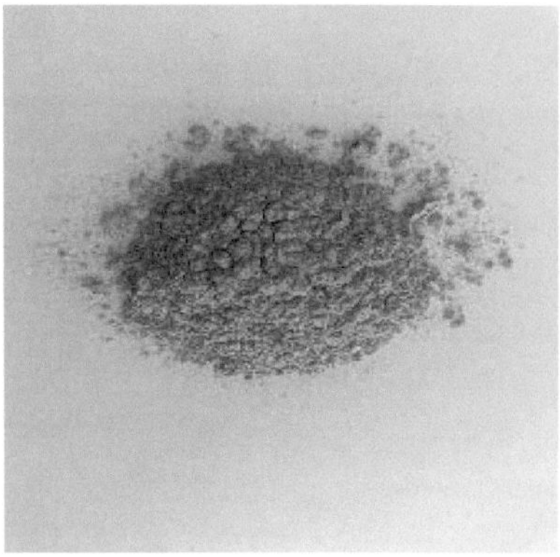

primary chemical constituents of sulphoaluminate cement are CaO, Al_2O_3, SO_3, and SiO_2, with respective concentrations of 47.37, 11.85, 18.44, and 14.93%. Figure 8.9 confirms that the mineralogical components of sulphoaluminate cement include $C_4A_3\hat{S}$, C_2S, $CaSO_4$, and $CaCO_3$.

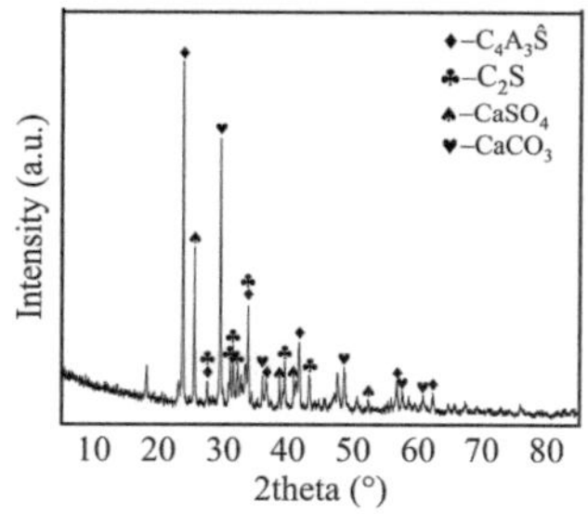

Fig. 8.9 XRD pattern of sulphoaluminate cement

Table 8.4 Chemical composition of sulphoaluminate cement (wt.%)

Component	CaO	SiO_2	Al_2O_3	SO_3	Fe_2O_3	MgO	Na_2O	TiO_2	Ignition loss
Proportion	47.37	11.85	18.44	14.93	1.41	1.36	1.98	0.45	10.05

Natural river sand was used as fine aggregate, with a particle size range of 0.282–0.425 mm. Coarse Aggregate: Natural crushed stone was used as coarse aggregate, with a particle size range of 5–20 mm. The experimental water used in this study is municipal tap water.

8.2.2 Preparation of LCCM

The preparation process of LCCM is illustrated in Fig. 8.10. The pre-processed coal gangue, desulfurization gypsum, and carbide slag were weighed according to the mix ratio. An appropriate amount of water was sprayed to facilitate pellet formation. The mixture was then placed into a mold and pressed into pellets with a diameter of 6 cm and a thickness of 3 mm using a pellet press. The pellets were dried in an oven for 3 h and then removed. Subsequently, the pellets were calcined in a high-temperature furnace at 1280 °C for 30 min. After calcination, the pellets were rapidly cooled. Once cooled to room temperature, the pellets were ground and sieved through a 200-mesh sieve to obtain the LCCM.

The XRD pattern of the LCCM is shown in Fig. 8.11, and its chemical composition is presented in Table 8.5. As shown in Table 8.5, the main chemical components of the LCCM are CaO, SiO_2, Al_2O_3, and SO_3, with contents of 44.79, 17.68, 15.28, and 17.50%, respectively. The primary mineral components are $C_4A_3\hat{S}$, $CaSO_4{\cdot}2H_2O$, C_2S, and $CaSO_4$. The compressive strength of the paste made from the LCCM is shown in Table 8.6. It can be observed that the compressive strength of the cementitious material at a water-to-binder ratio of 0.45 meets the standard for 42.5 rapid-hardening cement.

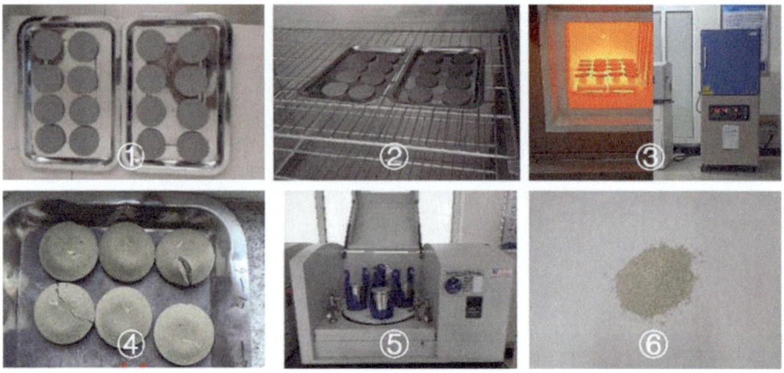

Fig. 8.10 Preparation process of LCCM

Fig. 8.11 XRD pattern of LCCM

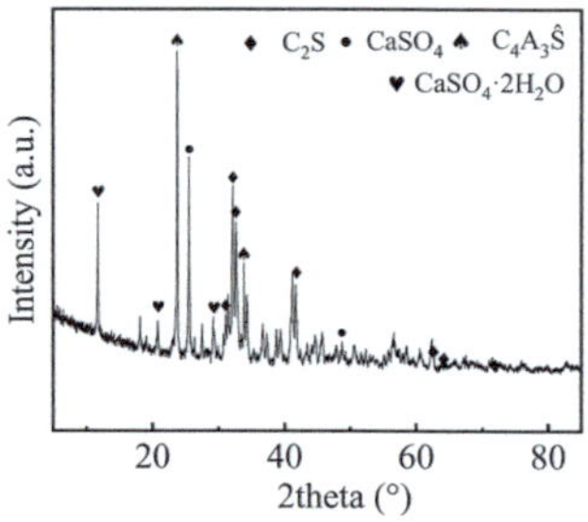

Table 8.5 Chemical composition of LCCM (wt.%)

Component	CaO	SiO_2	Al_2O_3	SO_3	Na_2O	MgO	TiO_2	P_2O_5	Ignition loss
Proportion	44.79	17.68	15.28	17.50	2.25	1.10	0.19	0.12	1.24%

Table 8.6 Compressive strength of LCCM (MPa)

Age	6 h	12 h	1 d	3 d	7 d	28 d
Compressive strength	19.57	30.18	35.33	39.42	40.00	41.67

8.2.3 Preparation and Mix Proportion of Concrete Containing LCCM

The cementitious material used in the experimental group was LCCM, while the control group used sulphoaluminate cement. The concrete mix proportions are shown in Table 8.7. Firstly, the mass of water, cementitious material, coarse aggregate, and fine aggregate was precisely weighed according to the mix ratio. Then, an appropriate amount of release agent was evenly applied to the surface of the 100 mm × 100 mm × 100 mm cubic mold. In accordance with the provisions of "Test Methods for Physical and Mechanical Properties of Concrete" (GB/T 50081-2019), a forced-disk

Table 8.7 Mix proportion of concrete (kg/m^3)

Samples	Water	Cementitious materials	Coarse aggregate	Fine aggregate	W/C
LCC	219.41	487.58	1072.67	585.09	0.45
SACC	219.41	487.58	1072.67	585.09	0.45

Note LCC denotes the concrete containing LCCM. SACC means the concrete containing sulphoaluminate cement

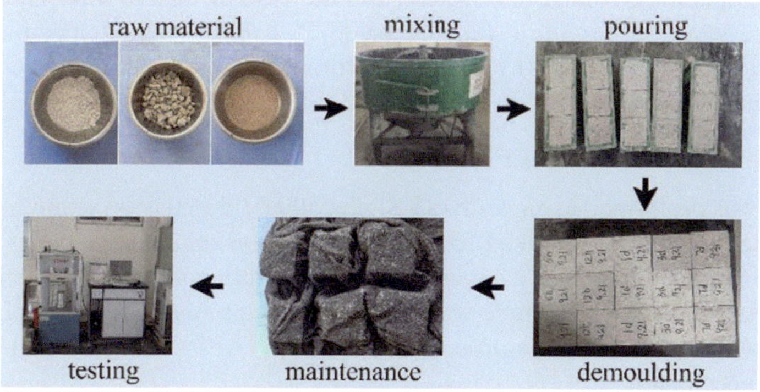

Fig. 8.12 Preparation process of concrete

mixer was used to add the weighed cementitious material, sand, coarse aggregate, and water in sequence, and the mixture was continuously stirred for at least two minutes until it reached a uniform state. The well-mixed concrete was quickly poured into the mold coated with release agent and placed on a vibrating table for vibration to ensure the density of the concrete. After 6 h of plastic film covering curing, the specimens were carefully demolded with an air pump. The curing conditions simulated the standard curing environment of concrete components at the actual construction site, that is, the specimens were placed on a wooden board and covered with black felt, and water was added at regular intervals every day to maintain a moist state. After the specimens were cured to the specified age, their mechanical properties were determined in accordance with relevant standards. The specific process of concrete preparation is shown in Fig. 8.12.

8.2.4 Test Methods

Thermogravimetric-differential scanning calorimetry (TG-DSC)

The thermogravimetric-differential scanning calorimetry analysis was performed using a NETZSCH STA 449 F5 simultaneous thermal analyzer. The measurement temperature range was 50–1000 °C, with a heating rate of 10 °C/min.

Fourier transform infrared spectroscopy (FTIR)

The FTIR was conducted using a Shimadzu IR Tracer-100 Fourier transform infrared spectrometer. The scanning wavelength range was 400–4000 cm^{-1}, with 45 scans. Samples were prepared using the potassium bromide (KBr) pellet method.

X-ray diffraction (XRD)

The XRD was performed using a Rigaku Ultima IV X-ray diffractometer with a Cu target. The operating voltage was 40 kV, and the current was 40 mA. The scanning angle range was 5–85°, with a step size of 0.02°, and a scanning speed of 8°/min. Samples weighing more than 1 g and sieved through a 200-mesh sieve were pressed into square shapes and placed in the instrument for testing.

Scanning electron microscopy (SEM)

The SEM was performed using a ZEISS Sigma 500 field-emission scanning electron microscope. The sample utilized for SEM testing was an aggregate-free fragment extracted from the interior of a concrete specimen. Following the completion of the concrete strength test, the sample was collected using tools such as a rubber mallet and steel pliers. The hydration process of the sample was terminated by immersion in anhydrous ethanol.

Nuclear magnetic resonance (NMR)

NMR technology was used to analyze the pore structure of the specimens. The tests were conducted using a MesoMR23-060H-I NMR imaging analyzer (Suzhou Niuman Analytical Instrument Corporation, China). Prior to testing, the specimens were subjected to vacuum saturation with water. To prevent the loss of surface moisture during testing, which could affect the results, the vacuum-saturated specimens were sealed with plastic film and then placed into the testing coil for pore structure analysis using the NMR analyzer.

X-ray fluorescence spectrometer (XRF)

The chemical composition of the sample was analyzed using an XRF analyzer, specifically the ScopeX 980CS model.

Compressive strength test

The compressive strength test was performed in accordance with the standard "Test Method for Mechanical Properties of Ordinary Concrete" (GB/T 50081-2002). Specimens were prepared as 100 mm × 100 mm × 100 mm cubes, with three specimens per group. These specimens were cured until they reached the specified age for testing. During the test, a loading rate of 5 kN/s was applied until specimen failure occurred, and the compressive strength data for concrete specimens at each age were recorded. The strength values for each group of three specimens were determined based on the following criteria:

(1) The arithmetic mean of the test results from the three specimens was taken as the representative strength value for the group.

(2) If the difference between the maximum or minimum value and the median value exceeded 15% of the median value, the maximum and minimum values were discarded, and the median value was adopted as the compressive strength value for the group.
(3) If both the maximum and minimum values deviated from the median by more than 15% of the median value, the test results for the group were considered invalid.

The curing ages were 6 h, 12 h, 1 day, 3 days, 7 days, and 28 days. The compressive strength of the concrete cubes was calculated using the Eq. 8.1:

$$f_a = \frac{F}{A} \tag{8.1}$$

where: f_a is the compressive strength of the sample, MPa. F is the failure load of the sample, N. A is the bearing area of the sample, mm^2.

8.3 Early Age Strength Formation Mechanism of Concrete Containing LCCM

8.3.1 Hydration Products at Early Age

Figure 8.13 presents the XRD pattern of the hydration products of sulphoaluminate cement at various curing ages, including 6 h, 12 h, 1 day, 3 days, 7 days, and 28 days. The results reveals that the diffraction peak angles for C_2S are located at 32.134°, 32.551°, and 39.403°, respectively. The diffraction peak angles for AFt are identified at 9.062°, 15.744°, 17.794°, 18.857°, 22.862°, 24.647°, and 27.437°. A diffraction peak corresponding to $C_4A_3\hat{S}$ is observed at 23.655°. The diffraction peaks of $CaCO_3$ are detected at 29.46°, 35.993°, and 48.52°. The diffraction peaks of $CaSO_4$ are located at 25.46° and 52.3°, respectively. Before hydration, the primary mineralogical constituents of the sample, as depicted in Fig. 8.13, were $C_4A_3\hat{S}$, $CaSO_4$, $CaCO_3$, and C_2S. From 6 h to 28 days, the sample contains unhydrated $C_4A_3\hat{S}$, C_2S, $CaCO_3$, as well as the hydration product AFt. Notably, the contents of AFt and $CaCO_3$ are relatively high, whereas the contents of $C_4A_3\hat{S}$ and C_2S are relatively low, suggesting incomplete hydration of the sample.

In contrast to the samples prior to hydration, AFt diffraction peaks were clearly observed in the XRD pattern after 6 h of hydration. Concurrently, the diffraction peaks of $C_4A_3\hat{S}$ were significantly reduced, and those of $CaSO_4$ were almost undetectable. These observations indicate that the sample began to hydrate at the 6-h mark. It is evident that the hydration reaction of $C_4A_3\hat{S}$ occurs relatively rapidly. During hydration, $C_4A_3\hat{S}$ reacts with $CaSO_4$ to generate a considerable amount of AFt, as described in Eq. 8.2 (Hargis et al., 2014). The diffraction peak intensity of C_2S remained largely stable, suggesting that the hydration of C_2S was minimal within the first 6 h.

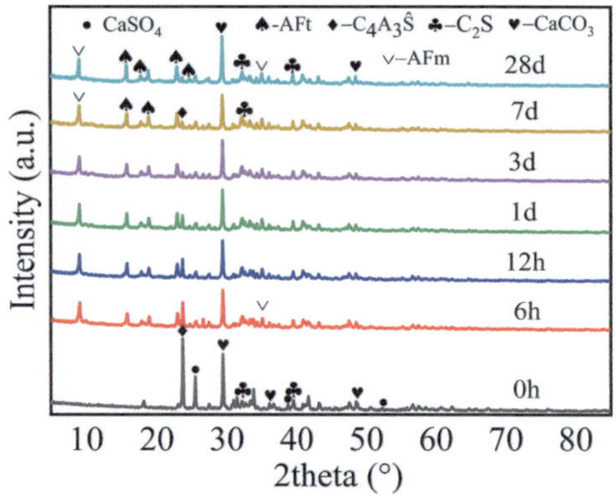

Fig. 8.13 XRD pattern of sulphoaluminate cement at different curing ages

$$C_4A_3\hat{S} + 2C\hat{S} + 38H \rightarrow C_3A \cdot 3C\hat{S} \cdot 32H + 2AH_3 \quad (8.2)$$

During the hydration period from 6 h to 28 days, the intensity of the diffraction peaks of $C_4A_3\hat{S}$ and $CaSO_4$ generally weakens, while the intensity of the AFt diffraction peaks strengthens. The diffraction intensity of $C_4A_3\hat{S}$ at 23.665° decreases most significantly at 6 h, becomes very weak at 7 days, and disappears by 28 days. Correspondingly, the intensity of the AFt diffraction peaks at 15.744°, 18.857°, and 22.862° increases most noticeably at 6 h, after which the rate of increase slows. From 6 h to 28 days, the intensity of the AFt diffraction peaks continues to increase. The intensity of the $CaSO_4$ diffraction peaks decreases most significantly at 6 h, indicating that $CaSO_4$ primarily participates in the hydration reaction generating AFt within the first 6 h.

Figure 8.14 shows the XRD patterns of hydration products of LCCM at curing ages of 6 h, 12 h, 1 day, 3 days, 7 days, and 28 days. As shown in Fig. 8.14, the LCCM consists of $C_4A_3\hat{S}$, C_2S, $CaSO_4$, and $CaSO_4{\cdot}2H_2O$. From 6 h to 28 days, the hydration products include AFt, AFm, and unhydrated C_2S. Compared to the XRD pattern before hydration, the diffraction peaks of AFm appear at 6 h, the diffraction peaks of $CaSO_4{\cdot}2H_2O$ disappear, and the diffraction peaks of $CaSO_4$ significantly decrease. The diffraction peaks of AFt appear at 15.744°, 18.857°, and 22.862°. The intensity of the $C_4A_3\hat{S}$ diffraction peak at 23.655° significantly decreases. The intensity of the C_2S diffraction peaks shows no significant change. This indicates that the hydration reactions of $C_4A_3\hat{S}$, $CaSO_4{\cdot}2H_2O$, and $CaSO_4$ dominate within the first 6 h. The hydration product AFt generated during this period is the primary source of strength for LCCM at 6 h.

$$C_4A_3\overline{S} + 18H \rightarrow C_3A \cdot C\overline{S} \cdot 12H + 4AH_3 \quad (8.3)$$

$$C_2S + AH_3 + 5H \rightarrow C_2ASH_8 \quad (8.4)$$

$$C_2S + AH_3 + C_2ASH_8 \rightarrow C_3ASH_4 + \text{C-S-H} \quad (8.5)$$

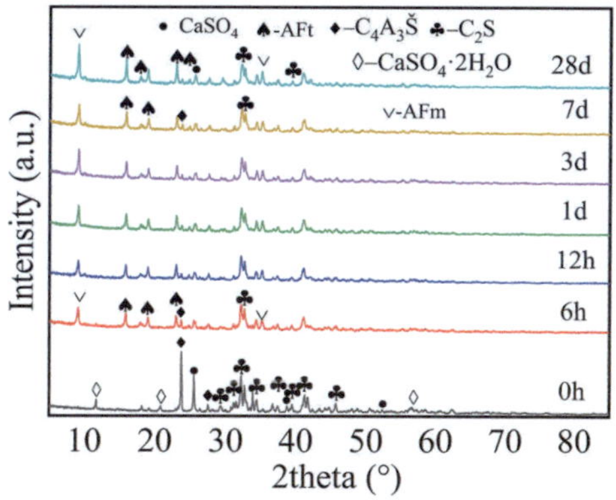

Fig. 8.14 XRD pattern of LCCM at different curing ages

$$3CH + AH_3 + 3CSH_2 + 20H \rightarrow C_6AS_3H_{32} \quad (8.6)$$

During the hydration period from 6 h to 28 days, the diffraction peak intensities of $C_4A_3\hat{S}$, C_2S, $CaSO_4$, and $CaSO_4·2H_2O$ generally decreased. In contrast, the diffraction peak intensity of AFt gradually increased. The variation in the diffraction peak intensity of $C_4A_3\hat{S}$ was consistent with that observed in sulfoaluminate cement, showing a significant decrease at 6 h, a weak intensity at 7 days, and a slight prominence at 28 days, indicating its imminent disappearance. This suggests that $C_4A_3\hat{S}$ reacts rapidly within the first 6 h and is nearly fully hydrated by 28 days. The diffraction peaks of $CaSO_4$ and $CaSO_4·2H_2O$ almost disappeared by the 6-h mark, indicating that their content was insufficient, and $C_4A_3\hat{S}$ hydration produced AFm, as shown in Eq. 8.3. The decrease in the diffraction peak intensity of C_2S was relatively slow, indicating that C_2S had begun to hydrate but at a slower rate.

Compared with sulphoaluminate cement, the mineral compositions of the two cementitious materials both contain $C_4A_3\hat{S}$, C_2S, and $CaSO_4$, while sulphoaluminate cement additionally contains $CaCO_3$. During the hydration process, the main hydration products of both materials are AFt and AFm. Moreover, the hydration behaviors of $C_4A_3\hat{S}$ and C_2S in these two cementitious materials show a high degree of similarity. Through XRD analysis, it was found that no characteristic peaks of $Ca(OH)_2$, C-S-H, or AH_3 were detected in the hydration products of the two cementitious materials. The reasons why the characteristic peaks of C-S-H are difficult to detect can be attributed to the following three points: (1) The reaction rate of C_2S is relatively slow in the early hydration stage, resulting in a low generation amount of C-S-H. Meanwhile, some C-S-H exists in an amorphous gel state, making it difficult to accurately detect through XRD technology. (2) C_2S preferentially reacts with AH_3 to form the intermediate product C_2ASH_8 (Eq. 8.4), and then C_2ASH_8 further reacts to generate C_3ASH_4 and C-S-H (Eq. 8.5) (Su et al., 2019). The absence of the characteristic peak of $Ca(OH)_2$ is mainly due to its low generation amount and its easy reaction and consumption with other hydration products. For instance, $Ca(OH)_2$ can react with AH_3 and $CaSO_4$ to form $C_6AS_3H_{32}$ (Eq. 8.6), or react with $Al(OH)_3$

to form calcium aluminate hydrate. The absence of the characteristic peak of AH_3 may be attributed to two reasons: (1) Some AH_3 exists in an amorphous state and is difficult to detect by XRD technology. (2) AH_3 may have participated in the reaction processes as shown in Eqs. 8.4–8.6.

8.3.2 Thermal Analysis of Early Age Hydration Products

As shown in Fig. 8.15, the primary hydration products of LCCM are AFt, AFm, and AH_3. In sulfoaluminate cement, the hydration products also include AFm-CO_3, which is attributed to the presence of $CaCO_3$. Figures 8.15a–f represent the TG-DSC curves of sulfoaluminate cement paste at hydration ages of 6 h, 12 h, 1 day, 3 days, 7 days, and 28 days, respectively. Similarly, Fig. 8.15g–l represent the TG-DSC curves of LCCM paste at the same hydration ages. Figures 8.16 and 8.17 summarize the TG curves of sulfoaluminate cement and LCCM at different hydration ages, respectively. Table 8.8 lists the mass losses of hydration products calculated from the TG curves. The temperature ranges corresponding to the endothermic decomposition or dehydration of hydration products for both cementitious materials at various ages are as follows:

(1) The endothermic peak of AFm occurs in the range of 160–220 °C (Wang et al., 2017).
(2) The endothermic peak of $Al(OH)_3$ dehydration occurs in the range of 220–280 °C.
(3) The endothermic peak of AFt dehydration occurs in the range of 100–160 °C.
(4) The endothermic decomposition of $CaCO_3$ occurs in the range of 600–800 °C (Lan et al., 2022).
(5) The endothermic peak of AFm-CO_3 occurs in the range of 800–920 °C (Hargis et al., 2014).

From Fig. 8.15a to e, it can be observed that the hydration samples of sulfoaluminate cement exhibit endothermic peaks corresponding to AFt, AFm, AH_3, and AFm-CO_3 at the age of 6 h. This indicates that the cement sample has already begun to hydrate, and these hydration products are likely the source of early strength in sulfoaluminate cement concrete. The endothermic peak of AFm is very small, suggesting that the content of $CaSO_4$ is insufficient, and the hydration of $C_4A_3\hat{S}$ produces AFm through the reaction shown in Eq. 8.3. The appearance of the AFm-CO_3 endothermic peak indicates the occurrence of a hydration reaction between $C_4A_3\hat{S}$ and $CaCO_3$.

Furthermore, based on Table 8.8, the mass loss rate due to the dehydration of AFt increases with age from 6 h to 28 days, with values of 2.2677, 2.3510, 2.6498, 3.0290, 9.6485, and 11.8701%, respectively. Notably, from 3 to 7 days, the mass loss of AFt increases by 6.6195%, with a mass loss rate increase of 218.54%. The significant increase in AFt contributes to the rapid growth in compressive strength of sulfoaluminate cement concrete between 3 and 7 days. The mass loss rate of

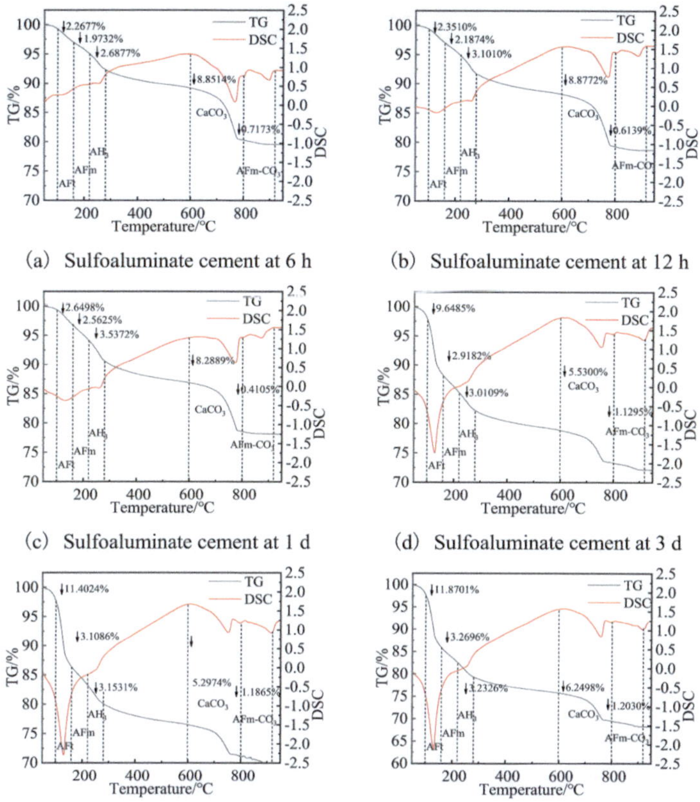

(a) Sulfoaluminate cement at 6 h (b) Sulfoaluminate cement at 12 h

(c) Sulfoaluminate cement at 1 d (d) Sulfoaluminate cement at 3 d

(e) Sulfoaluminate cement at 7 d (f) Sulfoaluminate cement at 28 d

Fig. 8.15 TG-DSC curve of sulphoaluminate cement and LCCM at different curing ages

AFm also increases with age from 6 h to 28 days, with values of 1.9732, 2.1874, 2.5625, 2.7027, 2.9182, and 3.2696%, respectively. This indicates that the formation of AFt consumes a large amount of $CaSO_4$ starting from 6 h, leading to insufficient $CaSO_4$ content, which is consistent with the XRD analysis. The rate of increase in the mass loss rate of AFm slows down with age, which is attributed to the presence of $CaCO_3$. The CO_3^{2-} ions can replace the SO_4^{2-} ions in AFm, forming AFm-CO_3 and AFt, as shown in Eq. 8.7. This ensures the continuous growth of AFt content, slows the growth rate of AFm, and maintains the sustained strength development of sulfoaluminate cement concrete. The slow increase in AFm may be due to its generation rate exceeding the dissolution rate of $CaCO_3$ in the paste. The mass loss rate of $CaCO_3$ generally decreases with age, with values of 8.8514, 8.8772, 8.2889, 8.3977, 5.5300, and 6.2498% from 6 h to 28 days. The increase in $CaCO_3$ content at 28 days may be due to insufficient removal of the carbonated surface layer during sample preparation. The mass loss rate of AH_3 increases from 6 h to 3 days but decreases by 1.1073% from 3 to 7 days. This is because AH_3 reacts with $Ca(OH)_2$

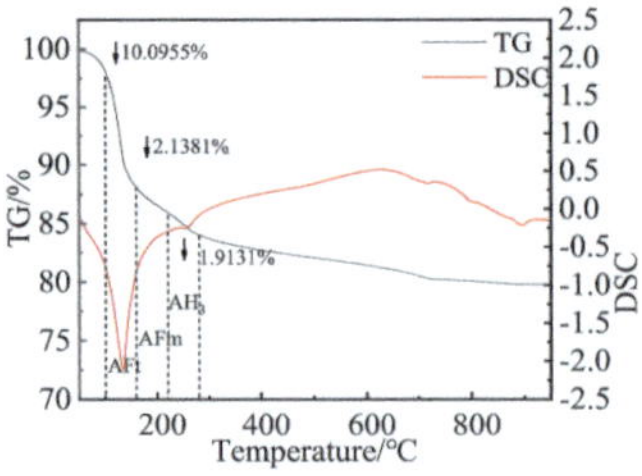

(g) LCCM at 6 h

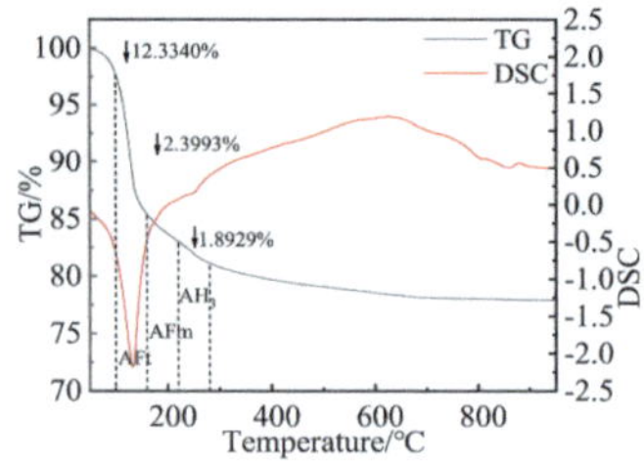

(h) LCCM at 12 h

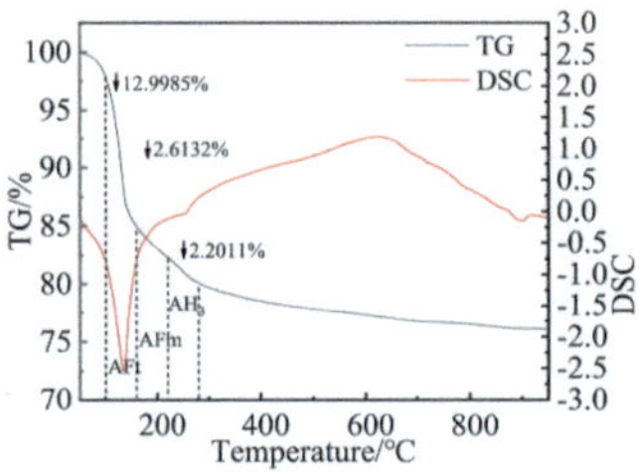

(i) LCCM at 1 d

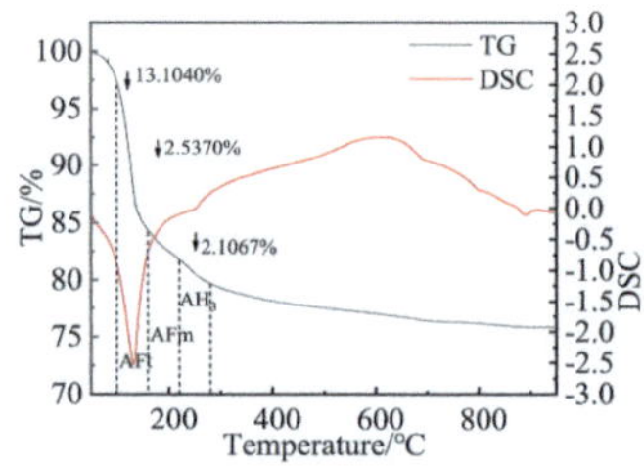

(j) LCCM at 3 d

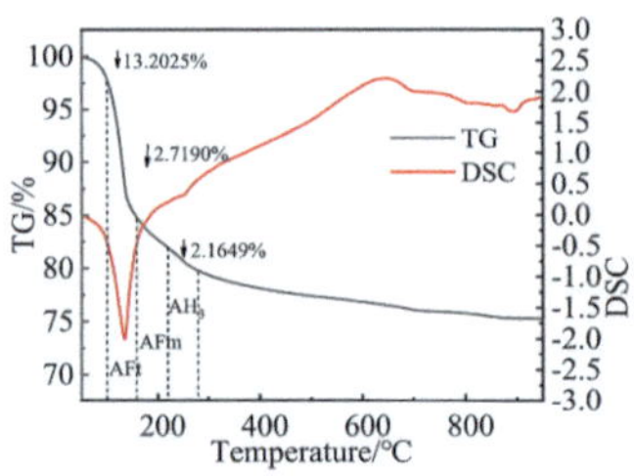

(k) LCCM at 7 d

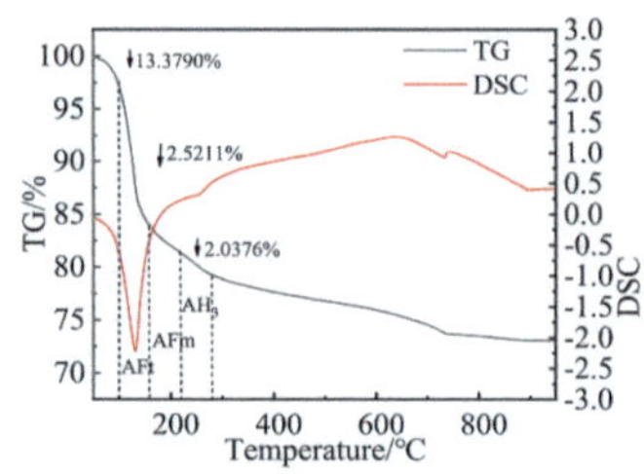

(l) LCCM at 28 d

Fig. 8.15 (continued)

Fig. 8.16 TG curve of sulphoaluminate cement

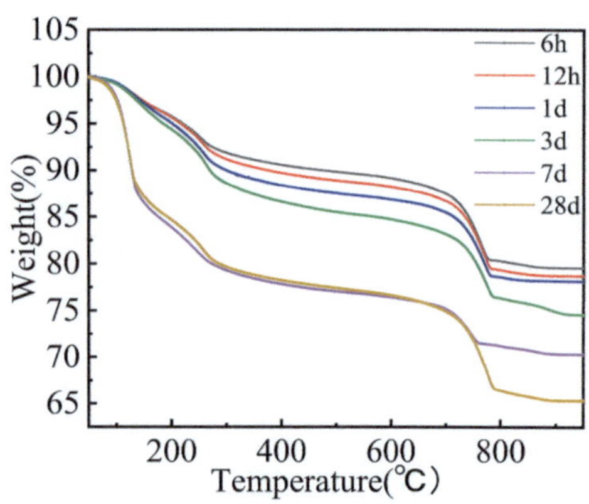

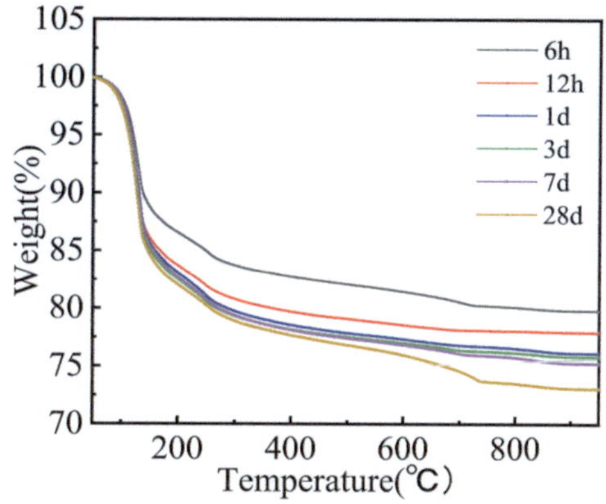

Fig. 8.17 TG curve of LCCM

Table 8.8 The mass loss rate of hydration product (%)

Samples	Phase	6 h	12 h	1 d	3 d	7 d	28 d
Sulphoaluminate cement	AFt	2.2677	2.3510	2.6498	3.0290	9.6485	11.8701
	AFm	1.9732	2.1874	2.5625	2.7027	2.9182	3.2696
	AH_3	2.6877	3.1010	3.5372	4.1182	3.0109	3.2326
	$CaCO_3$	8.8514	8.8772	8.2889	8.3977	5.5300	6.2498
	AFm-CO_3	0.7173	0.6139	0.4105	1.6818	1.1295	1.2030
LCCM	AFt	10.0955	12.334	12.9985	13.104	13.2025	13.379
	AFm	2.1381	2.3993	2.6132	2.537	2.719	2.5211
	AH_3	1.9131	1.8929	2.2011	2.1067	2.1649	2.0376

and $CaSO_4$ to form AFt. The mass loss rate of AFm-CO_3 also generally increases with age. Consistent with the XRD analysis, no endothermic peak of $Ca(OH)_2$ is observed in the 400–500 °C. This is because the early hydration of C_2S is slow, producing little $Ca(OH)_2$, which is consumed during the formation of AFt.

$$\begin{aligned}&3(3CaO \cdot Al_2O_3 \cdot CaSO_4 \cdot 12H_2O + 2CaCO_3 + 18H_2O \rightarrow \\ &2(3CaO \cdot Al_2O_3 \cdot CaCO_3 \cdot 11H_2O + 3CaO \cdot Al_2O_3 \cdot 3CaSO_4 \cdot 32H_2O\end{aligned} \quad (8.7)$$

From Fig. 8.15g to l, it can be observed that the main hydration products of LCCM are AFt, AFm, and AH_3, and the presence of AFt and AFm has also been confirmed by XRD analysis. Additionally, based on Table 8.8, it can be seen that at the age of 6 h, the mass loss rate of AFt reaches 75.46% of that at 28 days, and the mass loss rate of AFm reaches 84.80% of that at 28 days. After 6 h, the mass loss rates of AFt at 12 h to 28 days are 12.334, 12.9985, 13.1040, 13.2025, and 13.3790%, respectively, with a final increase of 3.2835%. The mass loss rates of AFm at 12 h to 28 days are 2.3993, 2.6132, 2.5370, 2.7190, and 2.5211%, respectively, with a final increase of 0.383%. The mass loss rates of AH_3 at 12 h to 28 days are 1.8929, 2.2011, 2.1067,

2.1649, and 2.0376%, respectively, with a final increase of 0.1245%. This indicates that the mass loss rates of AFt, AH_3, and AFm increase relatively slowly.

The mass loss rate in the temperature range of 280–720 °C is attributed to the dehydration of C-S-H. After 720 °C, the heat flow curve shows a significant decline, while the mass curve remains almost unchanged. This is likely due to the lattice transformation of unhydrated C_2S absorbing heat. The decreasing trend of the heat flow curve in the 720–950 °C range fluctuates with age, which qualitatively indicates a gradual reduction in C_2S content. From Fig. 8.16, it can be seen that the mass loss rate due to the dehydration of C-S-H in this temperature range increases slowly with age, indicating that C_2S is undergoing a slow hydration process. The mass loss rate of AH_3 fluctuates, and no endothermic peak of $Ca(OH)_2$ is observed in the 400–500 °C, as both are consumed during the formation of AFt. The fluctuation in the mass loss rate of AFm is due to its partial conversion to AFt.

Compared to sulfoaluminate cement, the heat flow curves of LCCM do not show endothermic peaks for $CaCO_3$ or AFm-CO_3 but do exhibit peaks for AFt, AFm, and AH_3. At the age of 6 h, the mass loss rates of AFt and AFm in LCCM are higher than those in sulfoaluminate cement, by 7.8278 and 0.1649%, respectively. This is likely the reason why the strength of LCCM at 6 h is higher than that of sulfoaluminate cement concrete. From Table 8.8, it can be seen that the content of AFt in sulfoaluminate cement increases rapidly from 3 to 7 days, while in LCCM, the content of AFt increases most rapidly after 6 h. This may be due to the presence of $CaCO_3$, which delays the period of significant AFt formation. The mass loss rates of AFm and AH_3 in both cementitious materials are similar at various ages from 6 h to 28 days, but the mass loss rate of AFt is significantly higher in LCCM, exceeding that in sulfoaluminate cement by 7.8278, 9.983, 10.3487, 10.075, 3.3765, and 1.5089%, respectively.

The quantitative calculation method for chemically bound water based on TG-DSC testing is as follows:

(1) For the mass curve of LCCM, the mass curve shows a decreasing trend from 50 to 950 °C, which is caused by the loss of chemically bound water from hydration products. Therefore, the chemically bound water content is equal to the mass loss rate at 950 °C.
(2) For the mass curve of sulfoaluminate cement, the decrease in the mass curve from 50 to 950 °C is due to the combined effect of the loss of chemically bound water from hydration products and the decomposition of $CaCO_3$ releasing CO_2. Therefore, when calculating the chemically bound water content, the influence of CO_2 release from $CaCO_3$ decomposition on the mass loss rate must be considered. Thus, the chemically bound water content of this cementitious material is equal to the mass loss rate at 950 °C minus the mass loss rate caused by the thermal decomposition of $CaCO_3$ releasing CO_2.

The results of chemically bound water content calculated using the above method are shown in Tables 8.9 and 8.10. From Fig. 8.18, it can be observed that the chemically bound water content of both LCCM and sulfoaluminate cement increases with age. Based on Tables 8.9 and 8.10, using the chemically bound water content at 6 h as a reference, the chemically bound water content increases by 0.8299, 2.0041, 5.4331, 12.5262, and 16.8363% at 12 h, 1 day, 3 days, 7 days, and 28 days, respectively. This indicates that the chemically bound water content grows slowly from 6 h to 1 day, while the growth rate is faster before 6 h and after 1 day. This is because, before 6 h, $C_4A_3\hat{S}$ and C_2S are rapidly consumed to form AFt. The hydration rate decreases as the content of $CaSO_4$ diminishes, leading to a slower increase in chemically bound water content after 6 h. With increasing age, the dissolution of $CaCO_3$ in the cement paste gradually increases, and the CO_3^{2-} ions gradually participate in the hydration reaction with AFm to form AFm-CO_3 and AFt. Moreover, for every 1 mol of AFm consumed, 6 mol of chemically bound water is added. Therefore, the growth rate of chemically bound water content increases again. This is consistent with the development pattern of the compressive strength curve of sulfoaluminate cement concrete in subsequent chapters.

Taking the chemically bound water content at 6 h as the baseline, the chemically bound water content of the LCCM increased by 1.8731, 3.682, 3.9666, 4.5522 and 6.7179% from 12 h to 28 days, respectively. This indicates that the significant increase in chemically bound water content mainly occurs within the first 6 h of hydration, after

Table 8.9 Quantitative calculation results of chemically bound water based on thermogravimetric mass curve at 6 h, 12 h, and 1 d

Curing age	6 h		12 h		1 d	
Material type	SCA	LCCM	SCA	LCCM	SCA	LCCM
Mass loss rate at 950 °C (%)	20.441	20.2132	21.2967	22.0863	21.8826	23.8952
Mass loss rate of $CaCO_3$ (%)	8.8514	–	8.8772	–	8.2889	–
Chemically bound water content (%)	11.5896	20.2132	12.4195	22.0863	13.5937	23.8952

Note SCA means the sulfoaluminate cement. LCCM represents the low-carbon cementitious materials

Table 8.10 Quantitative calculation results of chemically bound water based on thermogravimetric mass curve at 3 d, 7 d, and 28 d

Age	3 d		7 d		28 d	
Material type	SCA	LCCM	SCA	LCCM	SCA	LCCM
Mass loss rate at 950 °C (%)	25.4204	24.1798	29.6548	24.7654	34.6757	26.9311
Mass loss rate of $CaCO_3$ (%)	8.3977	–	5.53	–	6.2498	–
Chemically bound water content (%)	17.0227	24.1798	24.1158	24.7654	28.4259	26.9311

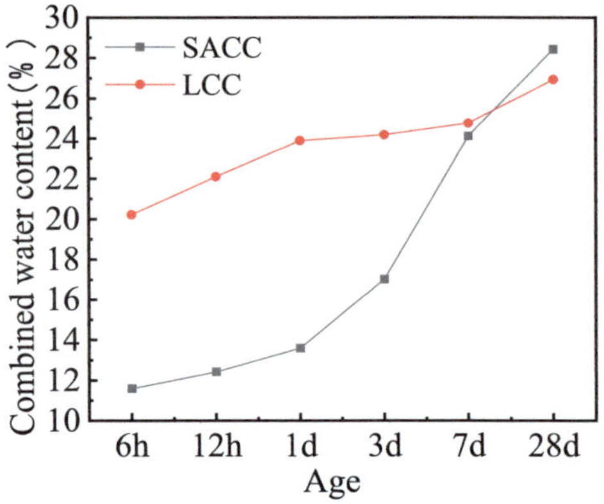

Fig. 8.18 Chemically bound water content based on TG-DSC curve

which the growth becomes gradual without any notable rapid increase. Compared to sulfoaluminate cement, the chemically bound water content of this cementitious material from 6 h to 28 days was higher by 8.6236, 9.6668, 10.3015, 7.1571, 0.6496, and −1.4948%, respectively. Additionally, the chemically bound water content of this cementitious material at 6 h was 8.6236% higher than that of sulfoaluminate cement, which is the fundamental reason why the 6-h strength of LCCM is higher than that of sulfoaluminate cement concrete. The lower chemically bound water content of this material at 28 days compared to sulfoaluminate cement may be attributed to the following reasons: (1) uneven distribution of hydration products in the test samples; (2) prolonged storage of the samples in a desiccator after preparation, which may have led to partial loss of chemically bound water.

8.3.3 Molecular Structure Analysis of Early Age Hydration Products

Figure 8.19a and b show the FTIR spectrum of LCCM at different hydration ages in the wavelength ranges of 400–1800 and 3000–4000 cm^{-1}, respectively. The hydration ages are 6 h, 12 h, 1 day, 3 days, 7 days, and 28 days. Table 8.11 lists the FTIR absorption characteristics of the minerals and hydration products. From the spectra, it can be observed that the characteristic peak of $CaSO_4$ appears at 679 cm^{-1}, which is caused by the bending vibration of the SO_4 group. The characteristic peaks of C_2S appear at 520 c and 847 cm^{-1}, resulting from the bending and asymmetric stretching vibrations of the SiO_4 tetrahedron. The main mineral phase in LCCM, $C_4A_3\hat{S}$, exhibits characteristic absorption peaks at 418, 616, 644, 877, and 1103 cm^{-1}. Among these, the peaks at 418 and 877 cm^{-1} are caused by the asymmetric contraction of the AlO_4 group, while the peaks at 616 and 644 cm^{-1} result from the coupled vibration of AlO_4 and SO_4. The peak at 1103 cm^{-1} is attributed to the asymmetric stretching vibration of the (SO_4) group. AFt shows characteristic peaks at 1119,

Table 8.11 Mineral composition and FTIR absorption characteristics peaks of hydration products

NO	Group	Reference frequency (cm^{-1})	Compound	Citation
679 cm^{-1} 667 cm^{-1}	$V_4(SO_4)$	679	$CaSO_4$	Yin et al., (2018)
847 cm^{-1}	SiO_4	847	C_2S	Carrasco and Vázquez (1996)
520 cm^{-1}	SiO_4	520	C_2S	Carrasco and Vázquez (1996)
418 cm^{-1} 413 cm^{-1}	$V_4(AlO_4)$	413	$C_4A_3\hat{S}$	Li (1984)
613 cm^{-1} 616 cm^{-1}	(AlO_4)	616	$C_4A_3\hat{S}$	Li (1984)
644 cm^{-1}	$V_3(AlO_4)$ $V_4(SO_4)$	644	$C_4A_3\hat{S}$	Li (1984)
877 cm^{-1}	(AlO_4)	882	$C_4A_3\hat{S}$	Carrasco and Vázquez (1996)
1105 cm^{-1} 1100 cm^{-1}	$V_3(SO_4)$	1103	$C_4A_3\hat{S}$	Li (1984)
1422 cm^{-1}	$V_3(CO_3)$	1437	$CaCO_3$	Trezza and Lavat (2001)
1104 cm^{-1} 1119 cm^{-1}	$V_3(SO_4)$	1110	AFt	Carrasco and Vázquez (1996)
3422 cm^{-1}	$V_2(H_2O)$	3422	AFt	Carrasco and Vázquez (1996)
3634 cm^{-1} 3627 cm^{-1}	OH	3635	AFt	Carrasco and Vázquez (1996)
520 cm^{-1}	OH	520	AH_3	Chi (2019)
3430 cm^{-1}	OH	3430	AH_3	Carrasco and Vázquez (1996)

3422, and 3634 cm^{-1}, caused by the asymmetric stretching of (SO_4), H_2O, and (OH), respectively. The characteristic peak of AH_3 at 3430 cm^{-1} is due to the stretching vibration of the (OH) group.

Before hydration, the FTIR spectrum confirms the presence of $CaSO_4$, $C_4A_3\hat{S}$, and C_2S in the LCCM, consistent with the XRD analysis results. At 6 h, the characteristic peaks of $C_4A_3\hat{S}$ at 418, 520, and 877 cm^{-1} significantly decrease in intensity and peak area. The characteristic peaks at 613 and 644 cm^{-1} disappear, while the characteristic peak at 1105 cm^{-1} shifts to 1119 cm^{-1} without significant changes in area, due to the overlapping characteristics peak of $C_4A_3\hat{S}$ and AFt. The characteristic peak of $CaSO_4$ at 679 cm^{-1} decreases in area and shifts to 667 cm^{-1}. Additionally, broad and flat absorption peaks of AH_3 and AFt appear in the 3300–3700 cm^{-1}. This indicates that the formation of AFt consumes a significant amount of $C_4A_3\hat{S}$ and $CaSO_4$. The rapid consumption of $C_4A_3\hat{S}$ and $CaSO_4$ within 6 h aligns with the XRD analysis results, and the substantial formation of ettringite and AH_3 is consistent with the thermogravimetric analysis. From the perspective of molecular functional groups, this further confirms the early formation of AFt, ensuring the high compressive strength of LCCM at 6 h.

As the hydration age increases, the absorption peaks of $C_4A_3\hat{S}$ at 418 and 877 cm^{-1} gradually decrease, becoming very weak by 28 days and nearly disappearing. The peak area at 520 cm^{-1} initially decreases and then slightly increases due to overlapping with the absorption peak of ettringite. The absorption peak of $CaSO_4$ at 679 cm^{-1} weakens rapidly before 6 h, after which the weakening slows down, nearly disappearing by 28 days. This indicates that the hydration reaction of LCCM is rapid within the first 6 h. The broad and flat absorption peaks of AH_3 and AFt in the 3300–3700 cm^{-1} also increase most significantly within the first 6 h, after which the rate of increase slows down. This further demonstrates that ettringite and AH_3 are predominantly formed within the first 6 h, with a slower formation rate thereafter. Additionally, no characteristic peak of $Ca(OH)_2$ was observed in the wavelength range of 3641–3644 cm^{-1}, which is consistent with the XRD and thermogravimetric analysis results of the LCCM.

Figure 8.20 shows the FTIR spectra of sulfoaluminate cement hydration. From the FTIR spectrum of sulfoaluminate cement, characteristic absorption peaks of $C_4A_3\hat{S}$ are observed at 413, 616, 644, 877, and 1100 cm^{-1}. The characteristic absorption peak of $CaSO_4$ appears at 679 cm^{-1}, while the characteristic peaks of C_2S are observed at 520 and 847 cm^{-1}. The FTIR spectrum of sulfoaluminate cement shows a characteristic absorption peak of $CaCO_3$ at 1422 cm^{-1}, which is caused by the stretching vibration of the CO_3 group. At the 6-h hydration age, the peak areas of the characteristic absorption peaks of $C_4A_3\hat{S}$, $CaSO_4$, and $CaCO_3$ all decrease. Meanwhile, broad and flat FTIR absorption peaks of AFt and AH_3 appear at 3422 and 3627 cm^{-1}, and a small characteristic peak of AFm-CO_3 appears at 716 cm^{-1}. This indicates that $C_4A_3\hat{S}$, $CaSO_4$, and $CaCO_3$ have begun to hydrate, forming hydration products such as AFt, AH_3, and AFm-CO_3. The peak area at 520 cm^{-1} does not decrease but instead increases, which is due to the overlap with the characteristic peak of AFt. As the hydration age increases, the characteristic peaks of $C_4A_3\hat{S}$ and $CaSO_4$ weaken rapidly, while the characteristic peak of $CaCO_3$ weakens more slowly. The characteristic peak of AFm-CO_3 does not increase significantly. The peak areas of AFt and AH_3 at 3422 and 3627 cm^{-1} gradually increase, which is consistent with the XRD results and aligns with the trends in ettringite and aluminum hydroxide content measured by TG.

From the FTIR spectrum in Figs. 8.19 and 8.20, it can be observed that both LCCM and sulfoaluminate cement contain $CaSO_4$, $C_4A_3\hat{S}$, and C_2S. Additionally, sulfoaluminate cement also contains $CaCO_3$. In terms of hydration products, the main hydration products of both cementitious materials are AFt and AH_3, while sulfoaluminate cement also produces AFm-CO_3. The FTIR analysis results are consistent with the findings from XRD and TG analyses.

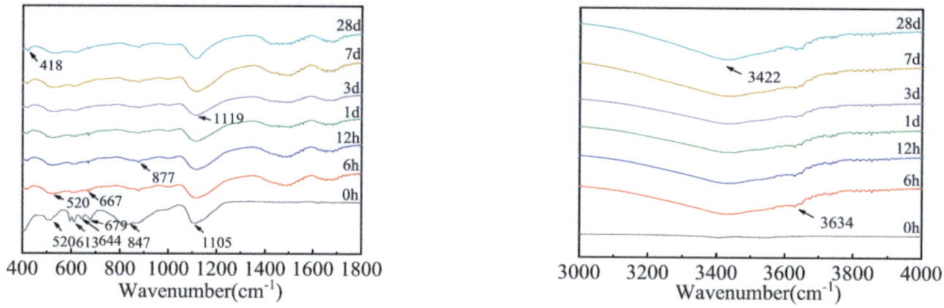

(a) Wavenumber between 400-1800 cm^{-1} (b) Wavenumber between 3000-4000 cm^{-1}

Fig. 8.19 FTIR spectrum of LCCM at different curing age

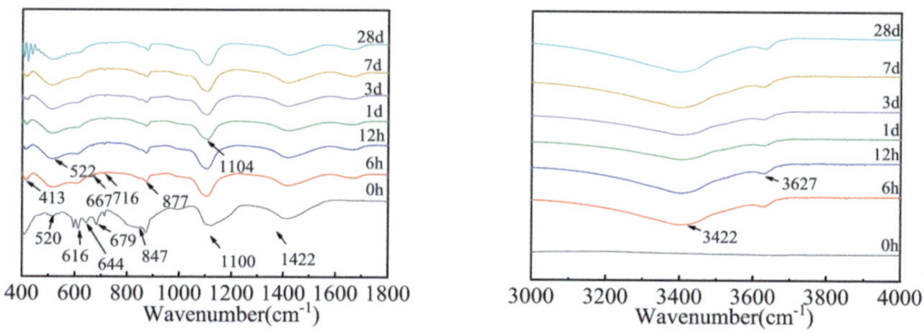

(a) Wavenumber between 400-1800 cm^{-1} (b) Wavenumber between 3000-4000 cm^{-1}

Fig. 8.20 FTIR spectrum of sulphoaluminate cement at different curing age

8.3.4 Micro-morphology of Early-Age Hydration Products

The performance of concrete is not only related to the hydration characteristics of the cementitious material but also closely associated with the microstructure of the hardened cement paste. By using high-magnification SEM, the types and composition of hydration products in hardened pastes of different cementitious materials at various ages can be determined. Generally, the AFt phase in cement hydration products appears as fine needles (Li et al., 2012), rods (Zhang et al., 2017; Hargis et al., 2013), blocks (Tosun & Baradan, 2010), columns, or tubes, while the AFm phase exhibits a layered structure, and AH_3 appears as flocculent, flaky, or honeycomb-like forms. C-S-H typically presents as needle-like, fibrous, or grass-like structures. Different hydration products form at varying rates and stages, influencing the performance of concrete at different periods. Therefore, observing the microstructure of hardened pastes at different ages provides a more intuitive analysis of the hydration degree of cementitious materials and helps understand the development of concrete strength.

Figures 8.21, 8.22 and 8.23 show the microstructures of hydration products in LCCM at 6 h, 12 h, 1 d, 3 d, 7 d, and 28 d, respectively. Figures 8.25, 8.26 and 8.27

depict the microstructures of hydration products in sulfoaluminate cement paste at the same ages. From the micrographs, it is evident that the main hydration products of both LCCM and sulfoaluminate cement are AFt, AFm, and AH_3. At the 6 h age, unhydrated $C_4A_3\hat{S}$, $CaCO_3$, and C_2S are also observed, which aligns with the results from XRD and TG-DSC analyses. Additionally, C-S-H is observed at the 28d age, complementing the limitations of XRD and TG-DSC tests and confirming that the mass loss curve between 280 and 600 °C is due to the dehydration of C-S-H. Compared to NMR pore structure analysis, NMR provides more accurate results, while SEM offers a more intuitive observation of the evolution of microcracks and pore structures over time

From Fig. 8.21, it can be seen that at the 6 h age, the LCCM exhibits numerous fine, randomly oriented AFt needles and a small amount of flocculent AH_3. This indicates rapid hydration of $C_4A_3\hat{S}$, consistent with XRD analysis. Figure 8.21 (6 h-2)–(6 h-3) reveal pore structures and microcracks. At the 12 h age, flaky AH_3 is observed in Fig. 8.21 12 **h-2**, and the width of microcracks is smaller than at 6 h. The fine AFt needles gradually thicken, transforming into coarse needles. Figure 8.21 12 **h-3** shows AFt interconnecting to form a spatial framework, with AH_3 filling the pores, leading to a denser microstructure.

As shown in Fig. 8.22, AH_3 in the LCCM exhibits a new morphology at the 1-day hydration age, specifically a honeycomb-like porous structure, filling the spaces between AFt and AFm. In Fig. 8.22 1d-2, well-developed prismatic AFt crystals begin to appear. In Fig. 8.22 1d-3, a significant amount of layered AFm can be clearly observed, indicating that the calcium aluminate in the hardened paste is in insufficient supply. AH_3 gradually grows into the micropores, leading to a reduction in pore volume. In Fig. 8.22 1d-1, a considerable amount of layered AFm is observed, showing an increasing trend compared to the 12-h age. This aligns with the intensity pattern of the AFm diffraction peak in XRD and matches the gradual increase in AFm

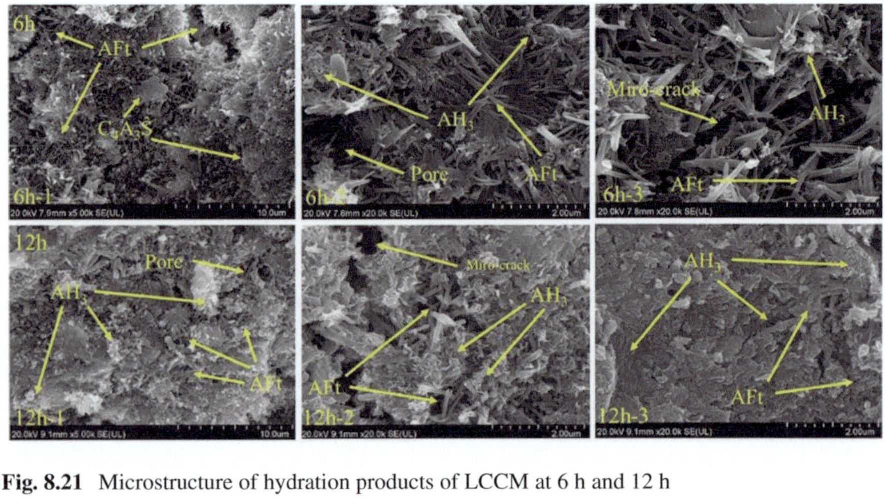

Fig. 8.21 Microstructure of hydration products of LCCM at 6 h and 12 h

Fig. 8.22 Microstructure of hydration products of LCCM at 1d and 3d

mass loss rate observed in thermogravimetric analysis. At the 3-day age, clusters of fine rod-like and blocky AFt crystals, as well as flaky AH_3 colloids, can be observed. Within the microcracks, multiple fine rod-like AFt crystals interlock, and layered AFm adheres to the rod-like AFt.

As shown in Fig. 8.23, at the 7-day hydration age of the LCCM, AFt crystals interlock and connect with each other, forming a denser structure compared to earlier ages. AH_3 fills the pores and gradually forms AFt crystal clusters or "crystal bridges" spanning the microcracks. In Fig. 8.23 7d-1, multiple AFt "crystal bridges" are observed within the microcracks, as illustrated in Fig. 8.23 7d-2. From Fig. 8.23 7d-3, it can be seen that the AH_3 colloid filling the spaces between AFm gradually occupies the entire pore. Due to the further growth of AFt crystals, their interlocking and bonding within the microcracks, along with the filling and binding effects of AH_3, lead to a gradual reduction in the width and length of the microcracks. The microstructure becomes increasingly denser, which aligns with the AFt content analysis results from TG-DSC. From Fig. 8.23 28d-1, it is observed that the hardened cement paste contains incompletely hydrated C_2S, as well as a significant amount of needle-like C-S-H. From Fig. 8.23 28d-3, it is evident that the needle-like C-S-H originates from the C_2S crystals, interlocking and filling the gaps between two C_2S crystals. This indicates that C_2S has undergone a certain degree of hydration by the 28-day age, which is one of the reasons for the continued increase in compressive strength of LCCM from 7 to 28 days. C-S-H gel is difficult to detect in XRD testing, but SEM analysis compensates for this limitation of XRD and provides further insight into the hydration process of the cementitious material.

Figure 8.24 shows the EDS spectra of the LCCM at different ages. From Fig. 8.24a, it can be observed that the crystal at point A primarily consists of S, O, and Ca, with elemental contents of 9.5, 57.79, and 24.88%, respectively. This crystal is identified as unhydrated $CaSO_4$. At point B, the crystal mainly contains O, Al, S, and Ca,

Fig. 8.23 Micro-morphology of hydration products of LCCM at 7 d and 28 d

with elemental contents of 67.34, 2.7, 7.9, and 20.47%, respectively. This crystal is identified as unhydrated $C_4A_3\hat{S}$. These results indicate that at the 7-day hydration age, unhydrated $CaSO_4$ and $C_4A_3\hat{S}$ still exist in the paste. From Fig. 8.24b, the fine needle-like crystal at Spectrum1 primarily consists of O, Si, and Ca, with contents of 67.28, 1.89, and 27.45%, respectively. This crystal is identified as C-S-H. This suggests that by the 28-day age, partial hydration of C_2S has occurred in the LCCM, which is the main reason for the continued strength development of the material at later stages.

Figure 8.25 shows the microstructural morphology of calcium sulfoaluminate cement at 6 h and 12 h of hydration. From the figure, it can be observed that the hydration products of the calcium sulfoaluminate cement paste at the 6 h age include AFt, AH_3, as well as unhydrated $CaCO_3$ and $C_4A_3\hat{S}$ crystals. AH_3 also exhibits a flaky morphology, forming clusters of aluminum gel. Additionally, numerous needle-like AFt crystals and interlocking AFt within microcracks are observed. From Fig. 8.25 12 **h-1**, it can be seen that at the 12 h age, flaky AH_3 appears in clusters in the cement paste. Notably, AFt exhibits a wide and flat elongated shape rather than a rod-like form. This may be due to the increase in pH of the paste caused by the dissolution of $CaCO_3$. The elongated AFt crystals interlock with each other, with their ends embedded in the aluminum gel clusters, leaving relatively large gaps between the elongated AFt crystals.

Figure 8.26 shows the microstructural morphology of calcium sulfoaluminate cement at 1 d and 3 d of hydration. At the 1 d age, the hydration products include flaky AFm, honeycomb-like aluminum gel, and fine rod-like AFt. From Fig. 8.26 1d-1, it can be observed that AFt, AFm, and AH_3 interlock and bond with each other, filling the spaces between microcracks. In Fig. 8.26 3d-1, unhydrated calcium sulfoaluminate crystals are observed, along with columnar AFt suspended in the cracks and flocculent AFt gradually growing toward the center of the pores. From

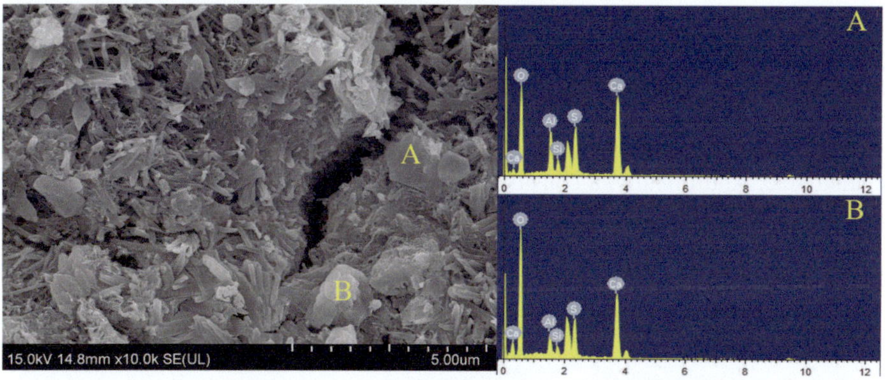

(a) 7 d

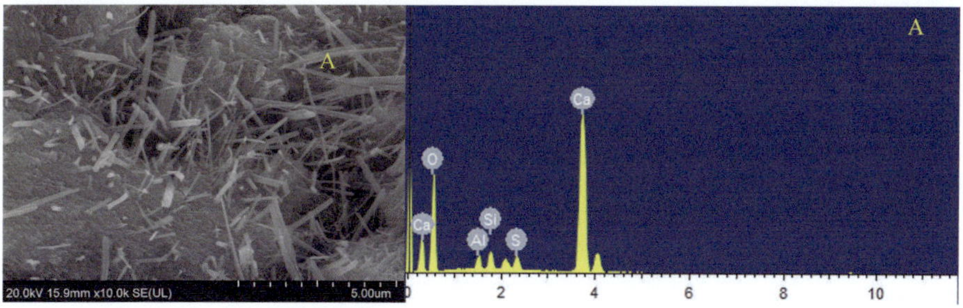

(b) 28d

Fig. 8.24 EDS energy spectrum of LCCM at 7 d and 28 d

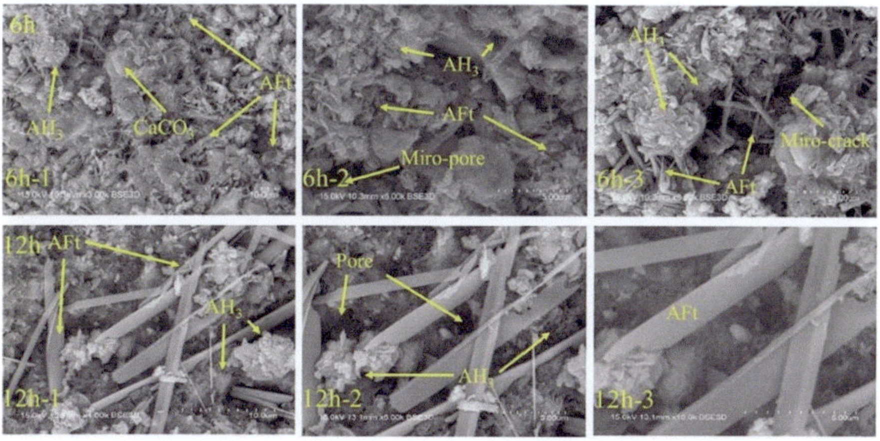

Fig. 8.25 Micro-morphology of hydration products of sulphoaluminate cement paste at 6 h and12 h

Fig. 8.26 3d-3, some larger-sized AFt crystals are observed, indicating that AFt continues to grow and develop as hydration progresses.

Figure 8.27 shows the microstructural morphology of calcium sulfoaluminate cement at 7 d and 28 d of hydration. At the 7 d age, a significant number of AFt crystals are observed in the calcium sulfoaluminate cement paste, with an overall smaller size. Additionally, a considerable amount of aluminum gel is found filling the spaces between the AFt crystals, forming a relatively dense matrix structure. However, the presence of some microcracks can still be observed. At the 28 d age, the cement paste contains numerous fine rod-like AFt crystals that interlock with

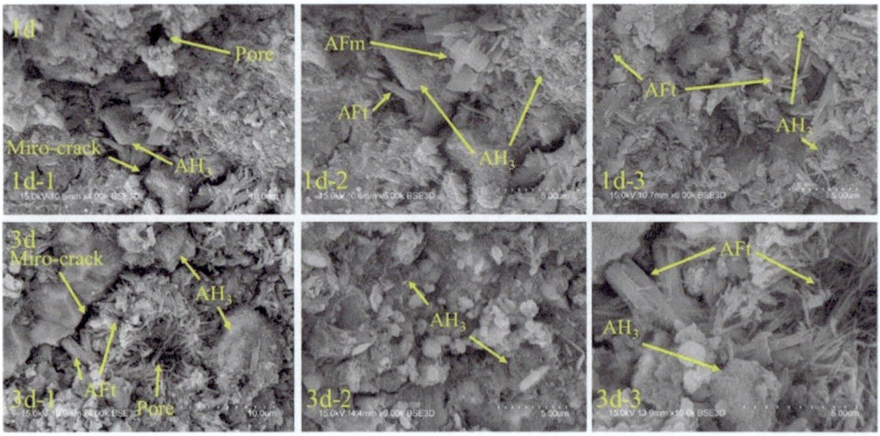

Fig. 8.26 Micro-morphology of hydration products of sulphoaluminate cement paste at 1 d and 3 d

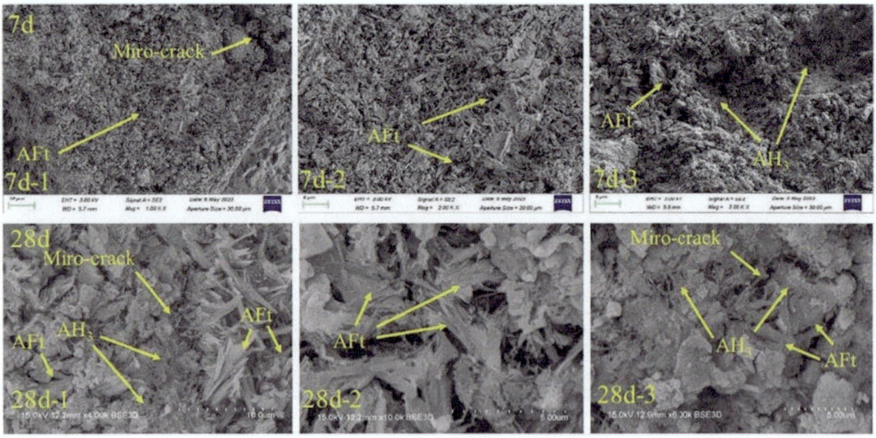

Fig. 8.27 Micro-morphology of hydration products of sulphoaluminate cement paste at 7 d and 28 d

each other, forming AFt "crystal bridges" as shown in Fig. 8.26 28d-2. Honeycomb-like aluminum gel fills the gaps between these AFt "crystal bridges" as illustrated in Fig. 8.26 28d-3. The width of the gaps gradually decreases due to the mutual bonding and filling effects of AH_3 and AFt.

Figure 8.28 shows the EDS spectra of calcium sulfoaluminate cement at 6 h of hydration. From the figure, it can be observed that the hydration products of the calcium sulfoaluminate cement paste at the 6 h age include AFt, AH_3, as well as unhydrated $CaCO_3$ and $C_4A_3\hat{S}$ crystals. AH_3 exhibits a flaky morphology, forming clusters of aluminum gel. Additionally, numerous needle-like AFt crystals and interlocking AFt within microcracks are observed. In the EDS spectrum of Fig. 8.28, the elongated crystal at point 1 primarily consists of O, Al, S, and Ca, with elemental contents of 57.78, 4.88, 6.40, and 28.37%, respectively. This elongated crystal is identified as AFt. The crystal at point 2 mainly contains O, Al, S, and Ca, with elemental contents of 60.52, 15.03, 2.67, and 20.96%, respectively. This crystal is identified as hydrated $C_4A_3\hat{S}$. These results indicate that at the 6 h age, the cement paste contains unhydrated $C_4A_3\hat{S}$ and AFt, which is consistent with the XRD analysis results.

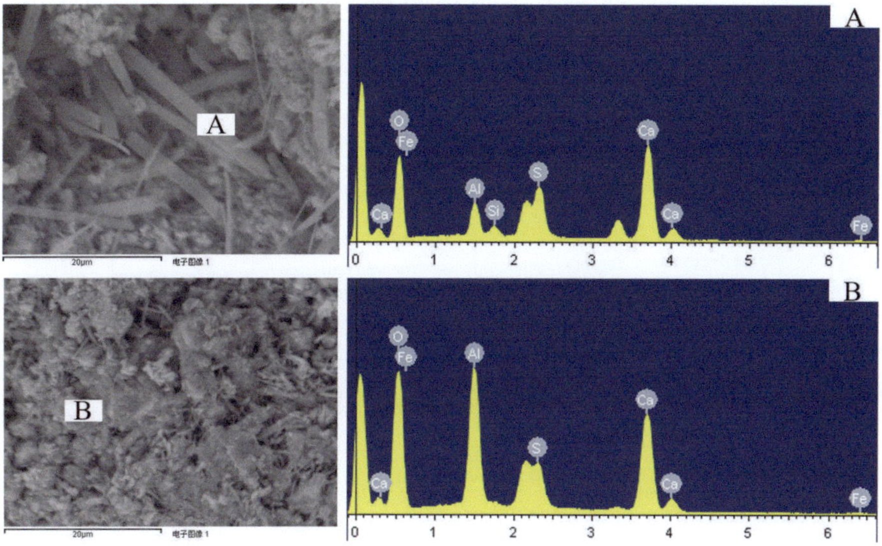

Fig. 8.28 EDS energy spectrum of sulphoaluminate cement

8.4 Pore Structure of Concrete Containing LCCM

8.4.1 Evolution Law of Pore Structure

Figure 8.29a and b show the T_2 spectrum distribution curves of SACC and LCC at ages of 12 h, 1 d, 3 d, 7 d, and 28 d, respectively. Table 8.12 presents the changes in the T_2 spectrum area of the two types of concrete with age. It can be intuitively observed that each T_2 spectrum has three peaks. From left to right, the first peak is Peak1, the second is Peak2, and the third is Peak3. Among them, the area of Peak1 is the largest, followed by Peak2, and Peak3 has the smallest area

From Fig. 8.29a, it can be seen that Peak1 is located between 0.02 and 6.58 ms. As the age increases, the area of Peak1 significantly decreases from 31,925.325 to 23,450.492, and the relaxation time corresponding to the maximum peak decreases from 1.16 ms to 0.56 ms. Peak2 is located between 6.58 and 127.27 ms. With increasing age, the area of Peak2 gradually increases from 630.680 to 1147.452, while the relaxation time corresponding to the maximum peak decreases from 53.36 ms to 21.07 ms. Peak3 is located between 119.83 and 2541.09 ms. As the age increases, the area of Peak3 generally shows an increasing trend, rising from 116.966 to 730.516,

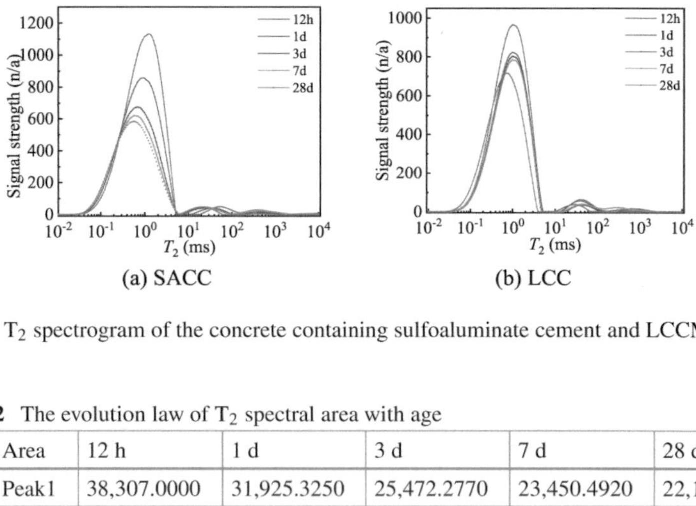

Fig. 8.29 T_2 spectrogram of the concrete containing sulfoaluminate cement and LCCM

Table 8.12 The evolution law of T_2 spectral area with age

Name	Area	12 h	1 d	3 d	7 d	28 d
SACC	Peak1	38,307.0000	31,925.3250	25,472.2770	23,450.4920	22,195.2720
	Peak2	01,615.4000	630.6800	1003.5130	1147.4520	990.4290
	Peak3	124.7630	116.9660	192.5040	730.5160	70.0150
	Total	40,043.5500	32,672.9710	26,668.2940	25,328.4600	23,255.7160
LCC	Peak1	30,031.0000	27,415.8050	26,273.0170	25,892.8870	24,185.8900
	Peak2	938.4000	1078.0040	599.7090	804.7230	648.6930
	Peak3	193.7500	198.8880	50.1390	558.5600	10.6380
	Total	31,163.3700	28,692.6970	26,922.8650	27,256.1700	24,845.2210

with the maximum peak corresponding to a relaxation time of 416.76 ms. The total peak area decreases from 32,672.971 at 1 d to 23,450.492 at 28 d.

From Fig. 8.29b, it can be observed that the three peaks in the T_2 spectrum of LCC are mainly located between 0.03 and 3000 ms. Peak1 is located between 0.03 and 5.75 ms. With increasing age, the area of Peak1 significantly decreases from 27,415.805 to 25,892.887, with the maximum peak corresponding to a relaxation time of 0.98 ms. Peak2 is located between 12.53 and 130.78 ms. As the age increases, the area of Peak2 decreases from 1078.004 to 804.723, with the maximum peak corresponding to a relaxation time of 39.77 ms. Peak3 is located between 145 and 2,677 ms. From 1 to 28 d, the total peak area decreases from 28,692.696 to 27,256.170, indicating that the internal structure of the concrete gradually becomes denser with increasing age.

Figure 8.30a and b present stacked bar charts showing the percentage changes in the area of each peak in the T_2 spectrum of SACC and LCC. From the bar charts in Fig. 8.30, it can be intuitively observed that the proportion of the Peak1 area generally increases with age. In contrast, the proportion of the Peak3 area generally decreases with age, reaching its lowest value at the 28 d

Figure 8.31 shows the pore size distribution diagrams of the two types of concrete. From the figure, it can be observed that the pore size distribution curves of both SACC and LCC exhibit three peaks. From left to right, these peaks are defined as Peak1, Peak2, and Peak3. The area of Peak1 is the largest, followed by Peak2, and Peak3 has the smallest area. Overall, the areas of all three peaks decrease with increasing age, and the peak pore sizes corresponding to each peak also decrease with age.

Based on the method proposed by Wu and Lian (1999), the internal pore structures of the two types of concrete are classified into four categories: harmless pores (pore size < 20 nm), Less harmful pores (20 nm $\leq$ pore size < 50 nm), Harmful pores (50 nm $\leq$ pore size < 200 nm), Highly harmful pores (pore size ≥ 200 nm).

The influence of different pore sizes on concrete performance varies. It can be intuitively observed that as the age increases, the overall pore size of the concrete gradually decreases. Highly harmful pores are gradually transformed into harmful pores, harmful pores are partially converted into less harmful pores, less harmful

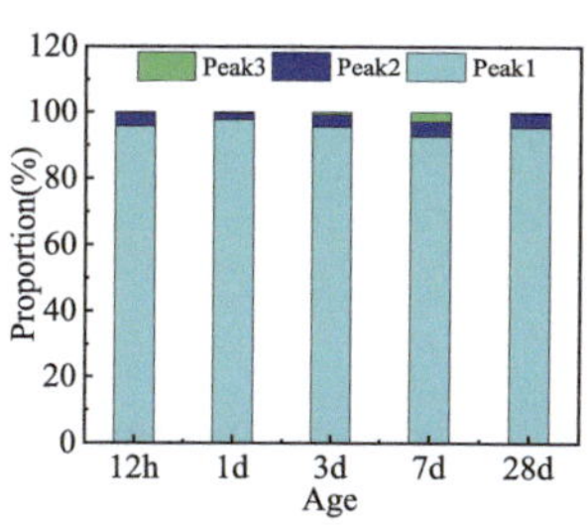

(a) Concrete containing sulfoaluminate cement

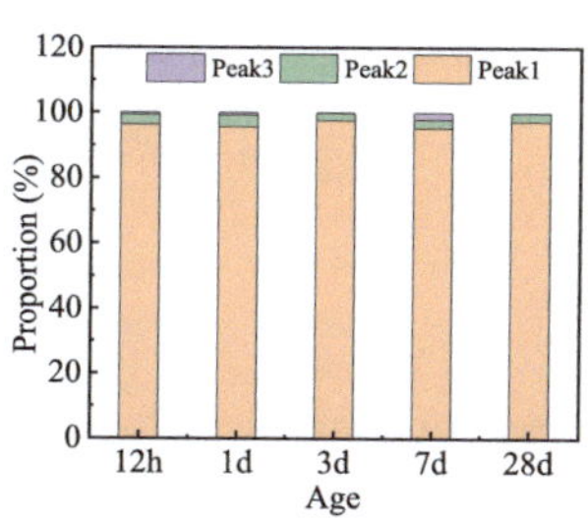

(b) Concrete containing LCCM

Fig. 8.30 T_2 spectrum area age evolution characteristic diagram

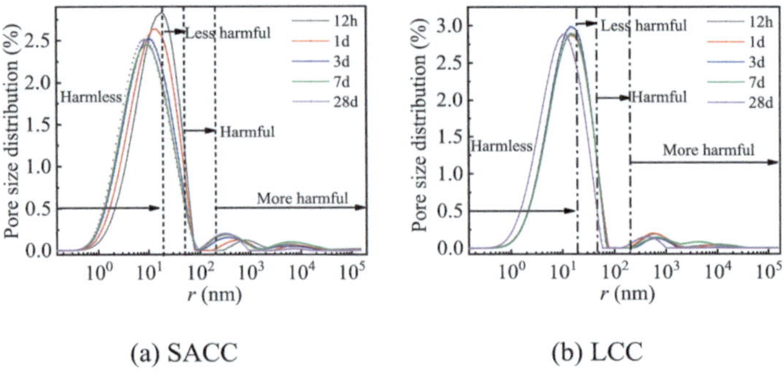

Fig. 8.31 Pore size distribution of the concrete containing sulfoaluminate cement and LCCM

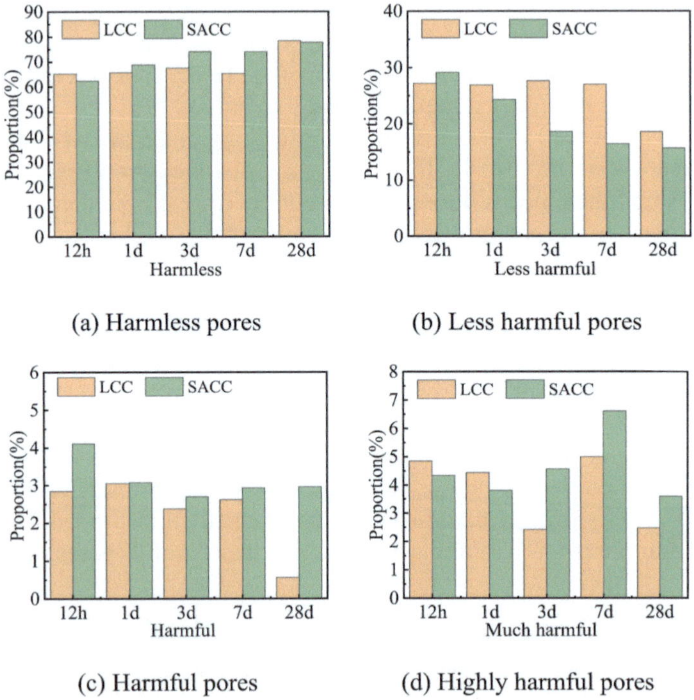

Fig. 8.32 Proportion of pore structure with different pore sizes varies with age

pores are partially transformed into harmless pores, and some harmless pores disappear. This is due to the gradual coarsening of hydration products such as AFt and the filling effects of gels like AH_3 and C-S-H.

8.4.2 Distribution of Pore Structure

Figure 8.32a–d illustrate the changes in the content of harmless pores, less harmful pores, harmful pores, and highly harmful pores in LCC and SACC with age, respectively.

Figure 8.32a: Harmless pores. The content of harmless pores in both types of concrete increases with age. At 1 d, 3 d, and 7 d, the content of harmless pores in sulfoaluminate cement is slightly higher than that in LCC. However, at the 28 d, the opposite is observed.

Figurc 8.32b: Less harmful pores. The content of less harmful pores in both types of concrete decreases with age. In LCC, the content fluctuates with age, while in SACC, it decreases steadily. After the 12 h, the content of less harmful pores in LCC is higher than that in sulfoaluminate cement. From 12 h to 28 d, the content of less harmful pores in LCC is higher by −2.00, 2.63, 9.02, 10.61, and 2.87%, respectively.

Figure 8.32c: Harmful pores. In LCC, the content of harmful pores decreases significantly from 12 h to 28 d, with values of 2.84, 3.06, 2.38, 2.62, and 0.56%, respectively, representing a decrease of 2.28 percentage points. In SACC, the content of harmful pores decreases from 4.11% at 12 h to 2.97% at 28 d, representing a decrease of 1.14%. At all ages, the content of harmful pores in LCC is lower than that in SACC by 1.71, 0.02, 0.32, 0.32, and 2.41%, respectively.

Figure 8.32d: Highly harmful pores. The content of highly harmful pores generally fluctuates and decreases with age. In LCC, the content of highly harmful pores at 12 h, 1 d, 3 d, 7 d, and 28 d is 4.85, 4.44, 2.42, 5.00, and 2.47%, respectively, representing a decrease of 2.38 percentage points from 12 h to 28 d. In SACC, the content of highly harmful pores decreases from 4.34% at 12 h to 3.60% at 28 d, representing a decrease of 0.74 percentage points. The content of highly harmful pores in both types of concrete gradually decreases with age, indicating that the internal structure of the concrete becomes denser and its performance improves. Compared to SACC, LCC has a higher content of highly harmful pores before the 1 d, but after 1d, its content is consistently lower, and the decrease in percentage points is significantly greater.

In summary, with increasing age, the content of harmless pores and less harmful pores in LCC gradually increases, while the content of harmful pores and highly harmful pores gradually decreases. This indicates that the internal microstructure of the concrete becomes denser over time, leading to improved concrete performance. Compared to SACC, LCC exhibits a higher proportion of less harmful pores and generally lower proportions of harmful and highly harmful pores. This further confirms that LCC has a denser microstructure and better overall performance, which aligns with the results of the concrete compressive strength tests.

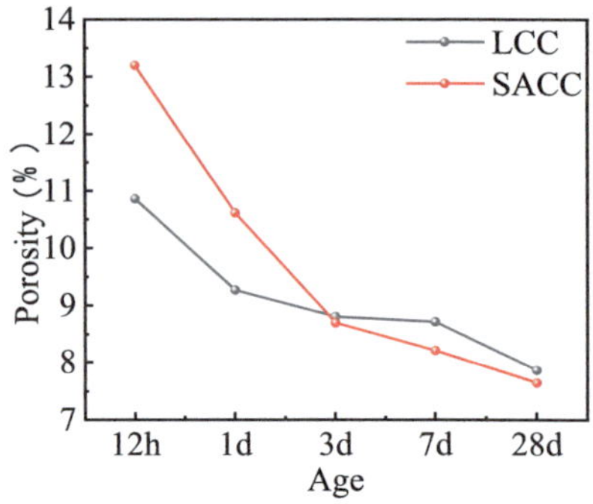

Fig. 8.33 Porosity of the concrete containing sulfoaluminate cement and LCCM

8.4.3 Porosity

Figure 8.33 shows the porosity changes of the two types of concrete. From the figure, it can be observed that the porosity of both SACC and LCC decreases with increasing age. At the 12 h, the porosity of LCC and SACC is 10.8581 and 13.1992%, respectively. By the 3 d, the porosity decreases to 8.8049 and 8.6976%, representing reductions of 2.0531 and 4.5016%, respectively. At the 28 d, the porosity further decreases to 7.8704 and 7.6469%, with reductions of 0.9345 and 1.0507% compared to the 3 d.

In summary, the porosity decreases more rapidly before the 3 d and slows down thereafter. This is because, before the 3d age, the hydration of $C_4A_3\hat{S}$ dominates, leading to the rapid formation of hydration products such as AFt and AH_3, which significantly reduce porosity. After the 3d, the hydration of $C_4A_3\hat{S}$ is largely complete, and the remaining C_2S hydrates at a slower rate. The slower formation of C-S-H gel results in a slower reduction in porosity.

8.4.4 Permeability Evolution Law Based on Pore Structure

The relationship between pore size r and relaxation time T_2 is expressed by Eq. 8.8. From this equation, it can be inferred that pore size r is directly proportional to relaxation time T_2; the larger the T_2, the larger the pore size r. When the pore size r reaches a certain threshold, the fluid inside the pore structure can flow freely, known as free fluid. When the pore size is smaller than this threshold, the fluid inside the pore structure cannot flow due to capillary action, adsorption, etc., and is referred to as bound fluid. In this study, the threshold is set at $T_2 = 10$ ms.

The free fluid saturation S_A is defined as the percentage of free fluid volume in the total pore volume of the concrete sample. The bound fluid saturation S_B is defined as the percentage of bound fluid volume in the total pore volume of the concrete sample. Generally, bound fluid exists in smaller micropores and microcracks, while free fluid exists in larger pores and wider cracks. Therefore, to some extent, the proportions of micropores, microcracks, larger pores, and cracks in concrete can be characterized by bound fluid saturation and free fluid saturation.

Table 8.13 presents the evolution of bound fluid saturation and free fluid saturation with age. From the analysis of Table 8.13, it can be observed that from 12 h to 28 d, bound fluid dominates in LCC, with bound fluid saturation consistently exceeding 92%. This indicates that LCC is primarily composed of micropores and microcracks, with relatively fewer larger pores and cracks.

$$r = 3\rho T_2 \tag{8.8}$$

Permeability is an important indicator for measuring the compactness of concrete. The Coates model is used to calculate permeability, as shown in the following formula:

$$K_{nmr} = \left(\frac{\varphi}{C}\right)^4 \left(\frac{S_A}{S_B}\right)^2 \tag{8.9}$$

where: K_{nmr} is permeability, mD. φ is porosity of concrete. S_A is free fluid saturation of concrete. S_B is bound fluid saturation of concrete. C is an undetermined coefficient.

According to Eq. 8.9, the variation patterns of permeability for the two types of concrete over time are calculated, as shown in Fig. 8.34. From the figure, it can be observed that the permeability of concrete generally exhibits a fluctuating decline with increasing age. For LCC, the permeability values from 12 h to 28 days are 0.0036, 0.0004, 0.0017, 0.0016, and 0.0006, respectively, showing an overall fluctuating downward trend. For sulfoaluminate cement, the permeability values from 12 h to 28 days are 0.0062, 0.0007, 0.0010, 0.0020, and 0.0006, respectively, also demonstrating a general downward trend.

In summary, as the concrete ages, the permeability gradually decreases. This, to some extent, indicates that the internal structure of the concrete becomes denser over time, and the porosity gradually reduces.

Table 8.13 Bound fluid saturation and mobile fluid saturation of samples at different ages

Sample	Saturation (%)	12 h	1 d	3 d	7 d	28 d
LCC	Bound fluid	95.1485	97.7169	94.8859	94.9983	96.185
	Mobile fluid	4.8515	2.2831	5.1141	5.0017	3.815
SACC	Bound fluid	95.663	97.7117	95.9993	93.7464	96.0502
	Mobile fluid	4.337	2.2883	4.0007	6.2536	3.9498

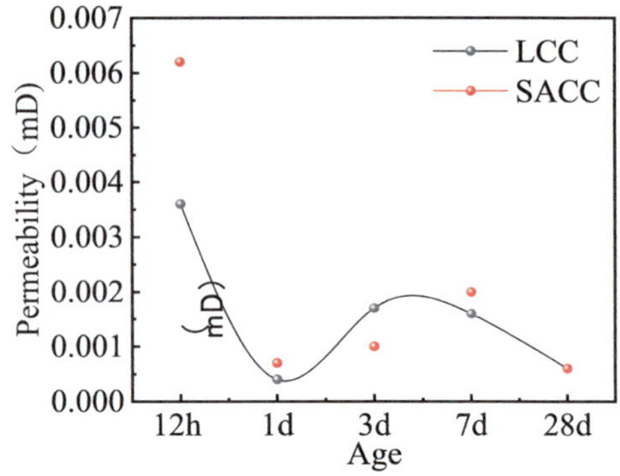

Fig. 8.34 Curve of concrete permeability changing with time

8.5 Early Strength Performance of Concrete Containing LCCM

8.5.1 Compressive Strength

The variation of concrete compressive strength with age is shown in Fig. 8.35. From the figure, it can be observed that at ages of 6 h, 12 h, 24 h, 72 h, 168 h, and 672 h, the compressive strengths of the concrete containing LCC in the experimental group are 35 MPa, 35.99 MPa, 39.79 MPa, 40.73 MPa, 46.72 MPa, and 54.84 MPa, respectively. For the control group, the compressive strengths of the concrete containing sulfoaluminate cement are 23.55 MPa, 24.31 MPa, 31.66 MPa, 38.88 MPa, 39.84 MPa, and 41.49 MPa, respectively.

The compressive strengths of both the concrete containing LCCM and sulfoaluminate cement increase with age. At different hydration ages, the compressive strength of the concrete containing LCCM is significantly higher than that of the concrete containing sulfoaluminate cement. At 6 h and 12 h, the strength of the concrete containing LCCM is 48% higher than that of the concrete containing sulfoaluminate cement. At 3 d, the strengths of the two types of concrete are the closest, with a

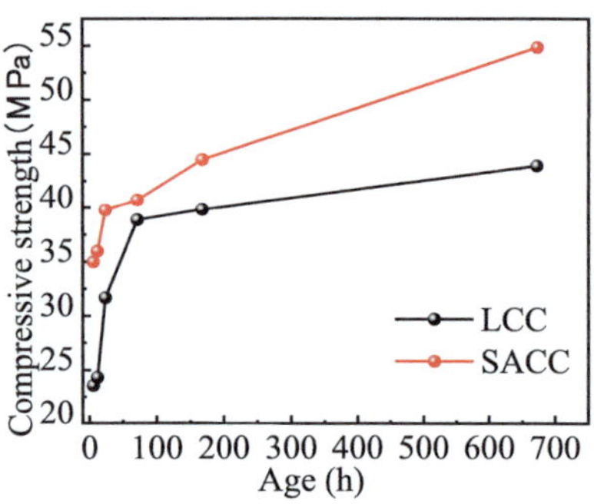

Fig. 8.35 Compressive strength of concrete containing LCCM and sulfoaluminate cement

difference of only 1.85 MPa. By 28 d, the strength of the concrete containing LCCM is 24% higher than that of the concrete containing sulfoaluminate cement. From these results, it can be concluded that this LCCM can be considered equivalent to R.C.42.5 calcium sulfoaluminate cement.

Table 8.14 shows the average growth rates of concrete compressive strength over different time periods. From Table 8.14, it can be observed that both the concrete containing LCCM and sulfoaluminate cement exhibit higher average strength growth rates during the 0–6 h and 12 h–1 d periods, with the 0–6 h period showing the highest growth rate. This indicates that both types of concrete possess early-strength properties.

Except for the 12 h–1 d and 1 d–3 d periods, the average strength growth rate of the concrete containing LCCM is consistently higher than that of R.C.42.5 sulfoaluminate cement. During the 0–6 h period, the average strength growth rate of the concrete containing LCCM is 1.91 MPa/h higher than that of sulfoaluminate cement, representing an increase of 48.7%. This demonstrates that, compared to sulfoaluminate cement, LCCM exhibits superior early-strength performance.

The early strength development of concrete is represented by the ratio of the compressive strength at different ages to the 28-day compressive strength, with a higher ratio indicating faster early-age strength development. The early strength characteristics of the two types of concrete are shown in Table 8.15. As can be seen from the table, the early strength development of both types of concrete increases with age, indicating that neither type exhibits strength regression. Within the first day of hydration, the early strength development of the concrete containing LCCM is higher than that of the concrete containing sulfoaluminate cement, demonstrating that the concrete containing LCCM has superior early-age strength properties. At 6 h of age, the compressive strength of both types of concrete exceeds 50% of their 28-day compressive strength. By 1 day of age, the strength reaches over 70% of the 28-day strength, and by 7 days of age, it exceeds 80% of the 28-day strength, indicating that the hydration phases are essentially fully hydrated by 7 days.

Table 8.14 The average growth rate (MPa/h) of concrete compressive strength in each time period

Type of concrete	Age					
	0–6 h	6–12 h	12 h–1 d	1–3 d	3–7 d	7–28 d
LCC	5.83	0.08	0.32	0.02	0.04	0.02
SACC	3.92	0.06	0.60	0.15	0.01	0.01

Table 8.15 Characterization table of early strength of concrete (%)

Type of concrete	Age					
	6 h	12 h	1 d	3 d	7 d	28 d
LCC	53.66	55.39	72.13	88.59	90.77	100
SACC	63.82	65.63	72.56	74.27	85.19	100

8.5.2 Failure Mode

Concrete is a typical heterogeneous material composed of hardened cementitious material surrounding aggregates. Its heterogeneity arises from the following reasons:

(1) It is a multiphase composite material consisting of cementitious materials, sand, and coarse aggregates in specific proportions.
(2) During the processes of pouring, vibration, and hardening, inherent defects such as microcracks, air bubbles, and micropores are introduced, and their distribution within the concrete is random.
(3) The shape, size, and orientation of coarse aggregates in the concrete are also random.

The strength and deformation of concrete are closely related to the development of inherent defects such as microcracks, micropores, and air bubbles. Cracks in concrete can generally be classified into two types: One type consists of randomly oriented microcracks that influence the macroscopic strength of concrete, such as tensile and compressive strength. The other type consists of aligned macroscopic cracks, which may lead to anisotropic mechanical properties in concrete.

The fracture of concrete is caused by the development of these inherent defects. The failure process essentially involves the initiation, propagation, and coalescence of microcracks and micropores, which eventually develop into macroscopic cracks, leading to concrete failure.

Concrete exhibits high compressive strength but low tensile strength. Therefore, when the principal tensile stress exceeds the tensile strength of concrete, a crack forms perpendicular to the direction of the principal tensile stress. As the tensile stress increases, additional cracks form, which gradually propagate and merge to form macroscopic cracks, ultimately causing concrete failure. The failure process of concrete can be divided into the following three stages.

The failure process of concrete can be divided into the following three stages:

(1) Initial defect damage stage

During the pouring and curing processes, water evaporation and hydration reactions occur in the cement paste, leading to hardening and shrinkage of the paste. Initial microcracks form at the interfaces between aggregates and cementitious materials, as well as within the cementitious materials themselves. These microcracks often appear at the interfaces between aggregates and the cement matrix, creating initial damage. Typically, such damage is randomly distributed within the concrete and remains relatively stable.

(2) Microcrack initiation and propagation stage

Under external loading, stress concentrations develop around various initial defects within the concrete, making these areas the most likely sites for microcrack initiation. As the load increases, relative sliding occurs between the mortar and aggregates along their interfaces, causing the cracks to propagate. During this stage, cracks extend

Fig. 8.36 Concrete containing LCCM compressive failure pattern diagram

steadily in a direction roughly aligned with the principal stress. If the loading is halted, the crack propagation ceases, hence this is referred to as the stable crack propagation stage. When these cracks continue to propagate and eventually interconnect, they form macroscopic cracks, entering the unstable propagation stage.

(3) Formation of interconnected macroscopic cracks

In this stage, both the number and width of cracks increase dramatically. Internal cracks within the paste and interfacial cracks between aggregates and paste interconnect, leading to the enlargement of gaps between aggregates and the paste and the loss of bonding. The formation of interconnected macroscopic cracks signifies the complete loss of load-bearing capacity of the concrete, marking its ultimate failure.

Figure 8.36 illustrates the compressive failure patterns of the concrete containing LCCM at ages ranging from 6 h to 28 days. Except for the 28-day age, the failure types of the concrete are predominantly tensile failures, with the final failure morphology characterized by two opposing conical shapes. In each case, a primary crack typically penetrates the concrete specimen.

During the initial period after pouring and vibrating, up to 1 day of age, water evaporation and hydration reactions lead to the formation of microcracks within the paste and at the interface between the paste and aggregates. Due to the short hydration period, the bonding strength between the hydration products of the LCCM and the aggregates is relatively weak, making the paste-aggregate interface a vulnerable region within the concrete. Consequently, the failure of the concrete specimens manifests as the interconnection of cracks within the paste and at the paste-aggregate interface.

From 3 days of age onward, the degree of hydration of the cementitious materials further increases. Hydration products gradually fill the micropores and microcracks at the paste-aggregate interface, enhancing the bonding strength between the paste and aggregates. As a result, the failure morphology evolves to include the interconnection of cracks within the paste, at the paste-aggregate interface, and through the aggregates themselves. Additionally, it can be observed from the figure that, starting from 3 days of age, apart from the primary failure crack, the paste-aggregate interfaces remain tightly bonded without significant fissures, which differs from the failure patterns observed at 1 day of age and earlier.

By 28 days of age, the hydration process has advanced further. In addition to the hydration products of $C_4A_3\hat{S}$, such as ettringite and AH_3 gel, the hydration of C_2S produces C-S-H, further strengthening the bond at the paste-aggregate interface. As a result, the failure is characterized by the appearance of fine microcracks on the sides

Fig. 8.37 Concrete containing sulfoaluminate cement compressive failure pattern diagram

of the concrete specimen. Due to the absence of restraint from the top and bottom surfaces in contact with the pressure plates, the outer middle section of the specimen fails and spalls off first.

Figure 8.37 depicts the compressive failure patterns of the concrete containing sulfoaluminate cement at ages ranging from 6 h to 28 days. As shown in the figure, the failure of samples is also primarily tensile in nature. The failure patterns at various ages indicate that microcracks initially appear on the surface, which then gradually propagate. The outer layer of the concrete in the middle section first bulges outward, and as the load increases, the bulging becomes more pronounced until it eventually detaches. Similar to the concrete containing LCCM, the failure patterns after 3 days of age also include the fracturing of aggregates.

For both types of concrete, at ages of 1 day or earlier, the cracks are numerous and of similar width, making it difficult to identify a primary crack. However, in the failure patterns at 3 days and beyond, one or two primary cracks penetrating the concrete can be easily observed, along with secondary cracks at the top and bottom that do not fully penetrate. This phenomenon may be attributed to the fact that, at ages of 1 day or earlier, the bond strength between the paste and aggregates is generally weak, resulting in multiple weak interfacial zones and thus numerous cracks without a distinct primary crack. At 3 days and beyond, the strength of the paste and the interfacial bond between the paste and aggregates are significantly stronger, so failure initiates at one or two relatively weaker interfaces, leading to the formation of primary cracks. Once the primary crack forms, stress redistribution occurs within the concrete specimen, causing further cracking at the next weakest interfacial zone in the separated section, resulting in the formation of a second primary crack or secondary cracks. However, a notable difference is that in the failure patterns at 3 days of age, the gaps between the paste and aggregates are larger in the concrete containing sulfoaluminate cement compared to the concrete containing LCCM.

8.5.3 Relationship Between Pore Structure and Compressive Strength

From the mesoscopic perspective of concrete failure, it is evident that the initial damage in concrete originates from microcracks and micropores formed during the processes of pouring, vibration, and curing. Under external loading, stress concentrations occur at these microcracks and micropores. As the load gradually increases, the concrete undergoes stages of crack initiation and propagation, followed by the formation of macroscopic main cracks, ultimately leading to complete failure. Therefore, the development and proportion of different types of pore structures are related to stress concentration, as well as the formation and propagation of microcracks. In summary, the pore structure of concrete has a significant influence on its compressive strength.

Based on the method proposed by Academician Wu Zhongwei, the internal pore structures of the two types of concrete can be classified into the following categories: harmless pores (pore diameter < 20 nm), less harmful pores (20 nm ≤ pore diameter < 50 nm), harmful pores (50 nm ≤ pore diameter < 200 nm), highly harmful pores (pore diameter ≥ 200 nm).

The influence of different pore sizes on concrete performance varies. Below, the impact of different pore types on compressive strength is studied using this classification method. The grey relational segment entropy analysis method is employed to quantitatively analyze the influence of harmless pores, less harmful pores, harmful pores, highly harmful pores, and porosity on the compressive strength of concrete. This method accurately identifies the pore types that have stronger or weaker effects on compressive strength. Using the content of harmless pores, less harmful pores, harmful pores, highly harmful pores, and total porosity at different ages as influencing factors, grey theory is applied to analyze the impact of different pore types and porosity on compressive strength. The compressive strength of concrete at different ages is taken as the reference sequence, which can be expressed as:

$$Y_n = [y_n(1), y_n(2), \ldots, y_n(k)] \tag{8.10}$$

where: Y_n is the compressive strength value of the samples at different ages. $y_n(k)$ is the compressive strength of the samples at the kth age.

Taking the harmless pore content (X_1), less harmful pore content (X_2), harmful pore content (X_3), more harmful pore content (X_4), and total porosity (X_5) of concrete as the comparison sequence, it can be set as:

$$X_n = [x_n(1), x_n(2), \ldots, x_n(k)] \tag{8.11}$$

where: X_n is the content of different types of pores or total pores. $x_n(k)$ is the content of different types of pores or total pores at the age of k.

The original sequence at different ages is shown in Table 8.16.

The grey entropy correlation degree between the different pore structure and porosity of the two kinds of concrete and their compressive strength can be obtained by averaging the original sequence as follows.

Table 8.16 Original sequence of pore structure and compressive strength of two types of concrete

Concrete	Age	X_1/%	X_2/%	X_3/%	X_4/%	X_5/%	Y/MPa
LCC	12 h	65.12	27.18	2.84	4.85	13.1992	35.99
	1d	65.56	26.93	3.06	4.44	10.6149	39.79
	3d	67.55	27.65	2.38	2.42	8.6976	40.73
	7d	65.33	27.05	2.62	5	8.2139	46.72
	28d	78.41	18.56	0.56	2.47	7.6469	54.84
SACC	12 h	62.37	29.18	4.11	4.34	13.1992	24.31
	1d	68.81	24.3	3.08	3.81	10.6149	31.66
	3d	74.11	18.63	2.7	4.56	8.6976	38.88
	7d	74.01	16.44	2.94	6.61	8.2139	39.84
	28d	77.74	15.69	2.97	3.6	7.6469	43.89

(1) Meaning processing:

$$\begin{cases} x'_n = \dfrac{X_n}{x_n(a)} = (x'_n(1), x'_n(2), ..., x'_n(k)) \\ y'_m = \dfrac{Y_m}{y_m(a)} = (y'_m(1), y'_m(2), \ldots, y'_m(k)) \end{cases} \tag{8.12}$$

where: $x'_n(k)$ is the mean value of $x_n(1), x_n(2),\ldots, x_n(k)$. $y'_m(k)$ is the mean value of $y_m(1), y_m(2),\ldots, y_m(k)$.

(2) The difference between reference sequence and comparison sequence:

$$\begin{cases} \Delta_{mn}(k) = \left|y'_m(k) - x'_n(k)\right| \\ \Delta_{mn} = (\Delta_{mn}(1), \Delta_{mn}(2), ..., \Delta_{mn}(k)) \\ (m = 1, 2, ..., l; n = 1, 2, ..., q; k = 1, 2, ..., p) \end{cases} \tag{8.13}$$

(3) Maximum difference and minimum difference:

$$\begin{cases} M = \max\limits_m \max\limits_n \Delta_m(k) \\ m = \min\limits_m \min\limits_n \Delta_m(k) \end{cases} \tag{8.14}$$

(4) Correlation coefficient:

$$\gamma_{nm} = \frac{m + \zeta M}{\Delta_n(k) + \zeta M} \tag{8.15}$$

where, γ_{nm} is the resolution coefficient, generally take as 0.5.

(5) Grey entropy correlation density

$$P_h = \frac{\gamma[y_n(h), x_n(h)]}{\sum_{k=1}^{n} \gamma[y_n(h), x_n(h)]}(h = 1, 2, ..., n) \tag{8.16}$$

(6) Grey entropy correlation density
Grey correlation entropy:

$$H = -\sum_{k=1}^{n} P_n \ln P_n \tag{8.17}$$

Grey entropy correlation degree:

$$E = \frac{1}{n}H \tag{8.18}$$

From Table 8.17, it can be observed that in the concrete containing LCCM, the pore type with the greatest influence on compressive strength is the harmless pore (pores with diameters < 20 nm), followed by multi-harmful pores (pores with diameters ≥ 200 nm), less-harmful pores (pores with diameters 20–50 nm), total porosity, and harmful pores (pores with diameters 50–200 nm). The gray entropy correlation degrees are 0.998972, 0.992869, 0.985133, 0.982167, and 0.966901, respectively.

From the analysis of the pore structure using NMR, it is evident that harmless pores constitute a significant proportion of the pore structure. A higher proportion of

Table 8.17 Grey entropy calculation results of pore structure and compressive strength

Type of concrete	Age	$-P_1\ln P_1$	$-P_2\ln P_2$	$-P_3\ln P_3$	$-P_4\ln P_4$	$-P_5\ln P_5$
LCC	12 h	0.314187	0.310906	0.305487	0.308776	0.2789
	1d	0.330701	0.327981	0.304242	0.336431	0.331817
	3d	0.329129	0.326771	0.35178	0.32799	0.35582
	7d	0.316373	0.352157	0.356287	0.338315	0.325675
	28d	0.317393	0.267696	0.238371	0.286449	0.288526
H	–	1.607783	1.58551	1.556167	1.597961	1.580736
E	–	0.998972	0.985133	0.966901	0.992869	0.982167
SACC	12 h	0.299164	0.268735	0.257484	0.308111	0.265039
	1d	0.328654	0.33577	0.353061	0.353622	0.341415
	3d	0.335762	0.350065	0.323414	0.345536	0.345254
	7d	0.328159	0.327525	0.333508	0.296051	0.331202
	28d	0.312143	0.302887	0.312448	0.274903	0.301013
H	–	1.603882	1.584983	1.579915	1.578224	1.583923
E	–	0.996548	0.984805	0.981656	0.980606	0.984146

harmless pores is beneficial for concrete strength. On the other hand, harmful pores have larger diameters, which increases the probability of stress concentration and the formation of microcracks, thereby directly affecting compressive strength.

In concrete containing sulfoaluminate cement, the pore type with the greatest influence on compressive strength is also the harmless pore, followed by less-harmful pores, harmful pores, total porosity, and multi-harmful pores. Through gray entropy correlation analysis, the influence of different pore types on the compressive strength of both types of concrete has been clearly quantified.

8.5.4 Relationship Between Hydration Product and Compressive Strength

When the coarse and fine aggregates meet the requirements for concrete sand and gravel, the compressive strength of concrete primarily depends on the strength of the cementitious material and the bond strength between the cementitious material and the aggregates. These strengths are derived from the content of various hydration products of the cementitious material. The gray theory system will be used to evaluate the influence of each hydration product on the concrete strength.

Taking the compressive strength of concrete at different ages as a reference sequence, the AFt content (Z_1), AFm content (Z_2), AH_3 content (Z_3), and AFt-CO_3 content (Z_4) can be set a:

$$Y_n = [y_n(1), y_n(2), \ldots, y_n(k)] \tag{8.19}$$

where, Y_n is the compressive strength value of low carbon concrete or SACC at different ages; $y_n(k)$ is the compressive strength of low carbon concrete or SACC at the kth age.

$$Z_n = [z_n(1), z_n(2), \ldots, z_n(k)] \tag{8.20}$$

where, Z_n is the content of different types of hydration products or total hydration products. the content of different types of hydration products or total hydration products at the kth age of $z_n(k)$.

The original sequence at different ages is shown in Table 8.18.

According to the Eqs. 8.12–8.18, the grey correlation entropy and grey entropy correlation degree of low carbon concrete can be calculated, as shown in Table 8.19.

From Table 8.19, it can be observed that in the concrete containing LCCM, the hydration product with the greatest influence on compressive strength is AH_3, followed by AFt and AFm, with gray entropy correlation degrees of 0.9693998, 0.963252, and 0.947682, respectively. The analysis results indicate that AH_3 has the most significant impact on the compressive strength of concrete, followed by AFt and AFm. This suggests that the cementitious effect of AH_3 plays a crucial role in the

Table 8.18 Original sequence of hydration products and compressive strength of two types of concrete

Type of concrete	Age	Z_1/%	Z_2/%	Z_3/%	Z_4/%	Y/MPa
LCC	12 h	12.334	2.3993	1.8929	–	35.99
	1d	12.9985	2.6132	2.2011	–	39.79
	3d	13.104	2.537	2.1067	–	40.73
	7d	13.2025	2.719	2.1649	–	46.72
	28d	13.379	2.5211	2.0376	–	54.84
SACC	12 h	2.351	2.1874	3.101	0.6139	24.31
	1d	2.6498	2.5625	3.5372	0.4105	31.66
	3d	3.029	2.7027	4.1182	1.6818	38.88
	7d	9.6485	2.9182	3.0109	1.1295	39.84
	28d	11.8701	3.2696	3.2326	1.203	43.89

Table 8.19 Grey entropy calculation results of hydration products and compressive strength

Type of concrete	Age	$-P_1 \ln P_1$	$-P_2 \ln P_2$	$-P_3 \ln P_3$	$-P_4 \ln P_4$
LCC	12 h	0.257945	0.259161	0.320604	–
	1d	0.285912	0.262339	0.272256	–
	3d	0.297927	0.311026	0.328165	–
	7d	0.324744	0.354256	0.335251	–
	28d	0.205016	0.173056	0.188525	–
H	–	1.725916	1.698017	1.736928	–
E	–	0.963252	0.947682	0.969398	–
SACC	12 h	0.332729	0.283477	0.288069	0.31386
	1d	0.303135	0.308019	0.310572	0.217209
	3d	0.280147	0.284541	0.321229	0.204746
	7d	0.272273	0.302327	0.278118	0.338852
	28d	0.235802	0.307503	0.26643	0.324438
H	–	1.788881	1.78013	1.741778	1.723601
E	–	0.998393	0.99351	0.972105	0.96196

concrete containing LCCM. Thermogravimetric analysis reveals that AFt is present in larger quantities and exists in a rod-like form, providing a skeletal support function. AFm, on the other hand, is present in smaller quantities, resulting in a lesser influence on strength.

In the concrete containing sulfoaluminate cement, the hydration products ranked by their influence on compressive strength from highest to lowest are AFt, AFm, AH_3, and AFm-CO_3, with gray entropy correlation degrees of 0.998393, 0.99351, 0.972105, and 0.96196, respectively. This differs from the results observed in the concrete containing LCCM. In the concrete containing sulfoaluminate cement, AFt has the greatest influence on strength, followed by AFm, AH_3, and AFm-CO_3.

In summary, the hydration products that have a significant impact on the compressive strength of concrete are AFt and AH_3, while AFm and AFm-CO_3 have a relatively weaker influence.

8.6 Conclusions

This chapter utilized coal gangue, desulfurized gypsum, and carbide slag as primary raw materials to prepare a multi-component solid waste-based low-carbon cementitious material (LCCM). The concrete containing LCCM was subsequently prepared by mixing LCCM with coarse/fine aggregates and water. For comparison, sulphoaluminate cement concrete with identical proportions was prepared as a control group. This approach not only achieved resource utilization of industrial solid wastes but also reduced carbon emissions and resource consumption in the cement industry. Through the study of the concrete containing LCCM, the following conclusions were drawn:

(1) Both LCCM and sulphoaluminate cement demonstrated AFt, AFm, and AH3 as primary hydration products, with their quantities increasing over curing time. Additionally, AFm-CO3 was identified in the hydration products of sulphoaluminate cement.
(2) Nuclear magnetic resonance analysis revealed that porosity in both the concrete containing LCCM and sulphoaluminate cement decreased with extended curing periods. The pore structure exhibited an overall reduction in pore size due to hydration product formation and filling effects, accompanied by increased harmless pores and decreased less harmful, harmful, and more harmful pores. The concrete containing sulphoaluminate cement displayed similar pore structure evolution patterns.
(3) Failure pattern analysis indicated that both the concrete containing LCCM and sulphoaluminate cement exhibited tensile failure modes. Within 1 day of curing, failure primarily occurred in the paste and paste-aggregate interface. From 3 days onward, aggregate failure emerged as an additional failure mode in both concrete types.
(4) Early strength development was characterized by the percentage of concrete strength relative to 28-day strength at various early ages. The concrete containing LCCM demonstrated superior early strength performance, achieving 63.82% at 6 h compared to 53.66% for the concrete containing sulphoaluminate cement.
(5) Grey theory analysis evaluated the influence of different pore types on compressive strength. For the concrete containing LCCM, the impact hierarchy was: harmless pores > more harmful pores > less harmful pores > harmful pores. For the concrete containing sulphoaluminate cement, the sequence was: harmless pores > less harmful pores > harmful pores > more harmful pores.

References

Carrasco, L. F., & Vázquez, T. (1996). Aplicación de la espectroscopia infrarroja al estudio de cemento aluminoso. *Materiales De Construcción, 46*(241), 53–66.

Chi, L. (2019). *Study on the activation and hydration mechanism of high belite sulphoaluminate cement* [Doctoral dissertation, Harbin Institute of Technology].

Hargis, C. W., Kirchheim, A. P., Monteiro, P. J., & Gartner, E. M. (2013). Early age hydration of calcium sulfoaluminate (synthetic ye'elimite, C_4A_3S) in the presence of gypsum and varying amounts of calcium hydroxide. *Cement and Concrete Research, 48*, 105–115.

Hargis, C. W., Telesca, A., & Monteiro, P. J. (2014). Calcium sulfoaluminate (Ye'elimite) hydration in the presence of gypsum, calcite, and vaterite. *Cement and Concrete Research, 65*, 15–20.

Lan, Y., Gu, Q., Peng, Y., Zeng, Q., & Zhang, Z. D. (2022). Early microstructure of sulphoaluminate and silicate cement pastes under different CO_2 curing pressures. *Journal of Zhejiang University (Engineering), 56*(12), 2454–2462.

Li, D. D. (1984). Infrared spectroscopic study of sulphoaluminate cement. *Journal of the Chinese Ceramic Society, 01*, 119–125.

Li, H., Sun, W., & Zuo, X. B. (2012). Microscopic analysis of mineral admixtures to improve the sulfate resistance of cement-based materials. *Journal of the Chinese Ceramic Society, 40*(08), 1119–1126.

Su, D., Li, K., Li, Q., & Zhao, G. (2019). Exploratory research on the synergistic preparation of high belite sulphoaluminate cement with various industrial solid wastes. *New Building Materials, 46*(12), 9–13.

Tosun, K., & Baradan, B. (2010). Effect of ettringite morphology on DEF-related expansion. *Cement and Concrete Composites, 32*(4), 271–280.

Trezza, M. A., & Lavat, A. E. (2001). Analysis of the system 3CaO· Al2O3–CaSO4· 2H2O–CaCO3–H2O by FT-IR spectroscopy. *Cement and Concrete Research, 31*(6), 869–872.

Wang, P., Li, N., & Xu, L. (2017). Hydration evolution and compressive strength of calcium sulphoaluminate cement constantly cured over the temperature range of 0 to 80 C. *Cement and Concrete Research, 100*, 203–213.

Wu, Z. W., & Lian, H. Z. (1999). *High Performance Concrete*. China Railway Publishing House.

Yin, Y., Yin, J., Zhang, W., Tian, H., Hu, Z., Ruan, M., ... & Chen, D. (2018). FT-IR and micro-Raman spectroscopic characterization of minerals in high-calcium coal ashes. *Journal of the Energy Institute, 91*(3), 389–396.

Zhang, W. S., Zhang, J. S., Ye, J. Y., Wang, H. X., Zhang, J. B., & Liu, J. C. (2017). Effect of synthesis conditions on the morphology of ettringite. *Journal of the Chinese Ceramic Society, 45*(05), 631–638.

Chapter 9
Impact Properties of Concrete Containing Low-Carbon Cementitious Materials

9.1 Introduction

Low-carbon concrete was developed by replacing traditional cement with low-carbon cementitious materials and coarse aggregates with coal gangue. A systematic investigation was conducted on the evolution of mechanical properties and failure modes of concrete at various hydration ages. The impact resistance, including the number of impacts, failure morphology, and impact energy, was analyzed through a series of impact tests. Microscopic testing techniques were employed to reveal the morphological features of hydration products and the structural characteristics of the interfacial transition zone (ITZ). Nuclear magnetic resonance (NMR) technology was utilized to quantitatively characterize the evolution of porosity and pore size distribution. Based on the Weibull distribution model, an impact life prediction formula was established, offering theoretical support for the durability design of low-carbon concrete under dynamic loading conditions.

9.2 Raw Material and Test Methods

9.2.1 Raw Material

Coal gangue was selected from the Ordos mining area in Inner Mongolia. Carbide slag was obtained from the Ulanqab region of Inner Mongolia. Desulfurization gypsum was sourced from a power plant in the Jungar region. For the control experiment, rapid-hardening sulfoaluminate cement (SAC-42.5) manufactured in Weifang, Shandong Province, China, was utilized. The chemical compositions of the raw materials are presented in Tables 9.1 and 9.2. The microscopic appearances of the coal gangue, carbide slag, and desulfurization gypsum are shown in Fig. 9.1.

C. Yan et al., *Low-Carbon Cementitious Materials with 100% Solid Wastes*,
https://doi.org/10.1007/978-981-96-8742-8_9

Table 9.1 Chemical composition of the raw materials (wt. %)

Materials	SiO_2	Al_2O_3	CaO	SO_3	TiO_2	Fe_2O_3	Na_2O	MgO	Loss
CG	33.74	31.67	1.81	1.59	1.21	0.74	/	0.64	28.52
CS	7.62	0.33	67.26	1.09	0.09	/	1.24	/	21.93
DG	7.21	0.31	25.62	47.62	/	/	3.91	/	14.62
Sulfoaluminate cement	11.85	18.44	47.37	14.93	0.45	1.41	1.98	1.36	10.05

CG means coal gangue, CS denotes carbide slag, DG represents desulfurization gypsum

Table 9.2 Performance indicators of the sulfoaluminate cement

Setting time (min)		Compressive strength (MPa)		Bending strength (MPa)	
Initial setting time	Final setting time	1 d	3 d	1 d	3 d
10	15	37	45	6	6.5

Fig. 9.1 Physical drawings of raw materials

Natural aggregates provided by a building materials factory in Ordos City were selected as the coarse aggregates, with a particle size range of 5–20 mm, as shown in Fig. 9.2a. For comparative analysis, coal gangue coarse aggregates were used as the control specimens. The specific preparation process is as follows: Firstly, the coal gangue was crushed by a jaw crusher; then, the particle size of the coal gangue coarse aggregates was controlled within the range of 5–20 mm, as shown in Fig. 9.2b. The performance indicators of the coarse aggregates are detailed in Table 9.3.

9.2.2 Preparation of LCCM

Mineral composition design

The design of LCCM takes coal gangue, calcium carbide slag, and desulfurization gypsum as the main raw materials. These materials respectively serve as the sources of Si, Al, Ca, and S. Through the synergistic effect of multiple solid wastes, LCCM

(a) Natural aggregate (b) Coal gangue aggregate

Fig. 9.2 Physical drawings of coarse aggregate

Table 9.3 Performance indicators of coarse aggregates

Types	Crush index (%)	Bulk density (kg/m^3)	Apparent density (kg/m^3)	Silt content (%)	Water content (%)	Water absorption (%)
Natural aggregate	12.4	1306	2659	3.56	0.93	5.32
Coal gangue aggregate	9.0	1401	2850	3.15	0.20	1.26

is prepared with C_2S and $C_4A_3\check{S}$ as the main mineral phases. The determination of the initial raw material ratio is based on the improved Bogue calculation method (Eqs. 9.1–9.5) (Zhang & Guolong, 2011). Combined with experimental data, the ratio was further optimized. The final optimal mix ratio is shown in Table 9.4.

$$W(CaO) = 0.6512W(C_2S) + 0.3762\,W\left(C_4A_3\check{S}\right) + 0.4616W(C_4AF) \tag{9.1}$$

$$W(SiO_2) = 0.3488W(C_2S) \tag{9.2}$$

$$W(Al_2O_3) = 0.5106W\left(C_4A_3\check{S}\right) + 0.2098W(C_4AF) \tag{9.3}$$

$$W(Fe_2O_3) = 0.33W(C_4AF) \tag{9.4}$$

$$W(SO_3) = 0.1311W\left(C_4A_3\check{S}\right) \tag{9.5}$$

Table 9.4 Raw material ratio and chemical composition of clinker (wt.%)

Group	Chemical components				Raw materials		
	Al_2O_3	SiO_2	CaO	SO_3	CG	CS	DG
Clinker	17.8	19.4	46.3	14.3	30	48	22

Raw material processing

The raw materials utilized in the experiment were first crushed and subsequently ground using a ball mill. The ground materials were then sieved through a 200-mesh square-hole sieve, ensuring that the sieve residue was controlled to less than 3%. Following this, the sieved materials were dried in a forced-draft drying oven at 105 °C. According to the batching scheme, the required quantities of materials were accurately weighed, mixed with 10% water by weight, and uniformly stirred. The homogeneously mixed raw meal was then loaded into steel molds and pressed into disc-shaped specimens with a diameter of 60 mm and a height of 5 mm under a pressure of 20 MPa. Finally, the specimens were dried in a forced-draft drying oven at 105 °C for 24 h and stored for subsequent use.

Clinker calcination

The dried samples were placed in a box-type high-temperature furnace. The heating rate of the furnace was set at 10 °C/min. The calcination temperature was set at 1260 °C. The calcination time was set at 40 min. After calcination, the samples were taken out of the furnace and naturally cooled to room temperature in a dry environment to obtain LCCM clinker.

Clinker grinding

The calcined clinker is ground by a rod mill, and then the ground powder is sieved through a 200-mesh square hole sieve, ensuring that the sieve residue is controlled below 3%. Finally, the obtained clinker is mixed with desulfurization gypsum at a mass ratio of 96:4 to complete the preparation of LCCM.

X-ray diffraction analysis indicates that the main mineral phases of LCCM are $C_4A_3\check{S}$ and C_2S, with a small amount of anhydrite and gypsum dihydrate, as shown in Fig. 9.3. The workability of LCCM was tested in accordance with GB/T 1346-2011 and GB/T 17,671-2021, and the results are presented in Tables 9.5 and 9.6.

9.2.3 *Mix Proportion Design of Concrete Containing LCCM*

The water-to-binder ratio (W/C) was set at 0.4. The reference mix proportion for the test concrete was established in accordance with the relevant standards outlined in the "Specification for Mix Proportion Design of Ordinary Concrete" (JGJ 55-2011). The parameters of the concrete mix proportion are presented in Table 9.7.

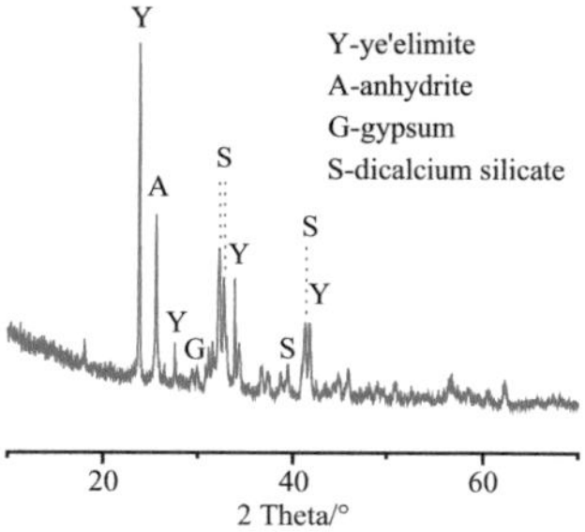

Fig. 9.3 XRD patterns of LCCM

Table 9.5 Setting time and water requirement for normal consistency of LCCM

Setting time (minutes)		Water requirement for normal consistency
Initial setting time	Final setting time	Water/Cement
25	38	0.28

Table 9.6 Mechanical properties of LCCM

Title	1 d	3 d	7 d	14 d	28 d
Compressive strength (MPa)	43.8	49.6	55.4	56.2	61.4
Flexural strength (MPa)	4.7	5.4	5.9	6.1	6.5

Table 9.7 Mix ratio of concrete containing LCCM (kg/m^3)

No	CG coarse aggregate	Natural coarse aggregate	Natural fine aggregate	LCCM	SAC-42.5	Water
A1	0	907.5	518.4	648	0	259.2
A2	0	907.5	518.4	0	648	259.2
A3	907.5	0	518.4	0	648	259.2

9.2.4 *Sample Preparation of Concrete Containing LCCM*

Compressive strength

The compressive strength test specimens of concrete were prepared as cubic blocks with dimensions of 100 mm × 100 mm × 100 mm. To ensure the uniformity and consistency of the specimens, vibration compaction was performed using a vibrating table during the preparation process. Demolding was carried out after curing at room temperature for 6 h. Subsequently, the specimens were transferred to a standard curing chamber maintained at a relative humidity of 95% and a temperature of 20 ± 2 °C for further curing.

Drop weight impact test

In accordance with the test method recommended by the ACI (American Concrete Institute) Committee 544, disc-shaped specimens measuring 153 mm in diameter and 63.5 mm in height were fabricated. The fabrication procedure and curing methodology were identical to those employed for the compressive strength specimens.

Scanning electron microscope

The sample utilized for SEM testing was an aggregate-free fragment extracted from the interior of a concrete specimen. Following the completion of the concrete strength test, the sample was collected using tools such as a rubber mallet and steel pliers. The hydration process of the sample was terminated by immersion in anhydrous ethanol.

Nuclear magnetic resonance

Cylindrical specimens, with a diameter and height of 50 mm each, were extracted from cubic concrete blocks (100 mm × 100 mm × 100 mm) using a concrete coring machine for subsequent NMR testing.

9.2.5 *Test Methods*

Compressive strength test

The compressive strength test was conducted in accordance with the relevant provisions of JTG 3420-2020 to evaluate the compressive and flexural strengths of concrete. Each testing group consisted of three samples. A YAS-600 type electro-hydraulic servo uniaxial pressure testing machine was employed to measure the compressive strength of samples, with a loading rate of 0.5 MPa/s. The four-point flexural strength of samples was assessed using a WDW-100E universal testing machine, with a loading rate set at 0.05 MPa/s.

Drop weight impact test

The impact resistance test of concrete is conducted in accordance with the methodology recommended by the American Concrete Institute (ACI) 544. Each group consists of six samples. Prior to testing, a layer of lubricant is applied to the base plate to prevent sample displacement during the impact process. The sample is then positioned at the center of the impact testing apparatus. Upon activation of the magnetic switch, the electromagnet lifts the weight to a predetermined height, allowing it to fall freely under gravity. Each hammer strike on the specimen is recorded as one cycle until the specimen fails due to the cumulative impact. Figure 9.4 illustrates the impact testing apparatus.

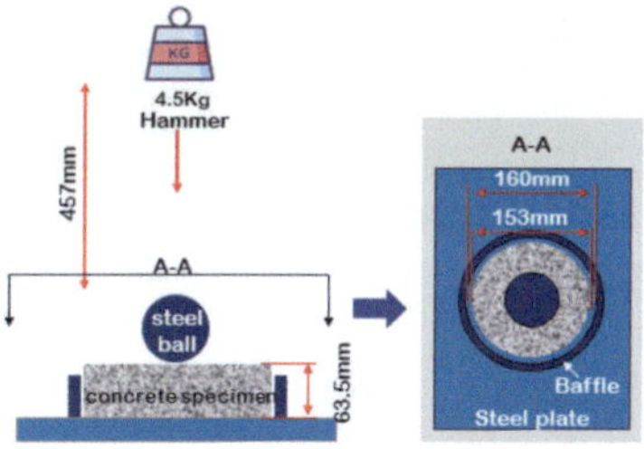

(a) Test diagram of impact resistance

(b) Impact resistance test device

Fig. 9.4 Impact resistance test diagram and experimental device diagram

Scanning electron microscope test

The micromorphology of the samples was observed using the Sigma SEM 300 field emission scanning electron microscope manufactured by Carl Zeiss. Gold coating was applied to the sample surface to enhance image quality. The operational parameters were as follows: an acceleration voltage range of 0.02–30 kV, a resolution of 1.0 nm, and a continuously adjustable step size of 10 V.

Nuclear magnetic resonance test

The MesoMR-60s nuclear magnetic resonance analyzer was employed to investigate the distribution characteristics of porosity and pore size in concrete specimens at various curing ages. The samples underwent pressurized water retention treatment to ensure complete saturation of the pores. The test conditions were as follows: a magnetic field strength of 0.23 T, a probe inner diameter of 50 mm, an operating frequency of 10.113 MHz, and the temperature of the magnet and measuring probe was precisely controlled at 35 ± 0.05 °C.

9.3 Mechanical Properties of Concrete Containing LCCM

9.3.1 Compressive Strength

Analysis of compressive failure modes of concrete

In the initial stage of the compressive strength test, the failure modes of A1, A2, and A3 concrete specimens exhibited a high degree of similarity. As the compressive load increased incrementally, the specimens initially underwent compressive deformation. During the loading process, localized bulging and spalling were observed on the specimen surfaces, accompanied by the formation of micro-cracks. These micro-cracks propagated from the surface toward the central region of the specimens and extended vertically, ultimately penetrating to the bottom of the specimens. Post-failure analysis revealed that all specimens displayed a characteristic quadrilateral pyramid shape with inverted and upright connections. However, upon comparing the failure modes of A1, A2, and A3 concrete specimens cured for 6 h, 1 day, and 3 days, respectively (as shown in Fig. 9.5), significant variations were noted in crack propagation paths, surface spalling extent, and ultimate failure characteristics. These differences highlight the evolution of mechanical properties in concrete specimens with varying mix proportions.

By comparing the failure modes of A1 and A2 concrete specimens, it is evident that the failure of the concrete specimens primarily involves the failure of the cement mortar layer and the interface failure between some natural coarse aggregates and the cement mortar. In comparison to A2, A1 concrete exhibits a more pronounced brittle sound upon damage. The cubic failure of A1 concrete can be encapsulated by "strong" mortar with "strong" aggregates, which possess high early-age strength, and some coarse aggregates on the damaged section may contribute to the failure. Conversely, the cubic failure of A2 concrete is encapsulated by "weak" mortar with "strong" aggregates; the natural gravel inherently possesses high strength, and under compressive loading, the interface between the natural gravel and the cement mortar

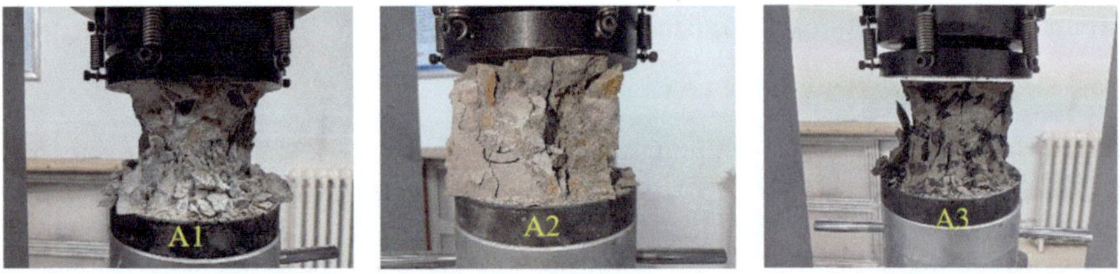

(a) Compression failure modes of concrete under compression at 6 h

(b) Compression failure modes of concrete under compression at 1 d

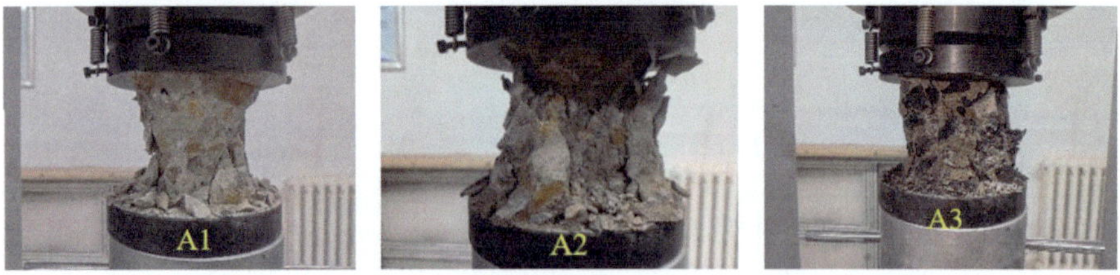

(c) Compression failure modes of concrete under compression at 3 d

Fig. 9.5 Compression failure modes of concrete in different age becomes the most vulnerable region. Upon damage, the natural gravel within A2 concrete remains largely intact, while the surface exhibits an uneven morphology. The failure shape of A2 concrete at 6 h and 1 day is predominantly attributed to incompletely hydrated cementitious materials. At 3 days, complete hydration of the cementitious materials leads to aggregate fracture, indicating that during the failure of A1 concrete, the bond strength between the aggregate and LCCM mortar surpasses that of sulphoaluminate cement mortar.

After the failure of A1 and A2 concrete specimens, the morphology of the specimens on the platen under the testing machine exhibited significant differences. The failure of the A1 concrete specimen under compression resulted in a distinct quadrangular conical shape. In contrast, the failure of the A2 concrete specimen under pressure led to fragments that remained adhered to the undamaged structure. Notably, the failure morphology of the A2 concrete specimen was most pronounced at 6 h and 1 day post-failure. Observations of the appearance of the A2 and A3 concrete specimens after failure revealed fewer fragments adhering to the A3 specimens. This phenomenon can be attributed to the short hydration time of the sulfoaluminate cement binder in the A2 concrete. Despite an improvement in strength, the ductility of the hardened gelled material was insufficient.

By comparing the failure modes of A2 and A3 concrete, it was observed that A3 concrete exhibits a faster failure rate compared to A2 concrete. In the axial compression test, the failure pattern of A3 concrete is analogous to that of A2 concrete. As the axial load increases, micro-cracks develop on the surface of A3 concrete and progressively expand until failure occurs. The cubic failure modes of A3 and A2 concrete differ significantly in terms of the failure section. For A2 concrete, damage primarily occurs in the interfacial transition zone between natural aggregate and cement mortar, with minimal damage to the natural aggregate itself. In contrast, the failure mode of A3 concrete involves not only the failure of the cement mortar layer but also the intrinsic failure of the non-spontaneous coal gangue coarse aggregate. This difference can be attributed to the lower strength of non-spontaneous combustion coal gangue aggregate compared to natural crushed stone, as well as the increased susceptibility of needle-like particles within the coal gangue aggregate to damage. Consequently, the failure section of A3 concrete appears relatively flat, and the cubic damage of A3 concrete can be characterized by a combination of "strong" mortar and "weak" aggregate.

Analysis of compressive strength test results

The compressive strength test results of A1, A2, and A3 concrete are detailed in Tables 9.8 and 9.9. Analysis of the data in Table 9.8 shows that the average compressive strengths of A1 concrete at 2 and 4 h reached 14.32 and 25.02 MPa, respectively. This indicates that the LCCM system exhibits excellent early-age strength development characteristics. From the data in Table 9.9, it can be seen that the compressive strengths of A2 concrete at 6 h, 1 day, and 3 days were 29.18, 32.84, and 41.01 MPa, respectively. In contrast, the compressive strengths of A1 concrete at the same curing ages were significantly higher, reaching 46.72, 51.43, and 53.41 MPa, respectively. Compared to A2 concrete, the compressive strengths of A1 concrete at 6 h, 1 day, and

3 days increased by 60.1, 56.6, and 30.2%, respectively. This significant difference fully demonstrates that the LCCM system exhibits a faster strength growth rate in the early hydration stage, highlighting its superior early-age mechanical properties.

Based on the comparative analysis of the compressive strengths of A1 and A2 concrete in Table 9.9, the compressive strength of A1 concrete showed a rapid increase within the first 6 h. This is primarily attributed to the presence of highly reactive $C_4A_3\check{S}$ components in its cementitious system. In contrast, A2 and A3 concrete, which use ordinary sulfoaluminate cement as the binder, exhibited slower early-age compressive strength development due to the relatively lower reactivity of $C_4A_3\check{S}$ in the early stages.

Studies indicate that the compressive strengths of all three types of concrete increased continuously with curing age. Among them, the compressive strengths of A1 concrete at 6 h, 1 day, and 3 days were significantly higher than those of A2 and A3 concrete. In terms of strength development patterns, the compressive strength of concrete showed rapid growth in the early stages, followed by a gradual slowdown, eventually reaching a stable strength value. This development pattern is closely related to the hydration process of cementitious materials. Hydration products accumulate and interlock according to specific patterns, forming a dense structure. However, as the curing age increases, the amount of hydration products gradually decreases, leading to a slowdown in strength growth. Notably, the strength growth rate of A1 concrete between 6 h and 1 day was significantly higher than that between 1 and 3 days, further highlighting its superior early-age strength development.

Compared to A2 concrete, the compressive strengths of A3 concrete at 6 h, 1 day, and 3 days decreased by 44.4, 46, and 80%, respectively. This significant strength

Table 9.8 Compressive strength of A1 at 2 and 4 h (MPa)

Number	2 h			4 h		
	1	2	3	1	2	3
A1	13.57	15.37	15.02	27.26	24.18	23.61
Mean	14.32			25.02		

Table 9.9 Compressive strength of samples at 6 h, 1 d, and 3 d (MPa)

Number	6 h			1 d			3 d		
	1	2	3	1	2	3	1	2	3
A1	45.49	47.53	47.14	51.19	50.27	52.85	52.49	54.56	53.19
Mean	46.72			51.43			53.41		
A2	29.47	27.66	30.41	33.27	34.45	30.80	42.97	38.34	41.72
Mean	29.18			32.84			41.01		
A3	19.40	21.34	19.87	23.22	24.02	20.26	22.76	21.71	23.79
Mean	20.20			22.50			22.75		

difference is mainly due to the material property deficiencies of coal gangue aggregates in A3 concrete. The rough surface and porous internal structure of coal gangue aggregates result in high water absorption, which negatively affects the early hydration of sulfoaluminate cement. Additionally, compared to natural crushed stone, coal gangue has lower intrinsic strength. As the curing age increases, although the strength of the cementitious system continues to improve, when the cementitious strength exceeds that of coal gangue, the coal gangue coarse aggregates fracture before the cement paste. This limits the later-stage strength growth of the concrete, ultimately leading to a significant reduction in the early-age compressive strength of A3 concrete.

9.3.2 Impact Resistance Analysis

Impact failure morphology

The morphological images of A1, A2, and A3 concrete before and after impact failure are shown in Figs. 9.6, 9.7, 9.8 and 9.9.

According to the observations in Fig. 9.6, the three groups of concrete specimens exhibited significant differences in color. The A1 group concrete appeared whitish, while the A2 and A3 groups, which used sulfoaluminate cement as the binder, displayed a typical brownish color.

(a) A1 (b) A2 (c) A3

Fig. 9.6 Morphological images of samples before impact

(a) A1 (b) A2 (c) A3

Fig. 9.7 Impact morphological images of the sample after curing for 6 h

(a) A1 (b) A2 (c) A3

Fig. 9.8 Impact morphological images of the sample after curing for 1 d

(a) A1 (b) A2 (c) A3

Fig. 9.9 Impact morphological images of the sample after curing for 3 d

From Fig. 9.7, it can be seen that at 6 h of curing, the concrete specimens in each group only showed localized damage on the central surface, with no significant deformation observed on the overall surface. As the curing time extended to 1 day and 3 days, the strength of the concrete increased accordingly, leading to enhanced impact resistance.

Figure 9.8 shows that at 1 day of curing, the specimens exhibited slight indentation on the central surface under multiple impacts, but the degree of deformation was not yet pronounced. In contrast, Fig. 9.9 reveals that the specimens at 3 days of curing displayed more significant deformation characteristics under impact. The central surface of the concrete specimens showed obvious indentation, accompanied by spalling of surface concrete. This is because the impact hammer rebounds after striking the specimen. To reduce the lateral stress exerted by the bottom device on the specimen, butter was applied to the contact area between the specimen and the impact device before testing. However, as the number of drop hammer impacts increased, the constraint of the butter on lateral stress gradually diminished, leading to concrete spalling on the specimens.

The failure morphologies of A1, A2, and A3 concrete under impact loading were similar. After the first crack appeared in the specimens, complete failure occurred with increasing impact cycles. The final failure mode of the specimens under drop hammer impact was characterized by a straight-line crack. All specimens exhibited a "crack-and-fail" behavior upon impact loading, often showing obvious brittle fracture characteristics. Numerous studies have shown that ordinary concrete specimens typically exhibit significant brittle failure after the formation of the first crack.

When subjected to a small impact force, the crack can easily propagate through the thickness of the specimen, causing the concrete to split in a brittle manner.

Figure 9.10a–i illustrate the morphology of impact craters on the concrete specimens of each group after being subjected to impact loading. In the initial stage, when the top surface of the concrete specimen remained uncracked, it could withstand significant impact forces. As the impact force concentrated on the central area beneath the steel ball, the surface in this region began to fail. Under repeated impact loading, the coarse aggregates in the concrete gradually became exposed, and the thin matrix layer was compressed and crushed. The exposed coarse aggregates formed a shielding zone behind them. These aggregates absorbed part of the impact force, thereby protecting the cement matrix and slowing down the damage rate in this region. As shown in the Fig. 9.10, all groups of concrete specimens exhibited localized stress concentration under impact loading. This stress concentration caused localized failure in the area beneath the steel ball, with cracks gradually propagating toward the surface and bottom of the specimen, ultimately leading to complete failure.

As shown in Figs. 9.11 and 9.12, a micro-crack observation instrument was used during the test to accurately measure the crack width of A1 and A2 concrete at the onset of initial cracking. Under continuous impact loading, the specimens typically developed one main crack first, followed by multiple secondary cracks. The size of the main crack was significantly larger than that of the other cracks. As the number of impacts increased, the cracks continued to propagate, eventually causing the concrete specimens to fail. These cracks expanded in arbitrary directions, with the main crack playing a dominant role in the failure process. The propagation path and width variation of the main crack directly reflected the impact resistance of the concrete.

Impact energy analysis

Impact resistance tests were conducted on concrete specimens prepared with three mix proportions. The impact energy of each group of concrete specimens was calculated using Eq. 9.6.

$$W = N \cdot mgh \tag{9.6}$$

where: N: the corresponding number of impact times when the concrete specimen cracks at the beginning or the end. m: the mass of the impact hammer, 4.5 kg. g: gravitational acceleration, 9.8 N/kg. h: the height of the impact hammer's drop, 457 mm.

When cracks appear in the specimen, its energy dissipation capacity also changes. Referring to literature (Haruna et al., 2021), the post-cracking energy dissipation ΔW is introduced. The formula for calculating the final cracking energy dissipation is as follows:

$$\Delta W = W_2 - W_1 \tag{9.7}$$

where: W_1: initial cracking impact energy. W_2: final crack impact energy.

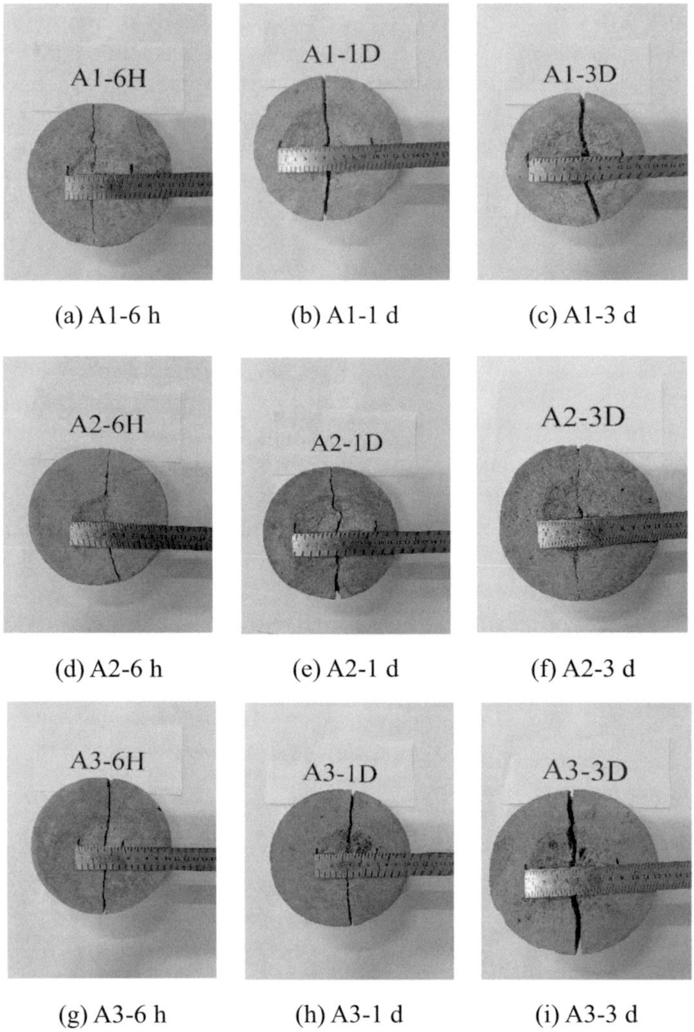

Fig. 9.10 Impact crater morphology of the sample at different ages

In impact tests, measurement results are susceptible to subjective human factors and material heterogeneity, leading to significant variability in concrete impact life data (Haruna et al., 2021). To reduce experimental errors and improve data reliability, this study employs statistical analysis methods to process the test data. The specific approach is as follows: from the test data of six specimens in each group, the maximum and minimum values are excluded, and the average of the remaining four

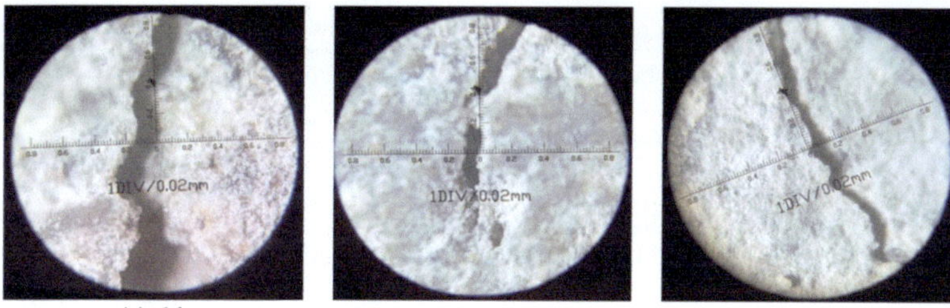

(a) 6 h (b) 1 d (c) 3d

Fig. 9.11 Crack width of A1 at different curing ages

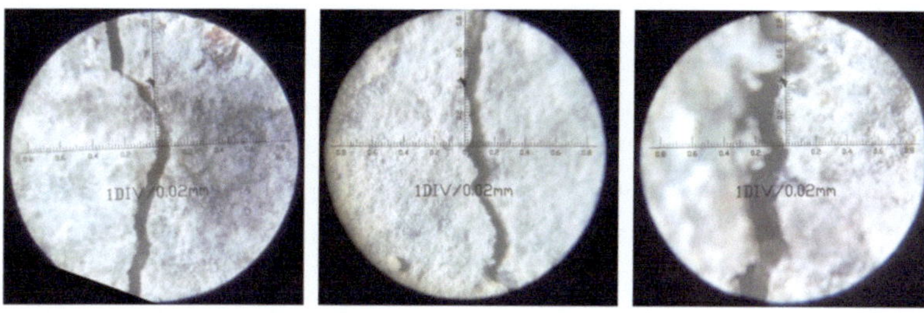

(a) 6 h (b) 1 d (c) 3d

Fig. 9.12 Crack width of A2 at different curing ages

data points is taken as the representative value of concrete impact energy consumption. Tables 9.10, 9.11 and 9.12 presents the statistical analysis results of this drop-weight impact test. This method effectively reduces data variability and provides a reliable basis for the objective study of concrete impact performance.

Based on the concrete impact life results in Tables 9.10, 9.11 and 9.12 (underlined data are excluded values), a histogram of impact life was plotted (Fig. 9.13). Analysis shows that the initial crack impact count (N_1) and final crack impact count (N_2) of A1 concrete specimens are higher than those of the other two types of concrete. Compared to A2 concrete, the average initial crack impact count of A1 concrete significantly increased. At 6 h, 1 day, and 3 days of curing, A1 concrete showed increases of 953, 122, and 5.3%, respectively. A1 concrete exhibited superior kinetic energy absorption capacity under repeated impact loading, primarily due to the following reasons. First, in terms of hydration degree, A1 concrete hydrates rapidly at early ages, generating a large number of hydration products such as ettringite (AFt) and aluminium gel (AH_3). These hydration products form a dense hardened cement paste, thereby enhancing impact resistance. Second, from the perspective of age-dependent strength, the strength of A1 concrete at all curing ages is superior to that of A2 concrete. Strength is a critical indicator of impact resistance. When concrete is subjected to impact loading, microcracks and pores form internally. If the strength is insufficient, these defects can easily expand, leading to failure. As the

Table 9.10 Impact resistance test results of A1 at different ages

Number		Impact number		Difference value	Impact power (J)		ΔW (J)
		N_1	N_2		W_2	W_1	
A1-6H	1	373	374	1	7537.48	7517.33	20.15
	2	339	340	1	6852.26	6832.10	20.16
	3	432	433	1	8726.55	8706.40	20.15
	4	396	397	1	8001.02	7980.87	20.15
	5	401	402	1	8101.79	8081.63	20.16
	6	468	469	1	9452.09	9431.93	20.16
A1-1D	1	1357	1358	1	27,368.72	27,348.57	20.15
	2	1277	1278	1	25,756.43	25,736.27	20.16
	3	1396	1397	1	28,154.72	28,134.57	20.15
	4	1462	1463	1	29,484.86	29,464.71	20.15
	5	1197	1198	1	24,144.13	24,123.98	20.15
	6	1291	1292	1	26,038.58	26,018.43	20.15
A1-3D	1	1825	1826	1	36,800.66	36,780.50	20.16
	2	1677	1678	1	33,817.91	33,797.75	20.16
	3	1877	1878	1	37,848.65	37,828.49	20.16
	4	1764	1765	1	35,571.28	35,551.13	20.15
	5	1671	1672	1	33,696.99	33,676.83	20.16
	6	1525	1526	1	30,754.55	30,734.39	20.16

curing age increases, the strength of A2 concrete gradually improves, enabling it to better resist high-intensity impact loads.

Table 9.10 shows that when coal gangue completely replaces natural crushed stone as coarse aggregate, the N_1 and N_2 values of A3 concrete significantly decrease. Compared to A2, the N_1 values of A1 at 6 h, 1 day, and 3 days decrease by 58.33, 51.01, and 78.25%, respectively. This is mainly attributed to the lower intrinsic strength of coal gangue, resulting in a substantial decline in the impact performance of A3 concrete. Additionally, the differences in Table 9.10 indicate that the difference between N_1 and N_2 for all groups of concrete specimens is 1. This suggests that the specimens fail rapidly after the appearance of microcracks. As the curing age increases, the compressive strength of concrete improves, and the internal stress distribution becomes more uniform. Furthermore, the increased hardening degree of the cementitious materials allows the concrete to better distribute external impact loads. Therefore, the difference between N_1 and N_2 is small.

Figure 9.13 illustrates the relationship between impact count and curing age for A1, A2, and A3 concrete. The impact resistance of the three types of concrete at different curing ages follows the order A1 > A2 > A3. Moreover, the impact count of concrete increases with the progression of hydration age, indicating that the impact resistance of concrete improves with age.

Table 9.11 Impact resistance test results of A2 at different ages

Number		Impact number		Difference value	Impact power (J)		ΔW (J)
		N_1	N_2		W_2	W_1	
A2-6H	1	39	40	1	806.15	785.99	20.16
	2	42	43	1	866.61	846.46	20.15
	3	50	51	1	1027.84	1007.69	20.15
	4	36	37	1	745.69	725.53	20.16
	5	36	37	1	745.69	725.53	20.16
	6	31	32	1	644.92	624.76	20.16
A2-1D	1	573	574	1	11,568.22	11,548.07	20.15
	2	594	595	1	11,991.45	11,971.30	20.15
	3	615	616	1	12,414.68	12,394.53	20.15
	4	722	723	1	14,571.13	14,550.97	20.16
	5	583	584	1	11,769.76	11,749.61	20.15
	6	601	602	1	12,132.53	12,112.37	20.16
A2-3D	1	1661	1662	1	33,495.45	33,475.30	20.15
	2	1552	1553	1	31,298.70	31,278.54	20.15
	3	1620	1621	1	32,669.15	32,648.99	20.16
	4	1762	1763	1	35,530.97	35,510.82	20.15
	5	1611	1612	1	32,467.61	32,487.76	20.15
	6	1694	1695	1	34,160.52	34,140.37	20.15

The impact energy at initial cracking and the impact energy at failure are two key parameters for evaluating the failure performance of concrete under repeated impact loading. The impact energy at initial cracking refers to the energy consumed when concrete develops microcracks under load or impact. These cracks can absorb and disperse impact forces. The impact energy at failure, on the other hand, refers to the energy consumed when the concrete specimen completely fails under a larger load or impact force. In the experiment, the difference between the impact energy at initial cracking and the impact energy at failure primarily stems from the different failure modes of concrete.

In the early stages of the test, microcracks form in the concrete specimen under load or impact. At this stage, the concrete still retains a certain level of strength and toughness, enabling it to effectively absorb and disperse impact forces. Therefore, the impact energy at initial cracking is generally lower than the impact energy at failure. This section of the experiment analyzed the impact energy absorption performance of A1, A2, and A3 concrete, using the impact energy at initial cracking and the impact energy at final cracking as references.

The impact energy at initial cracking ($\overline{W_1}$), impact energy at failure ($\overline{W_2}$), and post-cracking energy ($\overline{\Delta W}$) of the concrete are shown in Table 9.13.

Table 9.12 Impact resistance test results of A3 at different ages

Number		Impact number		Difference value	Impact power (J)		ΔW (J)
		N_1	N_2		W_2	W_1	
A3-6H	1	21	22	1	443.38	423.23	20.15
	2	26	27	1	544.15	524.00	20.15
	3	18	19	1	382.92	362.77	20.15
	4	20	21	1	423.23	403.07	20.16
	5	30	31	1	624.76	604.61	20.15
	6	29	30	1	604.61	584.46	20.15
A3-1D	1	419	420	1	8464.55	8444.40	20.15
	2	341	342	1	6892.57	6872.41	20.16
	3	416	417	1	8404.09	8383.94	20.15
	4	394	395	1	7960.71	7940.56	20.15
	5	355	356	1	7174.72	7154.56	20.16
	6	434	435	1	8766.86	8746.71	20.15
A3-3D	1	1021	1022	1	20,597.08	20,576.93	20.15
	2	862	863	1	17,392.64	17,372.49	20.15
	3	935	936	1	18,863.86	18,843.71	20.15
	4	877	878	1	17,694.95	17,674.79	20.16
	5	819	820	1	16,526.03	16,505.88	20.15
	6	1068	1069	1	21,544.31	21,524.15	20.16

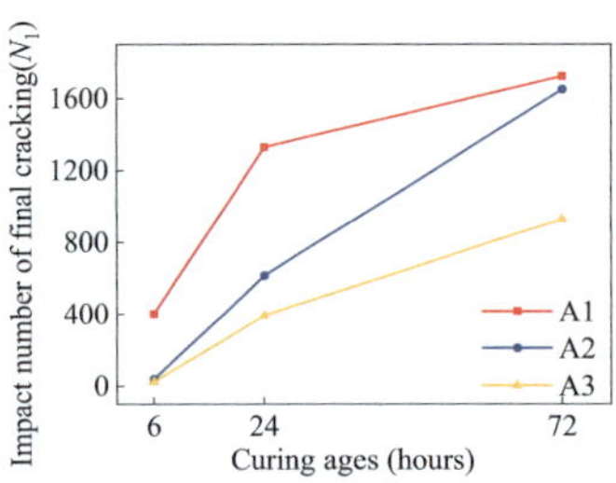

(a) Relationship between the initial cracking impacts number and curing age

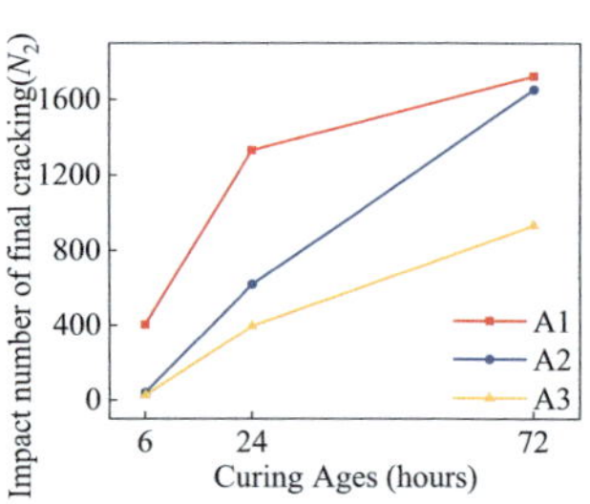

(b) Relationship between the final cracking impacts number and age

Fig. 9.13 Relationship between impact resistance numbers and curing age of samples

At curing ages of 6 h, 1 day, and 3 days, the order of impact energy at initial cracking and impact energy at failure is consistently: A1 > A2 > A3. Among them, A3 concrete exhibits the lowest impact energy at both initial cracking and failure, while A1 concrete shows the highest. Compared to A2 and A3 concrete, the impact energy at initial cracking of A1 concrete increases by 10 times and 16.1 times, respectively, at 6 h; by 2.16 times and 3.37 times, respectively, at 1 day; and by 1.45

Table 9.13 The average impact energy in initial cracking and failure of samples

Number	Curing age	$\overline{W_2}$ (J)	$\overline{W_1}$ (J)	$\overline{\Delta W}$ (J)
A1	6 h	8111.86	8091.71	20.15
	1 d	26,824.57	26,804.42	20.15
	3 d	34,748.34	34,728.18	20.15
A2	6 h	806.15	785.99	20.15
	1 d	12,407.96	12,387.81	20.15
	3 d	23,903.51	23,883.36	20.15
A3	6 h	503.84	483.69	20.15
	1 d	7943.92	7923.76	20.15
	3 d	18,769.81	18,749.66	20.15

times and 1.85 times, respectively, at 3 days. This indicates that the impact energy at initial cracking of A1 concrete improves most significantly before 6 h of curing. As the curing age increases, the impact energy at failure of A1 concrete shows a trend of rapid growth in the early stages and slower growth in the later stages. This trend is attributed to the generation of a large amount of hydration products in A1 concrete at early ages, enhancing its mechanical properties. As the hydration reaction approaches saturation, internal defects and damage in the concrete gradually reach equilibrium, leading to a slowdown and stabilization in the growth rate of impact energy at initial cracking.

Specific data show that the impact energy at initial cracking of A1 concrete at 6 h, 1 day, and 3 days is 8091.71 J, 26,804.42 J, and 34,728.18 J, respectively. The energy consumption at 3 days is the highest, increasing by 328 and 29% compared to 6 h and 1 day, respectively. For A2 concrete, the impact energy at initial cracking at the same curing ages is 785.99, 12,387.81, and 23,883.36 J, respectively. Compared to 6 h and 1 day, the energy consumption at 3 days increases by 2938 and 92%, respectively. For A3 concrete, the impact energy at initial cracking is 483.69, 7923.76, and 18,749.66 J, respectively. Compared to 6 h and 1 day, the energy consumption at 3 days increases by 3776 and 136%, respectively. The order of growth in impact energy at initial cracking at 3 days for the three types of concrete is: A1 < A2 < A3.

Under the same mix proportion conditions, A2 concrete, which uses natural crushed stone as coarse aggregate, exhibits superior impact resistance. Compared to A3 concrete, the impact energy at initial cracking of A2 concrete increases by 1.6 times, 1.5 times, and 1.2 times at 6 h, 1 day, and 3 days, respectively. Among these, the increase is the largest at 6 h and the smallest at 3 days.

The impact energy after initial cracking reflects the toughness of concrete. Analysis reveals that the average post-cracking energy of all three types of concrete is 20.15 J. The post-cracking energy is calculated as the difference between the impact energy at final cracking and the impact energy at initial cracking. It can also be determined by the difference in energy absorption between the plastic and elastic stages. The average post-cracking energy is the same for all groups of concrete.

This is because the cracks generated in concrete under impact represent a weakening phenomenon. When the concrete continues to bear excessive external loads after initial cracking, its overall strength and stability are difficult to maintain. This is also the reason why concrete fractures after only one drop hammer impact following initial cracking. The differences in impact energy at initial cracking among the groups are mainly influenced by their microstructures and material properties. Under load, changes in internal pores of concrete absorb, disperse, and delay the transmission of impact forces. Due to the poor mechanical properties of coal gangue, its ability to absorb impact forces is weaker, resulting in lower impact energy at initial cracking for A3 concrete.

9.4 Microstructure of Concrete Containing LCCM

9.4.1 Microstructure Characteristics of Concrete

Figure 9.14 presents SEM images of A1 concrete at curing ages of 6 h, 1 day, and 3 days. By examining these images, the densification of the cementitious materials in concrete at different ages can be clearly understood. Below is a detailed analysis of the micro-morphology of the cementitious materials at each curing age.

From Fig. 9.14a, it can be observed that during the 6-h hydration process of A1 concrete, a large amount of needle-like AFt and AH_3 were generated. Numerous small AFt crystals appeared at the interface between the cement and aggregates. The image shows that AFt and AH_3 envelop the aggregates, indicating that a certain degree of hydration reaction has occurred in A1 concrete at the 6-h curing age, producing a significant amount of hydration products. The visible cracks in the image are damage cracks caused by impact on the concrete specimen. Microcracks are hardly observed in other areas, with only a few visible pores at the interfaces. This is because the columnar AFt and AH_3 gels are interconnected, filling most of the pores. The hydration products densely and uniformly cover the aggregate surfaces, resulting in a compact interfacial transition zone and demonstrating excellent mechanical properties.

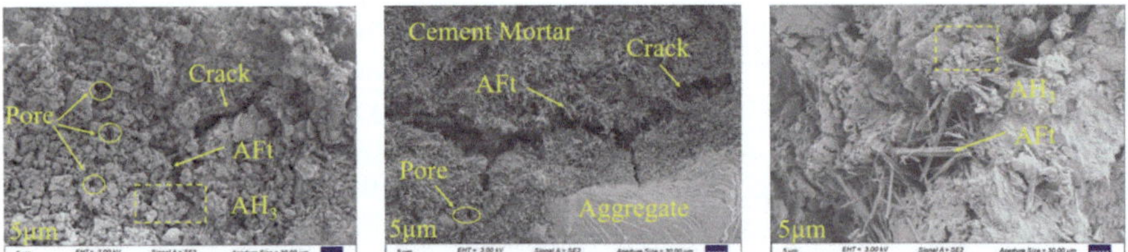

(a) 6 h (b) 1 d (c) 3 d

Fig. 9.14 SEM images of A1 sample at different curing ages

From Fig. 9.14b, it is evident that as hydration progresses, the amount of AFt generated gradually increases, and the AFt crystals become more robust. This is also the reason for the gradual increase in the strength of A1 concrete. Therefore, the high generation amount and distribution density of AFt crystals can reflect the degree of cement hydration and the hardening process of concrete, serving as one of the indicators for evaluating concrete strength. At the 1-day curing age, the AFt in the hydration products is not fully developed, resulting in short rod-like or flake-like morphologies in the image.

From Fig. 9.14c, it can be seen that at the 3-day curing age, spherical substances are generated after hydration, and the gaps are bridged by some columnar AFt crystals. As the LCCM paste hydrates, a large number of needle-like substances form on the microstructure surface, interconnecting with each other. AFt fills the pores in the LCCM, and no significant large pores are observed in the microstructure. AFt and AH_3 gels interconnect and fill the spaces, forming a dense hardened cement paste. The aggregates are also tightly enveloped by the hydration products. Macroscopically, as the cementitious materials fully hydrate, the strength development of A1 concrete is excellent.

The micro-morphology of A2 concrete is observed through Fig. 9.15. AFt, a hydration product of sulfoaluminate cement, exhibits morphologies dependent on formation conditions. At 6 h, the microstructure of A2 concrete contains numerous pores. The connectivity between particles is poor, and the filling by fine aggregates is not evident. Unhydrated cement particles are visible on the surface, and the overall structure is loose. This is because sulfoaluminate cement does not fully hydrate within 6 h, and its hydration rate is lower than that of LCCM. The amount of AFt generated is limited. The strength of A2 concrete at this stage is primarily provided by the interweaving of a small amount of AFt and AH_3 gel. Due to the high porosity, its strength at 6 h is lower than that of A1 concrete.

Figure 9.15b shows the SEM image of A2 concrete at a curing age of 1 day. Compared to A1 and A3 concrete, the microstructure of A2 concrete exhibits significant differences. The AFt distributed on the surface of A2 concrete mostly appears as long columnar crystals, covered with a large amount of gel-like material. At 1 day, sulfoaluminate cement begins to hydrate fully. The structure of the cementitious paste becomes denser. The combination of AFt and gel fills most of the pores in the aggregates, further enhancing the concrete's strength.

(a) 6 h (b) 1 d (c) 3 d

Fig. 9.15 SEM images of A2 sample at different curing ages

Figure 9.15c presents the SEM image of A2 concrete at a curing age of 3 days. The interface between natural aggregates and the cement paste is relatively dense, with fewer pores and smaller pore sizes. The cracks visible in the image are caused by impact on A2 concrete. The interface between aggregates and the cement matrix is generally smooth. As the curing age increases, the surface of the cementitious material becomes more compact. Microcracks on the surface of the cementitious material adhere to some of the cementitious phases. With ongoing hydration, pores are gradually filled, a process akin to "wound healing." The amount of AFt increases with the curing age of sulfoaluminate cement. AFt is the primary source of early strength in A2 concrete. AFt forms through both solid-phase and liquid-phase reactions. During hydration, pores between the cement paste and aggregates are filled with AFt. Some AFt remains in the matrix pores. When sufficient reactants and moisture are present in the pores, liquid-phase reactions continue. The continuous growth of AFt crystals enhances the strength of A2 concrete over time.

9.4.2 *Interface Transition Zone of Concrete Aggregate*

To gain a more intuitive understanding of the microstructure at the aggregate-cementitious material interface in concrete, SEM tests were conducted on concrete specimens at different curing ages. The test results are shown in Fig. 9.16.

From Fig. 9.16a, a distinct crack can be observed at the interface between the concrete aggregate and the cementitious material. These cracks represent weak regions within the concrete. Hydration products are present in the crack shown in the lower part of the image. This indicates that pores at the aggregate-cementitious material interface existed during the formation of the concrete. During loading, these cracks are prone to causing stress concentration. Stress concentration accelerates crack propagation and interconnection, adversely affecting the strength and toughness of the concrete. The crack in the upper part of the image shows no hydration products. These cracks result from uneven internal stress distribution in the

(a) 6 h (b) 6 h

Fig. 9.16 Interface transition zone of A1 sample at cuing age of 6 h concrete. Under external forces, stress gradually accumulates, eventually leading to crack formation.

From Fig. 9.16b, it is observed that LCCM is not fully hydrated in the early stages. The hydration products mainly consist of AFt and a small amount of gel. The aggregates enveloped by the hydration products exhibit poor compactness, resulting in a loose interfacial transition zone. This suggests that the hydration products did not sufficiently fill the interfacial pores. Consequently, the bonding strength at the aggregate interface is relatively low.

From Fig. 9.17, it can be observed that the hydration products in the cementitious material of A1 concrete have significantly increased. Compared to the microstructure at 6 h, Fig. 9.17a shows that the interfacial transition zone between aggregates and cementitious material is densely filled with AFt and AH_3 gel. No obvious cracks are observed between the aggregates and the cementitious material. The hydration products gradually extend toward the edges of the aggregates. At this stage, the hydration products in the interfacial transition zone mainly consist of AFt and AH_3 gel, which are interconnected. Although the number of pores in the transition zone increases, their sizes are small.

From Fig. 9.17b, it is evident that the cement particles in the interfacial transition zone undergo hydration reactions to form hydration products. These products interact with unhydrated cement particles, creating a tightly connected structure. This structure improves the interface between aggregates and cementitious material, significantly enhancing the strength of the concrete.

From Fig. 9.18a, it can be observed that the aggregate surface is enveloped by hydration products, resulting in an uneven structural surface. At this stage, the gaps in the cementitious material are significantly reduced. The interface between the concrete aggregate and the cementitious material is clearly visible in the image. This phenomenon occurs due to the large difference in elastic modulus between the aggregate and the cement paste, leading to weaker bonding between them. When the concrete is subjected to stress, the compactness of the cementitious material increases, and internal gaps decrease. Many amorphous AH_3 formations appear at the interface. Some AH_3 fills the interfacial pores, while the rest adheres to the aggregate

(a) 1 d (b) 1 d

Fig. 9.17 Interface transition zone of A1 sample at curing age of 1 d

Fig. 9.18 Interface transition zone of A1 sample at curing age of 3 d

surface. The presence of AH_3 gel enhances the bonding between the aggregate and the cementitious material, improving the overall performance of the concrete.

Figure 9.18b shows that as the concrete curing age increases, the cementitious material continues to hydrate, generating an increasing amount of hydration products. These products gradually interconnect at the microscopic level, making the cementitious material denser. Simultaneously, the interaction between the aggregate and the cementitious material strengthens, microscopic defects gradually decrease, and the overall compressive strength improves.

From Fig. 9.19a, it can be observed that there is a distinct interface between the aggregate and the cementitious material in A2 concrete. This is because A2 concrete has fewer hydration products at 6 h. These hydration products are mainly located at the interface between the aggregate and the cementitious material. Some of the gel fills the interfacial pores, while the rest adheres to the aggregate surface. At this stage, the cement hydration reaction is in its initial phase, and a large number of unhydrated cement particles have not yet participated in the reaction. This results in the lower compressive strength of A2 concrete at 6 h.

Figure 9.19b shows that the hydration products generated at the interface of A2 concrete at 1 day are more abundant than those at 6 h. As the curing age increases, the microstructure of the cementitious material becomes denser, and micro-cracks and pores are significantly reduced. This enhances the compressive strength of the concrete. At the same time, the bonding between the cementitious material and the

Fig. 9.19 Interface transition zone of A2 sample at different curing age

aggregate interface strengthens, improving the mechanical properties and durability of the concrete. Additionally, the self-healing ability of A2 concrete improves with increasing curing age, enabling it to automatically fill micro-cracks and fractures, ensuring overall performance and safety.

From Fig. 9.19c, it can be seen that after 3 days of hydration, needle-like hydration products are rarely observed in the interfacial transition zone of A2 concrete. A small amount of flocculent crystals grows along the aggregate surface, gradually forming an enveloping trend. The cracks at the interface are clear, and no hydration products are found at the junctions. This is caused by mechanical damage. The continuous contact and accumulation of aggregates and cement hydration products make the interfacial structure denser. As the curing age increases, the cementitious material gradually completes hydration, and the bonding between hydration products and aggregates strengthens, making the interfacial transition zone more compact. These factors collectively contribute to the continuous improvement in the compressive strength of sulfoaluminate cement concrete.

Figure 9.20 shows the micro-morphology of the aggregate-cementitious material interface in concrete with coal gangue as the coarse aggregate at curing ages of 6 h, 1 day, and 3 days. From Fig. 9.20a, it can be observed that the irregular surface shape of coal gangue leads to a mismatch at the interface between the aggregate and the cementitious material, resulting in cracks at the interface. The presence of hydration products in these cracks indicates that the cracks did not form during concrete preparation. At 6 h of curing, the surface of the cementitious material contains a large number of unhydrated cement particles. Due to the short curing time, cement hydration is insufficient. This results in the interfacial transition zone being filled with a small amount of hydration products and unhydrated cement clinker, leaving numerous voids within the cementitious material. The strength of the interfacial transition zone is provided by the bonding between hydration products and unhydrated cement particles. The low density of coal gangue causes it to absorb some water during mixing, reducing the water-to-binder ratio. At early ages, A3 concrete exhibits insufficient hydration, generating only a small amount of hydration products that fill the space between the aggregate and the cementitious material. The limited hydration products result in a less dense surface of the cementitious material, leading to lower early-age strength in A3 concrete.

(a) 6 h (b) 1 d (c) 3 d

Fig. 9.20 Interface transition zone of A3 sample at different curing age

Figure 9.20b shows that as the curing age increases, the hydration reaction intensifies, and the amount of hydration products significantly increases. Hydration products such as AFt fill the pores, reducing the porosity visible in the SEM images. However, large unhydrated cement particles still remain on the surface of the cementitious material. This is because the addition of coal gangue reduces the water content in the sulfoaluminate cement mortar. Some mineral particles in coal gangue absorb water and expand, increasing the porosity of the mortar and directly affecting the performance of A3 concrete.

From Fig. 9.20c, it can be seen that at 3 days of curing, the interface between coal gangue aggregate and the cementitious material is clearly visible. Cracks are present at the interface, with hydration products accumulated within them. The coal gangue coarse aggregate is enveloped by hydration products, and a small number of unhydrated cement particles remain on the surface of the cementitious material. Some unhydrated cement particles and hydration products are present in the layered cracks of coal gangue. The micro-morphology of the cementitious material is irregular, with a small amount of hydration product gel adhering to the aggregate surface and extending outward. At the cracks in the interfacial transition zone, the hydration products of the cement paste continue to expand along the cracks. This is due to the poor quality of coal gangue, while the hardened cementitious material and other aggregates have stronger mechanical properties. Under external forces, the coal gangue aggregate, with its weaker mechanical properties, is the first to fail.

9.4.3 Evolution of Micro-pore Structure in Concrete

T_2 spectrum

In the T_2 spectrum, small pores correspond to the range of the first peak, medium pores correspond to the range of the second peak, and large pores correspond to the range of the third peak. T_2 spectrum can be used to analyze the pore distribution of concrete at different curing ages. Figure 9.21 presents the T_2 spectrum of A1, A2, and A3 concrete at curing ages of 6 h, 1 day, and 3 days.

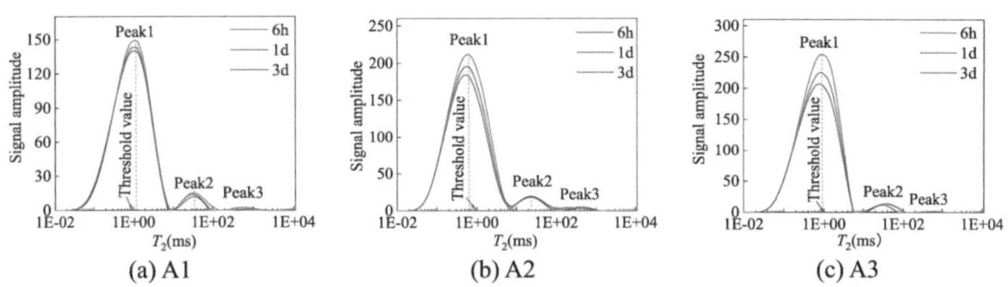

Fig. 9.21 T_2 spectrum of the sample of A1, A2, and A3 at different curing age

Analysis reveals that at the same temperature, the T_2 spectra curves of A1, A2, and A3 concrete shift along the positive direction of the time axis, and the distribution curves exhibit similar trends. Based on the number and magnitude of peaks, the T_2 spectra of each concrete show three peaks. The main peak corresponds to small pores, while the two secondary peaks correspond to medium and large pores, respectively. The main peak appears at shorter relaxation times, and Peak1 is significantly higher than Peak2 and Peak3. The main peak amplitudes of A1, A2, and A3 concrete are much larger than those of the secondary peaks, indicating that small pores account for the largest volume fraction, followed by medium pores. The T_2 spectrum signal amplitude of A1 concrete differs significantly from that of A2 and A3 concrete. The main peak amplitudes are 149, 212, and 255, respectively, while the secondary peak amplitudes are around 15 and 16. The secondary peak amplitudes of all concrete groups are much smaller than the main peak amplitudes. This further confirms that small pores occupy the largest volume fraction, followed by medium pores, with large pores being the smallest.

The calculated peak area results are shown in Table 9.14. The total area under the T_2 spectrum curve and the area of each peak can be calculated. The peak area reflects the number of pores within a specific size range. From the table, it can be observed that during the curing periods of 6 h to 1 day and 1 day to 3 days, the main peak areas of A1, A2, and A3 concrete decreased by an average of 0.62, 8.75, and 3.41%, respectively. The first secondary peak areas decreased by an average of 11.11, 20.12, and 29.80%, respectively. The second secondary peak areas decreased by an average of 53.01, 56.16, and 68.44%, respectively. The reduction in the main peak area is significantly smaller than that of the secondary peaks. This indicates that by 3 days of curing, the volumes of large, medium, and small pores in A1, A2, and A3 concrete have decreased, with the reduction being most pronounced for large pores.

Table 9.15 shows the changes in the T_2 spectrum area of each concrete specimen at 6 h, 1 day, and 3 days. From the table, it can be observed that during the 1–3 day curing period, the order of T_2 spectrum area for the concrete is A1 $<$ A2 $<$

Table 9.14 Peak area of T_2 spectrum of the sample at different curing ages

Number	Peak	Curing age		
		6 h	1 d	3 d
A1	Peak1	5417.05	5364.89	5350.43
	Peak2	271.44	264.39	212.52
	Peak3	18.00	13.30	2.68
A2	Peak1	7650.89	7145.32	6366.51
	Peak2	468.61	411.50	296.10
	Peak3	105.42	54.24	19.66
A3	Peak1	8833.32	8493.82	8240.71
	Peak2	523.80	521.73	212.88
	Peak3	50.30	27.48	2.33

Table 9.15 Total peak area of T_2 spectrum of the sample at different curing ages

Number	Ageing		
	6 h	1 d	3 d
A1	5706.50	5642.57	5565.63
A2	8224.92	7611.06	6682.26
A3	9407.42	9043.03	8455.92

A3. Concrete with lower strength exhibits the largest T_2 spectrum area. Under the same water-to-cement ratio, the porous nature of coal gangue increases the difficulty of material integration. The incomplete early hydration of sulfoaluminate cement results in an increased internal pore volume in A3 concrete. At 6 h of curing, the T_2 spectrum areas of A1, A2, and A3 concrete are 5706.49, 8224.92, and 9407.42, respectively.

Compared to A1 and A2 concrete, A3 concrete has the largest T_2 spectrum area. This phenomenon can be attributed to two reasons. First, incomplete vibration during the mixing process creates pores inside the coal gangue concrete specimens. Second, the high water absorption of coal gangue coarse aggregate generates bubbles during water absorption, leading to pore formation in the cementitious material. During the initial setting stage of the cementitious material, the number of internal pores in the concrete is the highest, resulting in the largest T_2 spectrum area and pore volume. As the curing age increases, the T_2 spectrum area of A1, A2, and A3 concrete decreases by an average of 4.15% during the 6-h to 1-day curing period and by an average of 6.69% during the 1-day to 3-day curing period. By 3 days of curing, the T_2 spectrum areas of A1, A2, and A3 concrete decrease to 556.63, 6682.26, and 8455.92, respectively.

In summary, at curing ages of 6 h, 1 day, and 3 days, the pore variation trends of A1, A2, and A3 concrete are similar. As the curing age increases, the T_2 spectrum area of each group of concrete continues to decrease. The T_2 spectrum areas of A1, A2, and A3 concrete decrease by 140.866, 1542.661, and 951.497, respectively. The reduction rates are 2.47, 18.76, and 10.11%, respectively. This indicates that A1 concrete undergoes more complete hydration at early ages, with hydration products filling some of the pores, resulting in higher early-age strength compared to A2 concrete. The reduction rates of the T_2 spectrum areas for A2 and A3 concrete are higher than that of A1 concrete. This is because sulfoaluminate cement gradually completes hydration as the curing age increases, and the generated hydration products fill the pores, reducing the pore volume. The decrease in internal pores enhances the compactness of the concrete.

Pore size distribution

Figure 9.22 displays the pore distribution curve after T_2 spectrum conversion. The longer the relaxation time of pore water, the larger the pore radius; the shorter the relaxation time, the smaller the pore radius. Pores with a diameter less than 0.02 μm are considered harmless. Pores with diameters between 0.02 and 0.05 μm are classified as less harmful. Pores with diameters between 0.05 and 0.20 μm are deemed

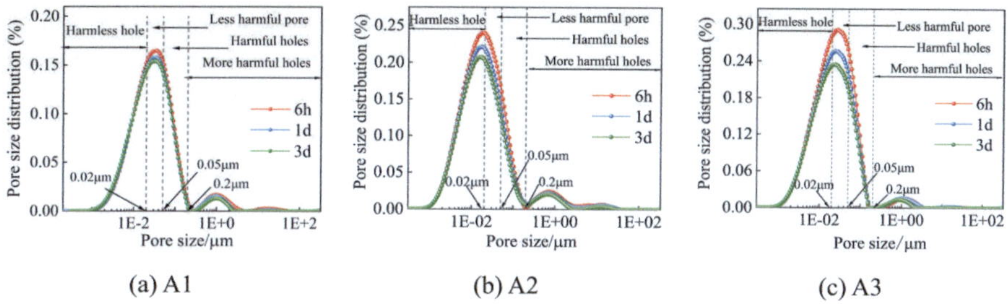

Fig. 9.22 Pore size distribution curve of samples at different curing ages

harmful. Pores with diameters greater than 0.20 μm are categorized as highly harmful. According to the pore size distribution diagram, the first peak mainly includes harmless pores, less harmful pores, and a small number of harmful pores. Highly harmful pores are primarily distributed within the second and third peaks.

Figure 9.23 displays the pore size distribution percentages of each group of concrete at curing ages of 6 h, 1 day, and 3 days. The pore size range for A1 concrete is 0.488–32,530.980 nm. For A2 concrete, the range is 0.488–34,869.670 nm. For A3 concrete, the range is 0.523–56,689.570 nm. At 6 h, 1 day, and 3 days, the percentages of harmless pores in A1 are 45.5, 54.1, and 56.3%, respectively. The percentages of less harmful pores are 32, 26, and 25.7%. The percentages of harmful pores are 19.6, 12.7, and 11.8%. For A2, the percentages of harmless pores are 52.1, 54.1, and 56.3%. The percentages of less harmful pores are 26.9, 26, and 25.7%. The percentages of harmful pores are 13.8, 12.7, and 11.8%. The percentages of more harmful pores are 7.31, 7.21, and 6.26%. For A3, the percentages of harmless pores are 45.5, 47, and 49.4%. The percentages of less harmful pores are 32, 30.9, and 30.3%. The percentages of harmful pores are 19.6, 18.7, and 17.8%. The percentages of more harmful pores are 2.85, 3.49, and 2.58%.

Fig. 9.23 Proportion of pore size of samples at different curing age

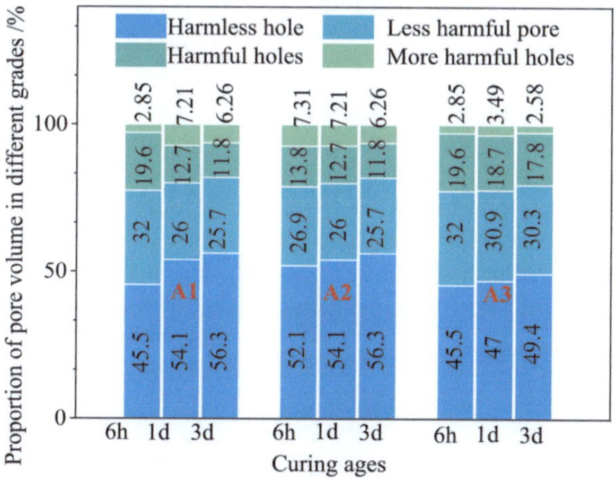

The proportion of harmful pores in A1, A2, and A3 concrete follows the order: A3 > A1 > A2. This indicates that, under the same water-cement ratio, concrete with lower strength has a higher proportion of more harmful pores. Coal gangue, used as coarse aggregate, has numerous pores, low density, and high water demand. This directly reduces the water-binder ratio of the concrete, leading to a decrease in hydration products. Consequently, the pores between coarse and fine aggregates are insufficiently filled, resulting in the formation of more harmful and more harmful pores.

The proportions of harmless and less harmful pores increase with higher strength. This suggests that as the curing age progresses, harmless and less harmful pores replace some of the harmful and more harmful pores, thereby enhancing the strength. Therefore, the key to increasing concrete strength lies in reducing the proportions of harmful and more harmful pores.

Based on the statistical analysis of Figs. 9.22 and 9.23, the proportions of harmless and less harmful pores in A1, A2, and A3 concrete increased by 3.2, 1.46, and 0.38%, respectively, during the 6 h–1 d curing period. During the 1–3 d curing period, they increased by 2.32, 2.32, and 2.29%, respectively. By the 3d mark, the cumulative increases were 5.52, 3.78, and 2.67%, respectively. This indicates that the proportions of harmless and less harmful pores increase with longer curing ages. In contrast, the proportions of harmful and more harmful pores in A1, A2, and A3 concrete decreased by 11.41, 5.53, and 1.32%, respectively, during the 6 h–1 d curing period. During the 1–3 d curing period, they decreased by 9.56, 9.56, and 8.21%, respectively. By the 3d mark, the cumulative decreases were 20.97, 15.09, and 9.52%, respectively. This demonstrates that the proportions of harmful and more harmful pores decrease with longer curing ages.

Based on the above analysis, the following conclusions can be drawn:

(1) The pore size follows the order: harmless pores < less harmful pores < harmful pores < more harmful pores. When A1 and A2 are cured for 3 days, the total porosity decreases, and the proportion of more harmful pores reduces. This is due to the ongoing hydration process, where some more harmful pores are filled with hydration products such as AFt and AH_3, gradually transforming into smaller pores, including harmless, less harmful, or harmful pores.
(2) The pore size of harmful pores lies between that of harmless, less harmful, and more harmful pores. Under external influences, some harmful pores shrink in size and transform into harmless or less harmful pores. Another portion originates from the shrinkage of more harmful pores, leading to fluctuations in the proportion of harmful pores.
(3) Within 3 days of concrete curing, pores are continuously filled, and harmful and more harmful pores gradually transform into harmless and less harmful pores. As a result, the proportions of harmless and less harmful pores continue to increase.

In summary, the reduction in the proportion of harmful pores is related to the decrease in total porosity, while the improvement in compressive strength is closely associated with the reduction in the proportion of more harmful pores.

9.5 Prediction of Impact Life for Concrete Containing LCCM

9.5.1 Weibull Distribution Model

The Weibull distribution function is a probability density function that includes location, shape, and scale parameters. It is also known as the three-parameter Weibull distribution function. This function is not only suitable for small sample sizes but also effectively adapts to various types of experimental data. In material life analysis, the Weibull distribution function offers the advantage of improving the accuracy of material reliability assessments through the conceptual analysis of minimum safe life. The probability density function (f(t)) of the three-parameter Weibull distribution is shown in Eq. 9.8.

$$f(t) = \frac{\beta}{\eta}\left(\frac{t-\gamma}{\eta}\right)^{\alpha-1} \exp\left[-\left(\frac{t-\gamma}{\eta}\right)^{\beta}\right] \quad t \geq \gamma \tag{9.8}$$

The expression of the cumulative failure probability function $F(t)$ is as follows:

$$F(t) = P(T \leq t) = 1 - \exp\left[-\left(\frac{t-\gamma}{\eta}\right)^{\beta}\right] \quad t \geq \gamma \tag{9.9}$$

In the Eq. 9.8, $\beta > 0$ represents the shape parameter, $\eta > 0$ represents the scale parameter, and $\gamma \geq 0$ represents the location parameter, denoted as $T \sim W$ (β, η, γ). In this formula, the shape parameter β characterizes the shape of the Weibull distribution curve. When $\beta = 1$, it corresponds to an exponential distribution. When $\beta = 2$, it corresponds to a Rayleigh distribution. When $3 < \beta < 4$, it approximates a normal distribution. When $\beta = 3.313$, it represents a normal distribution. This versatility indicates that the Weibull distribution has a broader applicability and stronger coverage capabilities. When $\gamma = 0$, the three-parameter Weibull distribution reduces to a two-parameter Weibull distribution model. When $t = \gamma$, the cumulative failure probability function is zero, and the reliability function is 100%. Therefore, γ is also referred to as the minimum safe (guaranteed) lifespan.

9.5.2 Impact Times Analysis

Using the Weibull distribution theory, the impact resistance performance of each group of concrete is analyzed in terms of probability distribution. The expression for the impact life function (or reliability function) of each concrete is as follows:

$$R(t) = 1 - F(t) = \exp\left[-\left(\frac{t-\gamma}{\eta}\right)^{\beta}\right] \quad t \geq \gamma \tag{9.10}$$

Due to the discrete nature of concrete and to simplify the analysis process, this chapter takes the minimum survival lifespan, that is, $\gamma = 0$. Therefore, the simplified Weibull distribution function is:

$$f(t) = \frac{\beta}{\eta}\left[\frac{t}{\eta}\right]^{\alpha-1} \exp\left[-\left(\frac{t}{\eta}\right)^{\beta}\right] \tag{9.11}$$

$$R(t) = 1 - F(t) = \exp\left[-\left(\frac{t}{\eta}\right)^{\beta}\right] \tag{9.12}$$

$$\frac{1}{R(t)} = \exp\left[\left(\frac{t}{\eta}\right)^{\beta}\right] \tag{9.13}$$

Taking the logarithm of both sides of Eqs. 9.11–9.12 twice results in Eq. 9.14:

$$\ln\left[\ln\frac{1}{R(t)}\right] = \beta\ln(t) - \beta\ln(\eta) \tag{9.14}$$

If $Y = \ln[\ln(1/R(t))], X = \ln(t), C = \beta\ln(\eta)$, it would:

$$Y = \beta X - C \tag{9.15}$$

From Eq. 9.15, it can be inferred that an approximate linear relationship exists between X and Y through linear regression analysis. Additionally, the values of the parameters β, C, and the correlation coefficient R^2 can be determined. When the R^2 value is high, the Weibull distribution can effectively describe the impact life of the concrete specimens in the experiment. Before conducting the analysis and verification, the initial crack impact resistance count (N_1) and the final crack impact count (N_2) of the impact specimens are first arranged in ascending order. Then, the survival probability is calculated using the expected estimation formula of the survival function.

$$R(t) = 1 - \frac{i}{k+1} \tag{9.16}$$

where: i is the number of test serial numbers after being sorted in ascending order, $i = 1, 2, \ldots, n$. k is the total number of samples for impact test, the total number of samples for this test is 6.

Based on the data from Table 9.10, the initial crack impact number (N_1) and the ultimate crack impact number (N_2) are used to determine the Weibull parameters for each group of specimens. The Weibull parameters for each group of specimens are presented in Table 9.16.

According to Table 9.16, the scatter points are plotted with Y $Y = \ln[\ln(1/R(t))]$ as the vertical axis and $X = \ln(t)$ as the horizontal axis. The estimated initial crack and final crack failure counts for each group of specimens are fitted to obtain the regression line graphs and corresponding regression results for the impact resistance counts of A1, A2, and A3 concrete at different curing ages. The regression results are shown in Figs. 9.24, 9.25, 9.26, 9.27, 9.28 and 9.29 and Table 9.17. From the regression line graphs, the values of the parameters β, C, and the correlation coefficient R_2 can be determined. The slope of the fitted line represents the shape parameter β of the Weibull distribution function.

Table 9.17 presents the Weibull distribution regression results for each group of concrete at different curing ages. The analysis shows that for A1 concrete, the average R^2 values in the Weibull linear regression of impact counts at 6 h, 1 d, and 3 d are 0.9705, 0.977, and 0.971, respectively. For A2 concrete, the average R^2values are 0.93, 0.968, and 0.957, respectively. For A3 concrete, the average R^2 values are 0.934, 0.945, and 0.932, respectively. These results indicate a linear correlation between Y and X, demonstrating that the impact counts of A1, A2, and A3 concrete conform to the two-parameter Weibull distribution for impact failure probability analysis.

9.5.3 Prediction of Impact Life of Concrete Under Different Failure Rates

According to Eqs. 9.12–9.15, the impact counts N for each group of concrete specimens A1, A2, and A3 under the corresponding failure probability $F(t)$ can be expressed as:

$$N = \exp\left\{\frac{\ln\left[\ln\left(\frac{1}{1-F(t)}\right)\right] + C}{\beta}\right\} \tag{9.17}$$

The parameters C and β in the Eq. 9.17 can be obtained from Table 9.17. The initial crack impact count (N_1) and failure impact count (N_2) for each concrete specimen under different failure probabilities $F(t)$ are calculated. The failure probability $F(t)$ is set to 0.10, 0.20, 0.30, 0.40, and 0.50 for estimation. By analyzing the impact life of each group of concrete, the estimated values of the impact life for each specimen are determined. The specific values are shown in Table 9.18.

Based on the data in Table 9.18, the natural logarithms of the estimated impact life values, $\ln(N_1)$ and $\ln(N_2)$, under different failure probabilities are selected to plot line graphs. These graphs illustrate the impact life of each concrete specimen group at various failure probabilities. As shown in Figs. 9.30 and 9.31, the estimated

Table 9.16 Weibull parameters of samples

Curing age	Specimen group	N	Ordinal number (i)	Survival probability $R(t)$	$X = \ln(t)$	$Y = \ln[\ln(1/R(t))]$
6 h	A1 initial cracking	339	1	0.8571	5.8260	−1.8698
		373	2	0.7143	5.9216	−1.0892
		396	3	0.5714	5.9814	−0.5805
		401	4	0.4286	5.9940	−0.1657
		432	5	0.2857	6.0684	0.2253
		468	6	0.1429	6.1485	0.6657
	A1 final cracking	340	1	0.8571	5.8289	−1.8698
		374	2	0.7143	5.9243	−1.0892
		397	3	0.5714	5.9839	−0.5805
		402	4	0.4286	5.9965	−0.1657
		433	5	0.2857	6.0707	0.2253
		469	6	0.1429	6.1506	0.6657
1 d	A1 initial cracking	1197	1	0.8571	7.0876	−1.8698
		1277	2	0.7143	7.1523	−1.0892
		1291	3	0.5714	7.1632	−0.5805
		1357	4	0.4286	7.2130	−0.1657
		1396	5	0.2857	7.2414	0.2253
		1462	6	0.1429	7.2876	0.6657
	A1 final cracking	1198	1	0.8571	7.0884	−1.8698
		1278	2	0.7143	7.1531	−1.0892
		1292	3	0.5714	7.1639	−0.5805
		1358	4	0.4286	7.2138	−0.1657
		1397	5	0.2857	7.2421	0.2253
		1463	6	0.1429	7.2882	0.6657
3 d	A1 initial cracking	1525	1	0.8571	7.3298	−1.8698
		1671	2	0.7143	7.4212	−1.0892
		1677	3	0.5714	7.4248	−0.5805
		1764	4	0.4286	7.4753	−0.1657
		1825	5	0.2857	7.5093	0.2253
		1877	6	0.1429	7.5374	0.6657
	A1 final cracking	1526	1	0.8571	7.3304	−1.8698
		1672	2	0.7143	7.4218	−1.0892
		1678	3	0.5714	7.4254	−0.5805
		1765	4	0.4286	7.4759	−0.1657
		1826	5	0.2857	7.5099	0.2253
		1878	6	0.1429	7.5380	0.6657

(continued)

Table 9.16 (continued)

Curing age	Specimen group	N	Ordinal number (i)	Survival probability $R(t)$	$X = \ln(t)$	$Y = \ln[\ln(1/R(t))]$
6 h	A2 initial cracking	31	1	0.8571	3.4340	−1.8698
		36	3	0.5714	3.5835	−0.5805
		36	2	0.7143	3.5835	−1.0892
		39	4	0.4286	3.6636	−0.1657
		42	5	0.2857	3.7377	0.2253
		50	6	0.1429	3.9120	0.6657
	A2 final cracking	32	1	0.8571	3.4657	−1.8698
		37	3	0.5714	3.6109	−0.5805
		37	2	0.7143	3.6109	−1.0892
		40	4	0.4286	3.6889	−0.1657
		43	5	0.2857	3.7612	0.2253
		51	6	0.1429	3.9318	0.6657
1 d	A2 initial cracking	573	1	0.8571	6.3509	−1.8698
		583	2	0.7143	6.3682	−1.0892
		594	3	0.5714	6.3869	−0.5805
		601	4	0.4286	6.3986	−0.1657
		615	5	0.2857	6.4216	0.2253
		722	6	0.1429	6.5820	0.6657
	A2 final cracking	574	1	0.8571	6.3526	−1.8698
		584	2	0.7143	6.3699	−1.0892
		595	3	0.5714	6.3886	−0.5805
		602	4	0.4286	6.4003	−0.1657
		616	5	0.2857	6.4232	0.2253
		723	6	0.1429	6.5834	0.6657
3 d	A2 initial cracking	1552	1	0.8571	7.3473	−1.8698
		1611	2	0.7143	7.3846	−1.0892
		1620	3	0.5714	7.3902	−0.5805
		1661	4	0.4286	7.4152	−0.1657
		1694	5	0.2857	7.4348	0.2253
		1762	6	0.1429	7.4742	0.6657
	A2 final cracking	1557	1	0.8571	7.3505	−1.8698
		1612	2	0.7143	7.3852	−1.0892
		1629	3	0.5714	7.3957	−0.5805
		1669	4	0.4286	7.4200	−0.1657
		1695	5	0.2857	7.4354	0.2253
		1763	6	0.1429	7.4748	0.6657

(continued)

Table 9.16 (continued)

Curing age	Specimen group	N	Ordinal number (i)	Survival probability $R(t)$	$X = \ln(t)$	$Y = \ln[\ln(1/R(t))]$
6 h	A3 initial cracking	18	1	0.8571	2.8904	−1.8698
		20	2	0.7143	2.9957	−1.0892
		21	3	0.5714	3.0445	−0.5805
		26	4	0.4286	3.2581	−0.1657
		29	5	0.2857	3.3673	0.2253
		30	6	0.1429	3.4012	0.6657
	A3 final cracking	19	1	0.8571	2.9444	−1.8698
		21	2	0.7143	3.0445	−1.0892
		22	3	0.5714	3.0910	−0.5805
		27	4	0.4286	3.2958	−0.1657
		30	5	0.2857	3.4012	0.2253
		31	6	0.1429	3.4340	0.6657
1 d	A3 initial cracking	341	1	0.8571	5.8319	−1.8698
		355	2	0.7143	5.8721	−1.0892
		394	3	0.5714	5.9764	−0.5805
		416	4	0.4286	6.0307	−0.1657
		419	5	0.2857	6.0379	0.2253
		434	6	0.1429	6.0730	0.6657
	A3 final cracking	342	1	0.8571	5.8348	−1.8698
		356	2	0.7143	5.8749	−1.0892
		395	3	0.5714	5.9789	−0.5805
		417	4	0.4286	6.0331	−0.1657
		420	5	0.2857	6.0403	0.2253
		435	6	0.1429	6.0753	0.6657
3 d	A3 initial cracking	819	1	0.8571	6.7081	−1.8698
		862	2	0.7143	6.7593	−1.0892
		877	3	0.5714	6.7765	−0.5805
		935	4	0.4286	6.8405	−0.1657
		1021	5	0.2857	6.9285	0.2253
		1068	6	0.1429	6.9735	0.6657
	A3 final cracking	863	1	0.8571	6.7604	−1.8698
		878	2	0.7143	6.7776	−1.0892
		900	3	0.5714	6.8024	−0.5805
		936	4	0.4286	6.8416	−0.1657
		1022	5	0.2857	6.9295	0.2253
		1069	6	0.1429	6.9745	0.6657

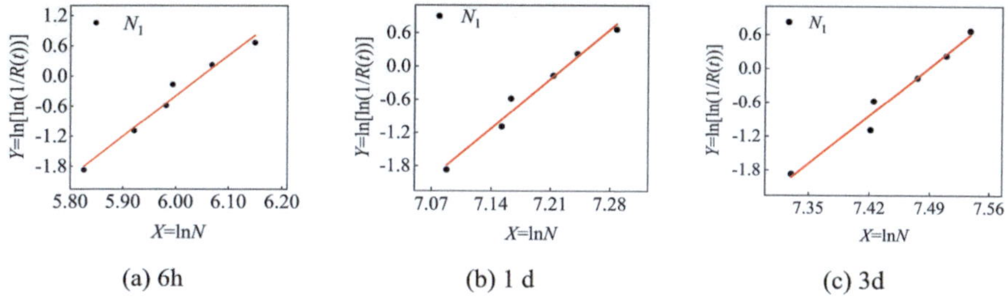

Fig. 9.24 Linear regression of initial crack impact times of A1

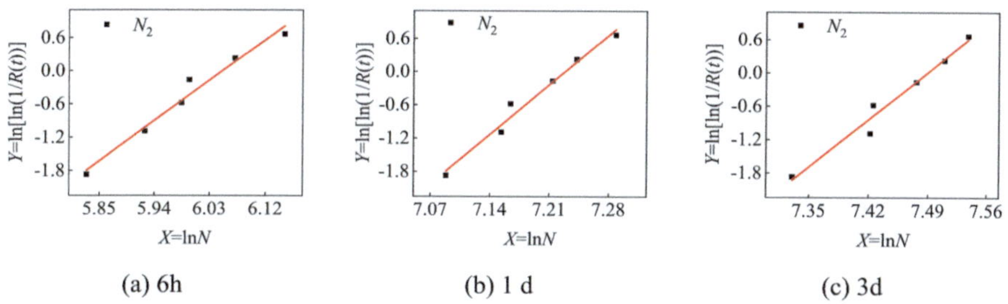

Fig. 9.25 Linear regression of final crack impact times of A1

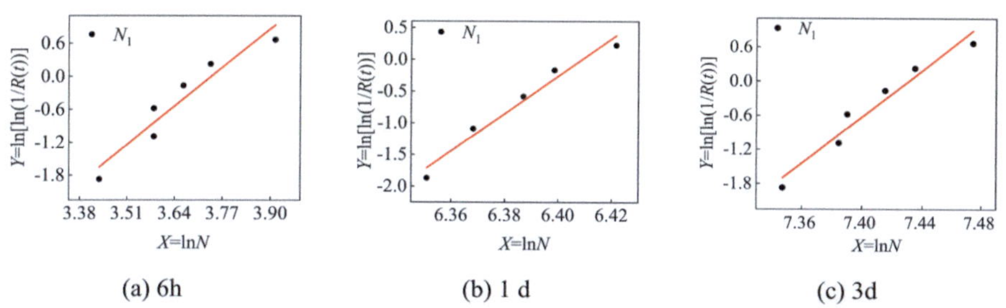

Fig. 9.26 Linear regression of initial crack impact times of A2

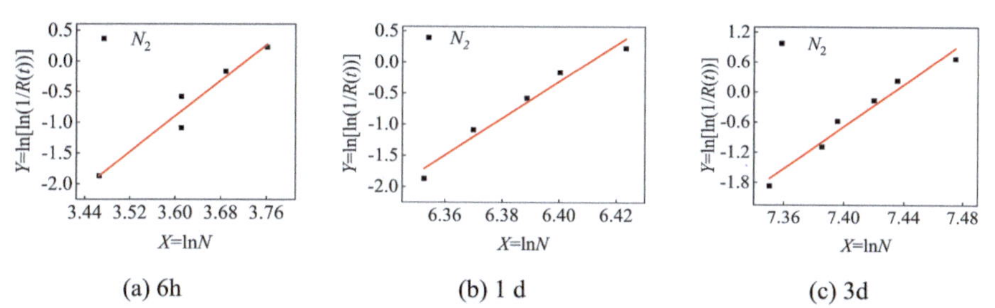

Fig. 9.27 Linear regression of final crack impact times of A2

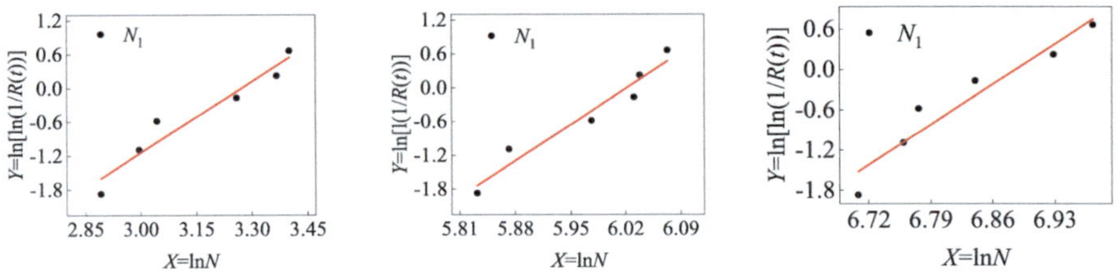

(a) 6h (b) 1 d (c) 3d

Fig. 9.28 Linear regression of initial crack impact times of A3

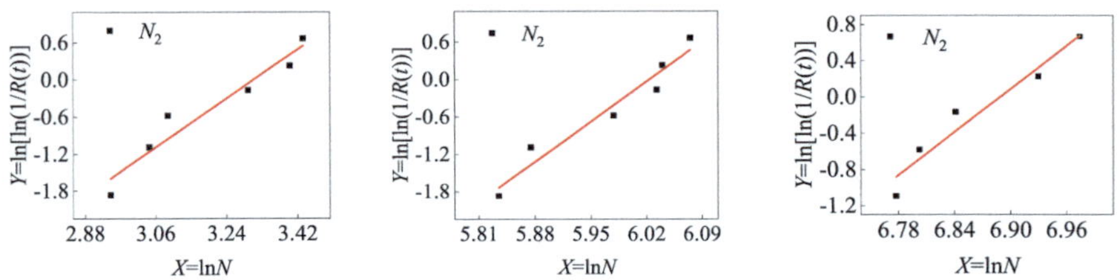

(a) 6h (b) 1 d (c) 3d

Fig. 9.29 Linear regression of final crack impact times of A3

impact life values align well with the experimental results. The estimated impact life values for A1 concrete are the highest among all groups, indicating its excellent impact resistance performance.

9.6 Conclusions

This thesis investigates the impact resistance performance of LCCM concrete. First, the basic cementitious materials and raw materials were determined, and the reference mix proportion of concrete was established. Subsequently, the macroscopic mechanical properties and microstructure of LCCM concrete, sulfoaluminate cement concrete, and coal gangue concrete were analyzed. Through microstructural analysis, the reasons for changes in the macroscopic properties of concrete were explored, and the impact life was estimated using the Weibull distribution and logarithmic distribution. The following conclusions were drawn:

(1) In compressive tests, with a water-to-cement ratio (W/C) of 0.4, the compressive strengths of A1 concrete at 6 h, 1 d, and 3 d were 46.72, 51.43, and 53.41 MPa, respectively. Compared to A2 concrete, these values increased by 60.1, 56.6, and 30.2%, respectively. Compared to the standard requirements, the 1d and 3 d compressive strengths of A1 concrete increased by 35.8 and 28.1%, respectively.

Table 9.17 Weibull distribution regression results of samples at different curing ages

Number	Curing age	Impact number	Regression coefficient		Coefficient of association R^2
			β	$\beta\ln(\eta)$	
A1	6 h	N_1	8.082	48.881	0.967
		N_2	8.102	49.022	0.974
	1 d	N_1	12.767	92.275	0.977
		N_2	12.776	92.354	0.977
	3 d	N_1	12.159	91.054	0.971
		N_2	12.166	91.115	0.971
A2	6 h	N_1	5.406	20.213	0.913
		N_2	7.241	26.961	0.947
	1 d	N_1	29.513	189.146	0.968
		N_2	29.563	189.513	0.968
	3 d	N_1	20.349	151.209	0.951
		N_2	20.916	155.467	0.963
A3	6 h	N_1	4.199	13.738	0.935
		N_2	4.378	14.487	0.934
	1 d	N_1	9.157	55.142	0.945
		N_2	9.181	55.309	0.945
	3 d	N_1	8.563	58.965	0.923
		N_2	7.892	54.369	0.942

The 3 d compressive strength of A1 concrete was 1.4 times the standard for heavy-duty cement pavements, indicating its excellent early-age compressive performance. A1 concrete fully meets the standard requirements and exhibits superior early-age performance compared to A2 concrete.

(2) In impact tests, an impact life analysis model was established using the Weibull distribution. The fitting correlation coefficient was close to 1, proving the model's effectiveness. Based on the relationship between impact life and failure probability, the impact life of A1, A2, and A3 specimens was estimated under different failure probabilities (0.10, 0.20, 0.30, 0.40, 0.50). The estimated values showed a consistent growth trend with the experimental results.

(3) The hydration progress of cementitious materials and the microstructure of the interfacial transition zone were observed through SEM testing. A1 concrete exhibited the highest density in the ITZ at early ages. Its strength was primarily provided by AFt and AH_3. As hydration progressed, AFt crystals in the pores continued to grow, further enhancing the density of the hardened cement paste and increasing strength.

(4) Nuclear magnetic resonance T_2 spectra revealed the pore distribution patterns in the concrete. A1 concrete had the fewest pores. Differences in pore size distribution were observed among A1, A2, and A3 concrete at 6 h, 1 d, and 3 d.

Table 9.18 The predicted values of impact times under different failure probability levels

Impact number	Group indication	Failure probability $R(t)$				
		0.1	0.2	0.3	0.4	0.5
N_1	A1-6H	320	352	373	390	405
	A1-1D	1,154	1,224	1,270	1,306	1,338
	A1-3D	1,486	1,580	1,642	1,691	1,734
	A2-6H	28	32	35	37	39
	A2-1D	563	577	586	594	600
	A2-3D	1,510	1,567	1,604	1,632	1,657
	A3-6H	15	18	21	22	24
	A4-1D	322	350	368	383	396
	A3-3D	752	821	868	905	938
N_2	A1-6H	321	353	374	391	406
	A1-1D	1,156	1,226	1,272	1,308	1,339
	A1-3D	1,487	1,581	1,643	1,693	1,736
	A2-6H	30	34	36	38	39
	A2-1D	564	578	587	595	601
	A2-3D	1,518	1,574	1,609	1,637	1,661
	A3-6H	16	19	22	23	25
	A4-1D	323	351	369	384	397
	A3-3D	738	812	861	901	937

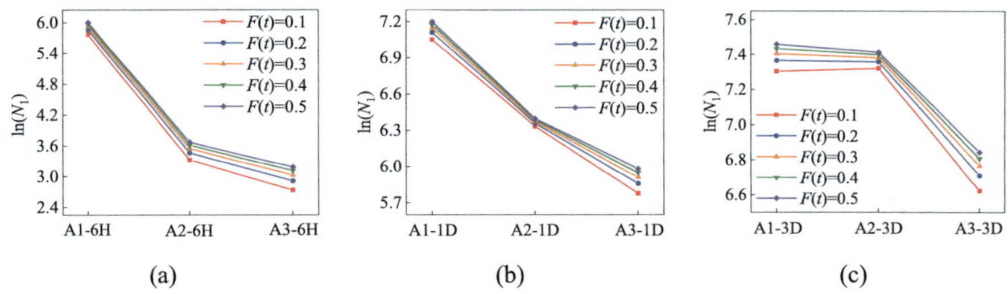

Fig. 9.30 The impact life $\ln(N_1)$ line chart of samples under different failure probabilities

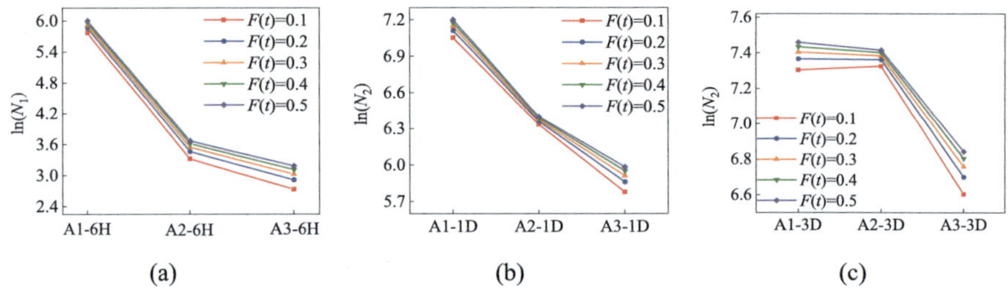

Fig. 9.31 The impact life $\ln(N_2)$ line chart of samples under different failure probabilities

Compressive strength was closely related to the proportions of harmful and more harmful pores. After 3 days of water-saturated curing, the proportions of harmful and more harmful pores in A1, A2, and A3 concrete decreased cumulatively by 20.97, 15.09, and 9.52%, respectively. A1 concrete showed a higher early reduction rate, indicating its higher internal density. This is also the reason for the higher early strength of A1 concrete.

References

Haruna, S. I., Zhu, H., Jiang, W., & Shao, J. (2021). Evaluation of impact resistance properties of polyurethane-based polymer concrete for the repair of runway subjected to repeated drop-weight impact test. *Construction and Building Materials, 309*, 125152.

JGJ 55-2011. (2011). Design Specification for Mix Proportion of Ordinary Concrete [S]. Beijing: Guangming Daily Press.

Test methods for water requirement of normal consistency, setting time and soundness of the Portland cement: GB/T1346-2011[S]. (2012). Beijing: General Administration of Quality Supervision, Inspection and Quarantine of the People's Republic of China.

Testing Methods of Cement and Concrete for Highway Engineering: JTG 3420-2020 [S]. (2011). Beijing: Ministry of Transport of the People's Republic of China.

Test method of cement mortar strength (ISO method): GB/T 17671-2021 [S]. (2022). Beijing: General Administration of Quality Supervision, Inspection and Quarantine of the People's Republic of China.

Zhang, S., & Guolong, S. (2011). Batching calculation of belite sulphoaluminate cement clinker. *Journal of Nanchang University, 33*(1), 30–32.

Chapter 10
Environmental Impact Assessment of Low-Carbon Cementitious Materials

10.1 Introduction

The resource-based industry in Ordos City, Inner Mongolia is mainly composed of five parts. They are coal-electricity industry chain, coal-to-oil industry chain, coal-to-gas industry chain, coal chemical industry chain and coal-electricity-aluminum industry chain. In the development of the above industries, a large amounts of solid wastes such as coal gangue, carbide slag and desulfurized gypsum will be discharged. It can be seen from the previous laboratory research that coal gangue, carbide slag and desulfurized gypsum can replace bauxite, limestone and gypsum respectively to produce all-solid waste low-carbon cementitious materials. However, at present, the products produced by cement plants in Ordos area are still mainly ordinary Portland cement. An all-solid waste low-carbon cementitious material production line with coal gangue, carbide slag and desulfurization gypsum as raw materials has not yet been built. Therefore, it is necessary to evaluate the environmental impact of each production link before the construction or upgrading of the production line.

Based on the theory of life cycle assessment, this chapter uses Simapro software to calculate. The production process was evaluated by using the measured data of laboratory preparation of all solid waste low carbon cementitious materials and natural resources preparation of sulphoaluminate cement as data sources. At the same time, the whole production process of the two cementitious materials is compared and analyzed. The evaluation results of energy consumption, climate change, environmental pollution and carbon emissions of the two cementitious materials in each link of the life cycle are obtained. The environmental impact advantages of all solid waste low carbon cementitious materials are clarified. In addition, by analyzing the environmental impact of the two cementitious materials in all aspects of production, the links with greater environmental impact are further improved. Achieve energy-saving and low-carbon production of all-solid waste low-carbon cementitious materials.

C. Yan et al., *Low-Carbon Cementitious Materials with 100% Solid Wastes*,
https://doi.org/10.1007/978-981-96-8742-8_10

10.2 Research Objectives and System Boundary

10.2.1 Research Objectives

The purpose of life cycle assessment theory is to quantify the energy consumption and environmental impact of products in all aspects of the life cycle. Then, the corresponding improvement suggestions are put forward for the source, production and processing of the product. In this study, coal gangue, carbide slag, and desulfurization gypsum were used as raw materials to prepare all-solid waste low-carbon cementitious materials. The energy consumption and environmental impact of each link in the whole life cycle are calculated from the whole production process of raw material treatment, raw material mixture preparation and cementitious material calcination. At the same time, the environmental impact categories and results of each link are analyzed, so as to put forward corresponding improvement measures or methods for the key links that cause serious environmental impact.

10.2.2 System Boundary

Usually, the system boundary of life cycle assessment theory is cradle to grave. Considering that the scene and scope of the use of all-solid waste low-carbon cementitious materials and traditional sulphoaluminate cement are consistent, the system boundary is selected from the cradle to the factory gate, that is, from industrial solid waste (or natural resources) to cementitious material formation. In this study, the production of 1 kg of cementitious material was used as the functional unit.

The system boundary of low-carbon cementitious materials prepared from all-solid waste is shown in Fig. 10.1. It can be seen that the main raw materials used by all-solid waste low-carbon cementitious materials are coal gangue, carbide slag and desulfurization gypsum. The preparation process includes raw material treatment (transportation, crushing, milling, and sieving), raw material mixture preparation (mixing, pressing, and drying) and cementitious material preparation (calcination, grinding). In the whole production process, the energy consumed mainly includes diesel and electricity, and the discharge mainly includes waste gas (CO_2, SO_2, etc.), waste liquid (sulfate) and waste solid (smoke, dust, etc.).

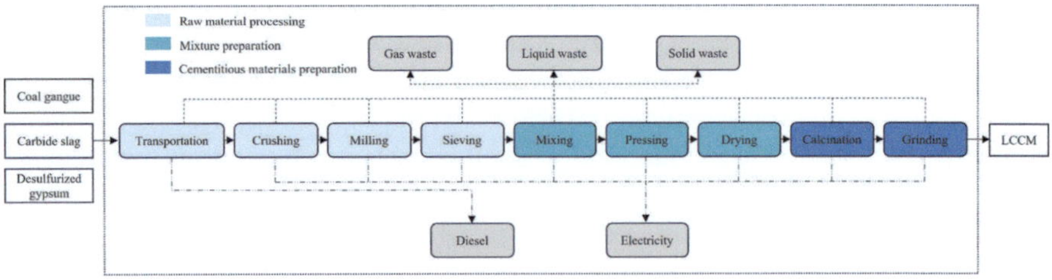

Fig. 10.1 System boundary for preparing LCCM with 100% solid wastes

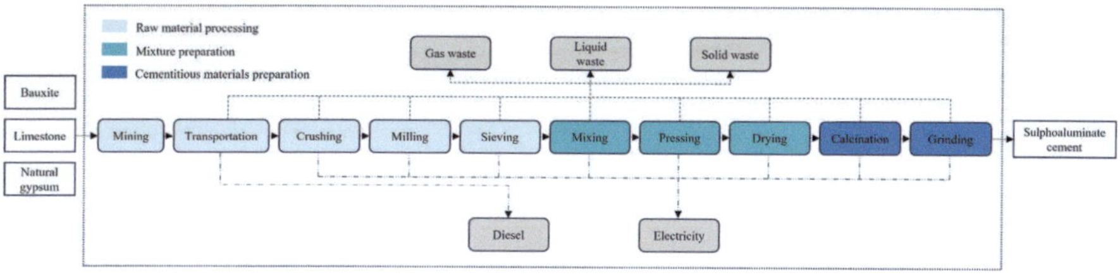

Fig. 10.2 System boundary for preparing traditional sulphoaluminate cement from natural resources

The system boundary of traditional sulphoaluminate cement prepared from natural resources is shown in Fig. 10.2. It can be seen that the main raw materials used in traditional sulphoaluminate cement are bauxite, natural gypsum and limestone. The preparation process includes: raw material treatment (mining, transportation, crushing, and grinding), raw material mixture preparation (mixing, pressing, and drying) and cementitious material preparation (calcination and grinding). In the whole production process, the energy consumed mainly includes diesel oil and electricity, and the discharge mainly includes waste gas, waste liquid and waste solid. Compared with the low-carbon cementitious material prepared from all solid waste in Fig. 10.1, the mining step appeared in the production process of traditional sulphoaluminate cement prepared from natural resources.

10.3 List Analysis

According to the system boundary, the input list and output list of all solid waste low carbon cementitious materials and traditional sulphoaluminate cement are obtained. The input list mainly includes resource input and energy input. The output list mainly includes pollutant emissions.

10.3.1 Resource Input List

The resource input of low-carbon cementitious materials prepared from all-solid waste mainly includes coal gangue, carbide slag and desulfurization gypsum. The resource input of traditional sulphoaluminate cement prepared from natural resources mainly includes bauxite, limestone and natural gypsum. According to the production data of traditional sulphoaluminate cement, the output rate of clinker is about 72% of the total amount of raw material mixture. Table 10.1 is the resource input list of the two cementitious materials.

Table 10.1 Resource input list (kg)

List type	Coal gangue	Carbide slag	Desulfurization gypsum	Bauxite	Limestone	Gypsum
LCCM	0.532	0.473	0.245	/	/	/
Sulphoaluminate cement	/	/	/	0.635	0.494	0.282

10.3.2 Energy Input List

In this study, combined with the resource situation and industrial layout of Ordos City, it is assumed that the cement plant is built in the Yili Industrial Park of Dalate Banner, Ordos City, Inner Mongolia. The selection of all raw material supply sites should follow the principle of proximity. Table 10.2 shows the energy input list in the production process of cementitious materials. The energy input of the two cementitious materials shows great differences in transportation, crushing and calcination, as described below.

The raw material transportation link of all solid waste low carbon cementitious material: coal gangue is taken from Dalate Yitai Coal Washing Co., Ltd., 27 km away from Yili Industrial Park; the calcium carbide slag is derived from the Yili resources of Inner Mongolia Yili Chemical Industry Park. The calcium carbide slag is mainly taken from the production line of the company's polyvinyl chloride, which is 2 km away from the Yili Industrial Park. Desulfurization gypsum is taken from Huaneng Power Plant in Dalate Banner, 36 km away from Yili Industrial Park.

The raw material transportation link of traditional sulphoaluminate cement: bauxite is located in Baode County, Shanxi Province, 280 km from Yili Industrial Park; limestone is taken from Zhungeer Banner, Ordos City, 160 km from Yili Industrial Park; the gypsum is taken from Hangjin Banner, Ordos City, 200 km from Yili Industrial Park.

The raw material crushing link of low-carbon cementitious material for all-solid waste: the particle size of coal gangue entering the plant is generally < 500 mm, and the hardness is about 3. Therefore, the two-stage crushing method is used to crush the coal gangue, that is, first coarse crushing (discharge particle size < 100 mm), then fine crushing (discharge particle size < 20 mm). The particle size of calcium carbide slag is concentrated below 75 μm, and the particle size of desulfurization gypsum is concentrated below 100 μm. Therefore, only carbide slag and desulfurization gypsum need to be ground and screened.

The raw material crushing link of traditional sulphoaluminate cement: the particle size of bauxite is generally less than 400 mm, and its hardness is the highest among the three raw materials. Therefore, the bauxite is crushed by two-stage crushing. The incoming particle size of limestone is generally less than 600 mm, which belongs to medium hardness. Therefore, the limestone is crushed by one-stage crushing. The particle size of gypsum is generally less than 400 mm, and the hardness is low. It is also broken by one-stage crushing.

Table 10.2 Energy input list

List type	Transport (t·km)	Crushing (kg)	Grinding (kg)	Screening (kg)	Homogenize (kg)	Pressing (kg)	Drying (kg)	Calcination (kg)	Grinding (kg)
Diesel oil (g)	53	/	/	/	/	/	/	/	/
Electric power (kW·h)	/	1.5	1.5	0.37	2.00	3.75	5.20	13	1.5

Calcination link: In the laboratory research stage, the heat source of the calcination link of the two cementitious materials is mainly electric power. So in the environmental impact assessment are the corresponding power input. Considering that coal gangue can provide a certain heat source, the electric power input can be reduced by 20 –30% in the calcination process of all-solid waste low-carbon cementitious materials.

10.3.3 Pollutant Output List

The pollutant emissions in the process of oil, electricity and coal acquisition refer to the SimaPro database. The output of pollutants mainly includes CO_2, NO_X, SO_2, CO and other emissions, COD, SS and oil and other wastewater emissions, as well as dust and noise pollution. Among them, the emission of SO_2 comes from fuel. The emission of CO_2 comes from the calcination process of fuel and cementitious materials. NO_X emission comes from the nitrogen oxides contained in the fuel itself.

10.4 Environmental Impact Analysis

10.4.1 Characterization Analysis

The ReCiPe 2016 Midpoint (H) method in SimaPro9.0 software was used to evaluate the environmental impact of the production process of the two cementitious materials. There are 18 assessment indicators, including Global Warming (GB), Ionizing Radiation (IR), Fine Particulate Matter Formation (FPMF), Terrestrial Acidification (TA), Land Use (LU), Ozone Formation-Human Health (OFHH), Mineral Resource Scarcity (MRS), Ozone Formation-Terrestrial Ecosystems (OFTE), Marine Eutrophication (MEA), Freshwater Eutrophication (FEA), Freshwater Ecotoxicity (FEB), Human Carcinogenic Toxicity (HCT), Terrestrial Ecotoxicity (TE), Marine Ecotoxicity (MEB), Human Non-carcinogenic Toxicity (HNT), Fossil Resource Scarcity (FRS), Stratospheric Ozone Depletion (SOD), Water Consumption (WC).

In the production process of two kinds of cementitious materials, the input of different types and quantities of resources and energy will have different effects on the environment. The resources and energy input in the list are converted into environmental impact assessment indicators of a unified unit. The calculation results of the environmental impact characterization of the production process of the two cementitious materials can be obtained, as shown in Tables 10.3 and 10.4. By analyzing the characteristic calculation results, the environmental impact proportion of the two cementitious materials in different production links under the same environmental impact type is obtained, as shown in Figs. 10.3 and 10.4. Figure 10.3 is the specific gravity diagram of each production link of all-solid waste low-carbon cementitious material under the same environmental impact type. It can be seen that in addition to the FRS environmental assessment indicators, the environmental impact of the calcination link in the remaining 17 environmental assessment indicators is the largest, accounting for about 31–43%. The second is the drying process of raw material mixture, accounting for about 11–17%; raw material mixture pressure cake link, accounting for about 8–13%. The environmental impact caused by crushing, ball milling, screening and homogenization of raw materials and ball milling of cementitious materials is the smallest, accounting for about 3–7%.

The analysis shows that in the process of preparing all-solid waste low-carbon cementitious materials in the laboratory, the equipment used in each production link is mainly driven by electric power. In all production and preparation processes, the calcination of cementitious materials takes the longest time, consumes the highest energy, and has the most significant impact on the environment. In addition, in the SOD and FRS environmental assessment indicators, the order of environmental impact of each production process from large to small is: raw material transportation > cementing material calcination > raw material mixture drying > raw material mixture pressing cake > raw material mixture homogenization > raw material crushing = raw material ball milling = cementing material ball milling > raw material screening. It can be seen that SOD and FRS have the most serious impact in the transportation of raw materials. The analysis shows that during the transportation of raw materials, the chlorofluorocarbons emitted by transport trucks rise to the stratosphere and are irradiated by ultraviolet light to decompose Cl-or Br-radicals. And a chain reaction with ozone, resulting in the destruction of the ozone layer. Therefore, the SOD index is the most significant in the process of raw material transportation. Similarly, for the FRS index, the transport truck is mainly powered by fuel oil, so the consumption of a large amount of fossil energy during transportation leads to the most serious impact of the FRS index.

In the environmental assessment indicators of IR, MRS and WC, the environmental impact of each production process is ranked from large to small: cementitious material calcination > raw material transportation > raw material mixture drying > raw material mixture pressing cake > raw material mixture homogenization > raw material crushing = raw material ball milling = cementitious material ball milling > raw material screening. Among the remaining 13 environmental assessment indicators, the environmental impact of each production process is ranked from large to small: cementitious material calcination > raw material mixture drying > raw material mixture pressing cake > raw material transportation > raw material mixture homogenization > raw material crushing = raw material ball milling = cementitious material ball milling > raw material screening.

It is assumed that the location of the traditional sulphoaluminate cement factory is also in the Yili Industrial Park of Dalate Banner, Ordos City. The proportion of each production link under the same environmental impact type is shown in Fig. 10.4. It can be seen that among the 18 environmental assessment indicators, the environmental impact caused by the transportation of raw materials is the largest, accounting for about 25–86%. The main reason is that the distribution of bauxite, limestone and gypsum mineral resources required for the preparation of traditional sulphoaluminate

Table 10.3 Characterization results for preparing LCCM with 100% solid wastes

Production link	GB	SOD	IR	OFHH	FPMF	OFTE	TA	FEA	MEA
Transport	1.86E+00	3.03E-06	1.29E-01	6.63E-03	5.38E-03	7.05E-03	1.55E-02	2.13E-04	2.79E-05
Crushing	1.56E+00	3.33E-07	2.42E-02	4.35E-03	2.39E-03	4.36E-03	5.34E-03	2.85E-04	1.78E-05
Grinding	1.56E+00	3.33E-07	2.42E-02	4.35E-03	2.39E-03	4.36E-03	5.34E-03	2.85E-04	1.78E-05
Screening	3.86E-01	8.22E-08	5.97E-03	1.07E-03	5.90E-04	1.07E-03	1.32E-03	7.04E-05	4.38E-06
Homogenization	2.09E+00	4.44E-07	3.23E-02	5.80E-03	3.19E-03	5.81E-03	7.12E-03	3.81E-04	2.37E-05
Pressuring	3.91E+00	8.33E-07	6.05E-02	1.09E-02	5.98E-03	1.09E-02	1.33E-02	7.14E-04	4.44E-05
Drying	5.42E+00	1.15E-06	8.39E-02	1.51E-02	8.30E-03	1.51E-02	1.85E-02	9.90E-04	6.16E-05
Calcination	1.36E+01	2.89E-06	2.10E-01	3.77E-02	2.07E-02	3.77E-02	4.63E-02	2.47E-03	1.54E-04
Grinding	1.56E+00	3.33E-07	2.42E-02	4.35E-03	2.39E-03	4.36E-03	5.34E-03	2.85E-04	1.78E-05
Production link	TE	FEB	MEB	HCT	HNT	LU	MRS	FRS	WC
Transport	2.46E+00	1.68E-02	2.95E-02	3.87E-02	5.17E-01	2.12E-02	3.15E-03	4.15E+00	2.00E-02
Crushing	7.70E-01	1.43E-02	2.00E-02	4.27E-02	3.31E-01	1.86E-02	4.55E-04	3.05E-01	3.81E-03
Grinding	7.70E-01	1.43E-02	2.00E-02	4.27E-02	3.31E-01	1.86E-02	4.55E-04	3.05E-01	3.81E-03
Screening	1.90E-01	3.53E-03	4.93E-03	1.05E-02	8.16E-02	4.59E-03	1.12E-04	7.53E-02	9.41E-04
Homogenization	1.03E+00	1.91E-02	2.66E-02	5.70E-02	4.41E-01	2.48E-02	6.07E-04	4.07E-01	5.08E-03
Pressuring	1.93E+00	3.58E-02	4.99E-02	1.07E-01	8.27E-01	4.65E-02	1.14E-03	7.64E-01	9.53E-03
Drying	2.67E+00	4.96E-02	6.92E-02	1.48E-01	1.15E+00	6.45E-02	1.58E-03	1.06E+00	1.32E-02
Calcination	6.68E+00	1.24E-01	1.73E-01	3.70E-01	2.87E+00	1.61E-01	3.94E-03	2.65E+00	3.30E-02
Grinding	7.70E-01	1.43E-02	2.00E-02	4.27E-02	3.31E-01	1.86E-02	4.55E-04	3.05E-01	3.81E-03

Note GB unit is kg CO_2 eq. IR unit is kBq Co-60 eq. FPMF unit is kg PM2.5 eq. TA unit is kg SO_2 eq. lU unit is m^2a crop eq. OFHH unit is kg NOx eq. MRS unit is kg Cu eq. OFTE unit is kg NOx eq. MEA unit is kg N eq. FEA unit is kg P eq. FEB, HCT, TE, and MEB units are kg 1,4-DCB. HNT unit is kg 1,4-DCB. FRS unit is kg oil eq. SOD unit is kg CFC11 eq. WC unit is m^3

Table 10.4 Characterization results for preparing traditional sulphoaluminate cement

Production link	GB	SOD	IR	OFHH	FPMF	OFTE	TA	FEA	MEA
Mining	0.00E+00	0.00E+00	0.00E+00	0.00E+00	0.00E+00	0.00E+00	0.00E+00	0.00E+00	0.00E+00
Transport	1.88E+01	3.07E-05	1.31E+00	6.73E-02	5.46E-02	7.15E-02	1.57E-01	2.16E-03	2.83E-04
Crushing	3.91E+00	8.33E-07	6.05E-02	1.09E-02	5.98E-03	1.09E-02	1.33E-02	7.14E-04	4.44E-05
Grinding	3.91E+00	8.33E-07	6.05E-02	1.09E-02	5.98E-03	1.09E-02	1.33E-02	7.14E-04	4.44E-05
Screening	3.86E-01	8.22E-08	5.97E-03	1.07E-03	5.90E-04	1.07E-03	1.32E-03	7.04E-05	4.38E-06
Homogenization	2.09E+00	4.44E-07	3.23E-02	5.80E-03	3.19E-03	5.81E-03	7.12E-03	3.81E-04	2.37E-05
Pressuring	3.91E+00	8.33E-07	6.05E-02	1.09E-02	5.98E-03	1.09E-02	1.33E-02	7.14E-04	4.44E-05
Drying	5.42E+00	1.15E-06	8.39E-02	1.51E-02	8.30E-03	1.51E-02	1.85E-02	9.90E-04	6.16E-05
Calcination	1.39E+01	2.89E-06	2.10E-01	3.77E-02	2.07E-02	3.77E-02	4.63E-02	2.47E-03	1.54E-04
Grinding	1.56E+00	3.33E-07	2.42E-02	4.35E-03	2.39E-03	4.36E-03	5.34E-03	2.85E-04	1.78E-05
Production link	TE	FEB	MEB	HCT	HNT	LU	MRS	FRS	WC
Mining	0.00E+00	0.00E+00	0.00E+00	0.00E+00	0.00E+00	0.00E+00	1.37E-02	0.00E+00	0.00E+00
Transport	2.49E+01	1.70E-01	2.99E-01	3.92E-01	5.24E+00	2.15E-01	3.20E-02	4.21E+01	2.02E-01
Crushing	1.93E+00	3.58E-02	4.99E-02	1.07E-01	8.27E-01	4.65E-02	1.14E-03	7.64E-01	9.53E-03
Grinding	1.93E+00	3.58E-02	4.99E-02	1.07E-01	8.27E-01	4.65E-02	1.14E-03	7.64E-01	9.53E-03
Screening	1.90E-01	3.53E-03	4.93E-03	1.05E-02	8.16E-02	4.59E-03	1.12E-04	7.53E-02	9.41E-04
Homogenization	1.03E+00	1.91E-02	2.66E-02	5.70E-02	4.41E-01	2.48E-02	6.07E-04	4.07E-01	5.08E-03
Pressuring	1.93E+00	3.58E-02	4.99E-02	1.07E-01	8.27E-01	4.65E-02	1.14E-03	7.64E-01	9.53E-03
Drying	2.67E+00	4.96E-02	6.92E-02	1.48E-01	1.15E+00	6.45E-02	1.58E-03	1.06E+00	1.32E-02
Calcination	6.68E+00	1.24E-01	1.73E-01	3.70E-01	2.87E+00	1.61E-01	3.94E-03	2.65E+00	3.30E-02
Grinding	7.70E-01	1.43E-02	2.00E-02	4.27E-02	3.31E-01	1.86E-02	4.55E-04	3.05E-01	3.81E-03

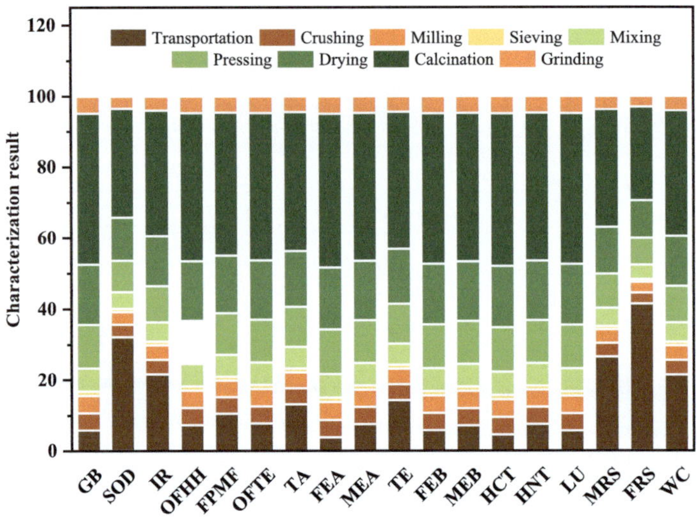

Fig. 10.3 Characterization analyses of environmental effect for LCCM with 100% solid wastes

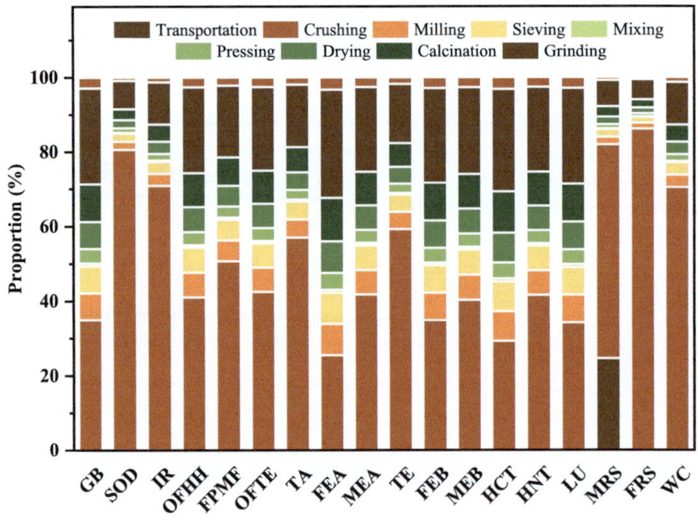

Fig. 10.4 Characterization analyses of environmental effect for traditional sulphoaluminate cement from natural resources

cement is far from the site. In the process of obtaining raw materials, a large amount of oil is consumed, and a large amount of pollutant gases such as CO_2, SO_2, and NO_X are discharged, resulting in a greater impact on the environment. In addition, the impact of other production processes on the environment is similar to that of all-solid waste low-carbon cementitious materials.

10.4.2 Standardization Analysis

By standardizing the characterization results obtained in Sect. 10.4.1, the relative size of different environmental assessment indicators can be compared. The standardized calculation results of the environmental impact of each production process of the two cementitious materials are shown in Tables 10.5 and 10.6. By analyzing the standardized calculation results, the contribution of different environmental assessment indicators of the two cementitious materials in each production process can be obtained, as shown in Figs. 10.5 and 10.6. Figure 10.5 is the environmental impact contribution ratio of all-solid waste low-carbon cementitious materials. It can be seen that in the whole production process, the contribution ratio of environmental impact from large to small is: MEB > HCT > FEB > HNT > TE > FRS > FEA > OFTE > OFHH > GB > TA > FPMF > IR, accounting for 38.14, 29.59, 22.64, 4.39, 1.59, 0.97, 0.84, 0.49, 0.42, 0.38, 0.27, 0.19 and 0.12% of the total environmental impact, respectively. Among them, MEB, HCT and FEB are the main indicators of environmental impact, contributing more than 90% to the total environmental impact. Similarly, it can be seen from Fig. 10.6 that the contribution of MEB, HCT and FEB to environmental impact is also the highest, accounting for 39.30, 26.51 and 21.74% of the total environmental impact, respectively.

10.5 Comparative Analysis of Products

The characterization and standardization comparison results of the two cementitious materials are shown in Table 10.7. It can be seen that among the 18 environmental impact assessment indicators, the environmental impact of preparing all-solid waste low-carbon cementitious materials is generally lower than that of traditional sulphoaluminate cement prepared from natural resources. Combined with the input and output list analysis of the life cycle, the main reason for the above results is that the mining, transportation and crushing of raw materials in the traditional sulphoaluminate cement production process consume a large amount of oil and power resources.

Comparing the GB index of the two products, it can be seen that the GB index of the all-solid waste low-carbon cementitious material is reduced by 40.82% compared with the traditional sulphoaluminate cement. According to the analysis, in the preparation process of all solid waste low carbon cementitious materials, due to the fragile and less grinding characteristics of coal gangue, carbide slag and desulfurization gypsum, a large amount of energy consumption and power consumption are saved in the process of raw material mining, transportation and crushing. In addition, in the calcination process, because the traditional sulphoaluminate cement uses limestone as the raw material, a part of carbon dioxide will also be emitted during the calcination process. This leads to its GB index is significantly higher than that of all solid waste low carbon cementitious materials.

Table 10.5 Standardization results for preparing LCCM with 100% solid wastes

Production link	GB	SOD	IR	OFHH	FPMF	OFTE	TA	FEA	MEA
Transport	2.33E-04	5.06E-05	2.68E-04	3.22E-04	2.10E-04	3.97E-04	3.78E-04	3.28E-04	6.05E-06
Crushing	1.96E-04	5.56E-06	5.03E-05	2.11E-04	9.36E-05	2.45E-04	1.30E-04	4.40E-04	3.85E-06
Grinding	1.96E-04	5.56E-06	5.03E-05	2.11E-04	9.36E-05	2.45E-04	1.30E-04	4.40E-04	3.85E-06
Screening	4.83E-05	1.37E-06	1.24E-05	5.21E-05	2.31E-05	6.05E-05	3.21E-05	1.08E-04	9.51E-07
Homogenization	2.61E-04	7.42E-06	6.71E-05	2.82E-04	1.25E-04	3.27E-04	1.74E-04	5.86E-04	5.14E-06
Pressuring	4.90E-04	1.39E-05	1.26E-04	5.28E-04	2.34E-04	6.13E-04	3.26E-04	1.10E-03	9.63E-06
Drying	6.79E-04	1.93E-05	1.75E-04	7.32E-04	3.24E-04	8.50E-04	4.52E-04	1.52E-03	1.34E-05
Calcination	1.70E-03	4.82E-05	4.36E-04	1.83E-03	8.11E-04	2.13E-03	1.13E-03	3.81E-03	3.34E-05
Grinding	1.96E-04	5.56E-06	5.03E-05	2.11E-04	9.36E-05	2.45E-04	1.30E-04	4.40E-04	3.85E-06
Production link	TE	FEB	MEB	HCT	HNT	LU	MRS	FRS	WC
Transport	2.37E-03	1.37E-02	2.85E-02	1.40E-02	3.47E-03	3.43E-06	2.62E-08	4.23E-03	7.48E-05
Crushing	7.43E-04	1.17E-02	1.93E-02	1.54E-02	2.22E-03	3.01E-06	3.79E-09	3.12E-04	1.43E-05
Grinding	7.43E-04	1.17E-02	1.93E-02	1.54E-02	2.22E-03	3.01E-06	3.79E-09	3.12E-04	1.43E-05
Screening	1.83E-04	2.88E-03	4.77E-03	3.81E-03	5.48E-04	7.43E-07	9.35E-10	7.68E-05	3.53E-06
Homogenization	9.91E-04	1.55E-02	2.58E-02	2.06E-02	2.96E-03	4.02E-06	5.05E-09	4.15E-04	1.91E-05
Pressuring	1.86E-03	2.91E-02	4.84E-02	3.86E-02	5.55E-03	7.53E-06	9.48E-09	7.79E-04	3.57E-05
Drying	2.58E-03	4.04E-02	6.71E-02	5.35E-02	7.70E-03	1.04E-05	1.31E-08	1.08E-03	4.96E-05
Calcination	6.44E-03	1.01E-01	1.68E-01	1.34E-01	1.92E-02	2.61E-05	3.29E-08	2.70E-03	1.24E-04
Grinding	7.43E-04	1.17E-02	1.93E-02	1.54E-02	2.22E-03	3.01E-06	3.79E-09	3.12E-04	1.43E-05

Table 10.6 Standardization results for preparing traditional sulphoaluminate cement

Production link	GB	SOD	IR	OFHH	FPMF	OFTE	TA	FEA	MEA
Mining	0.00E+00	0.00E+00	0.00E+00	0.00E+00	0.00E+00	0.00E+00	0.00E+00	0.00E+00	0.00E+00
Transport	2.36E-03	5.13E-04	2.72E-03	3.27E-03	2.13E-03	4.03E-03	3.83E-03	3.32E-03	6.14E-05
Crushing	4.90E-04	1.39E-05	1.26E-04	5.28E-04	2.34E-04	6.13E-04	3.26E-04	1.10E-03	9.63E-06
Grinding	4.90E-04	1.39E-05	1.26E-04	5.28E-04	2.34E-04	6.13E-04	3.26E-04	1.10E-03	9.63E-06
Screening	4.83E-05	1.37E-06	1.24E-05	5.21E-05	2.31E-05	6.05E-05	3.21E-05	1.08E-04	9.51E-07
Homogenization	2.61E-04	7.42E-06	6.71E-05	2.82E-04	1.25E-04	3.27E-04	1.74E-04	5.86E-04	5.14E-06
Pressuring	4.90E-04	1.39E-05	1.26E-04	5.28E-04	2.34E-04	6.13E-04	3.26E-04	1.10E-03	9.63E-06
Drying	6.79E-04	1.93E-05	1.75E-04	7.32E-04	3.24E-04	8.50E-04	4.52E-04	1.52E-03	1.34E-05
Calcination	1.75E-03	4.82E-05	4.36E-04	1.83E-03	8.11E-04	2.13E-03	1.13E-03	3.81E-03	3.34E-05
Grinding	1.96E-04	5.56E-06	5.03E-05	2.11E-04	9.36E-05	2.45E-04	1.30E-04	4.40E-04	3.85E-06
Production link	TE	FEB	MEB	HCT	HNT	LU	MRS	FRS	WC
Mining	0.00E+00	0.00E+00	0.00E+00	0.00E+00	0.00E+00	0.00E+00	1.14E-07	0.00E+00	0.00E+00
Transport	2.40E-02	1.39E-01	2.89E-01	1.42E-01	3.52E-02	3.48E-05	2.66E-07	4.29E-02	7.59E-04
Crushing	1.86E-03	2.91E-02	4.84E-02	3.86E-02	5.55E-03	7.53E-06	9.48E-09	7.79E-04	3.57E-05
Grinding	1.86E-03	2.91E-02	4.84E-02	3.86E-02	5.55E-03	7.53E-06	9.48E-09	7.79E-04	3.57E-05
Screening	1.83E-04	2.88E-03	4.77E-03	3.81E-03	5.48E-04	7.43E-07	9.35E-10	7.68E-05	3.53E-06
Homogenization	9.91E-04	1.55E-02	2.58E-02	2.06E-02	2.96E-03	4.02E-06	5.05E-09	4.15E-04	1.91E-05
Pressuring	1.86E-03	2.91E-02	4.84E-02	3.86E-02	5.55E-03	7.53E-06	9.48E-09	7.79E-04	3.57E-05
Drying	2.58E-03	4.04E-02	6.71E-02	5.35E-02	7.70E-03	1.04E-05	1.31E-08	1.08E-03	4.96E-05
Calcination	6.44E-03	1.01E-01	1.68E-01	1.34E-01	1.92E-02	2.61E-05	3.29E-08	2.70E-03	1.24E-04
Grinding	7.43E-04	1.17E-02	1.93E-02	1.54E-02	2.22E-03	3.01E-06	3.79E-09	3.12E-04	1.43E-05

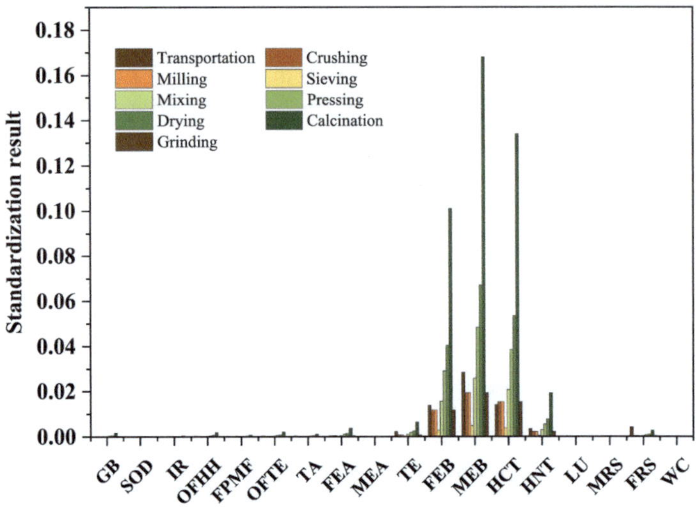

Fig. 10.5 Standardized environmental impact of LCCM with 100% solid wastes

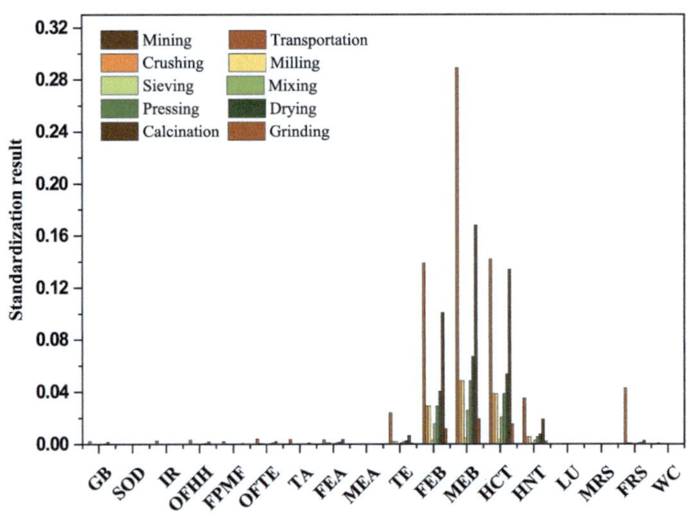

Fig. 10.6 Standardized environmental impact of traditional sulphoaluminate cement from natural resources

10.6 Conclusions

Based on the theory of life cycle assessment, this chapter compares and analyzes the whole production process of all-solid waste low-carbon cementitious materials and traditional sulphoaluminate cement. The evaluation results of energy consumption, climate change, environmental pollution and other evaluation categories of the two

Table 10.7 Comparison results of characterization and standardization for the two types of cementitious material

Category	Characterization			Standardization	
	Unit	LCCM	Sulphoaluminate cement	LCCM	Sulphoaluminate cement
GB	kg CO_2 eq	3.19E+01	5.39E+01	4.00E-03	6.76E-03
SOD	kg CFC11 eq	9.43E-06	3.81E-05	1.57E-04	6.37E-04
IR	kBq Co-60 eq	5.94E-01	1.85E+00	1.24E-03	3.84E-03
OFHH	kg NOx eq	9.03E-02	1.64E-01	4.38E-03	7.96E-03
FPMF	kg PM2.5 eq	5.13E-02	1.08E-01	2.01E-03	4.21E-03
OFTE	kg NOx eq	9.07E-02	1.68E-01	5.11E-03	9.48E-03
TA	kg SO_2 eq	1.18E-01	2.75E-01	2.88E-03	6.73E-03
FEA	kg P eq	5.69E-03	8.50E-03	8.77E-03	1.31E-02
MEA	kg N eq	3.69E-04	6.78E-04	8.01E-05	1.47E-04
TE	kg	1.73E+01	4.20E+01	1.67E-02	4.05E-02
FEB	kg	2.92E-01	4.88E-01	2.38E-01	3.98E-01
MEB	kg	4.13E-01	7.42E-01	4.00E-01	7.19E-01
HCT	kg	8.59E-01	1.34E+00	3.11E-01	4.85E-01
HNT	kg 1,4-DCB	6.88E+00	1.26E+01	4.61E-02	8.45E-02
LU	m^2a crop eq	3.78E-01	6.28E-01	6.13E-05	1.02E-04
MRS	kg Cu eq	1.19E-02	5.58E-02	9.90E-08	4.64E-07
FRS	kg oil eq	1.00E+01	4.89E+01	1.02E-02	4.98E-02
WC	m^3	9.32E-02	2.87E-01	3.50E-04	1.08E-03

cementitious materials in each link of the life cycle were obtained, and the environmental impact advantages of all-solid waste low-carbon cementitious materials were clarified. The main conclusions are as follows:

(1) Combined with the resource situation and industrial layout of Ordos City, it is assumed that the cement plant is built in Yili Industrial Park, Dalate Banner, Ordos City, Inner Mongolia. In the energy inventory input analysis, all-solid waste low-carbon cementitious materials and traditional sulphoaluminate cement showed significant differences. It is mainly manifested in raw material transportation, raw material crushing and calcination. The use of coal gangue, carbide slag and desulfurization gypsum to prepare all-solid waste low-carbon cementitious materials hardly needs to consider the energy consumption and environmental impact caused by the transportation of raw materials. In addition to the need for secondary crushing of coal gangue, carbide slag and desulfurization gypsum only need to be screened. The energy consumption and environmental impact of the crushing process are also significantly reduced. In addition, the calcination temperature of all-solid-waste low-carbon cementitious materials in the calcination process can be reduced by 150 °C

compared with traditional cementitious materials, and the energy consumption and environmental impact are further reduced.

(2) Among the 18 environmental impact assessment indicators, for all-solid waste low-carbon cementitious materials, except for the FRS indicator, the environmental impact caused by the calcination link in the remaining 17 environmental impact assessment indicators is the largest, accounting for about 31–43%. However, for traditional sulphoaluminate cement, considering the long distribution distance of mineral resources, more energy is consumed in the process of obtaining raw materials, and the emission of pollutants is large. Therefore, the environmental impact caused by this link is more serious. In the preparation process of the two cementitious materials, the three environmental impact indicators of MEB, HCT and FRB contributed the most, which were 38.14, 29.59 and 22.64%, respectively.

(3) Comparing the two cementitious materials, it can be seen that the all-solid waste low-carbon cementitious material has obvious environmental impact advantages in the process of raw material mining, transportation, crushing and calcination of cementitious materials. The 18 environmental impact indicators are less than those of traditional sulphoaluminate cement, and the greenhouse effect index can be reduced by 40.82%.

MIX
Papier aus verantwortungsvollen Quellen
Paper from responsible sources
FSC® C105338

If you have any concerns about our products,
you can contact us on
ProductSafety@springernature.com

In case Publisher is established outside the EU,
the EU authorized representative is:
Springer Nature Customer Service Center GmbH
Europaplatz 3, 69115 Heidelberg, Germany

Printed by Libri Plureos GmbH
in Hamburg, Germany